WATER DEFICITS
AND PLANT GROWTH

VOLUME III
Plant Responses and Control of Water Balance

CONTRIBUTORS TO THIS VOLUME

J. Gale

D. Hillel

Merrill R. Kaufmann

T. T. Kozlowski

Avinoam Livne

Aubrey W. Naylor

Johnson Parker

Alexandra Poljakoff-Mayber

E. Rawitz

Glenn W. Todd

Yoash Vaadia

Frank G. Viets, Jr.

WATER DEFICITS
AND PLANT GROWTH

EDITED BY

T. T. KOZLOWSKI

DEPARTMENT OF FORESTRY
THE UNIVERSITY OF WISCONSIN
MADISON, WISCONSIN

VOLUME III

Plant Responses and Control of Water Balance

1972

ACADEMIC PRESS   New York and London

ACADEMIC PRESS, INC.
111 Fifth Avenue, New York, New York 10003

United Kingdom Edition published by
ACADEMIC PRESS, INC. (LONDON) LTD.
24/28 Oval Road, London NW1

LIBRARY OF CONGRESS CATALOG CARD NUMBER: 68-14658

PRINTED IN THE UNITED STATES OF AMERICA

CONTENTS

4. PROTOPLASMIC RESISTANCE TO WATER DEFICITS
Johnson Parker

5. WATER DEFICITS AND ENZYMATIC ACTIVITY
Glenn W. Todd

6. WATER DEFICITS AND NUTRIENT AVAILABILITY
Frank G. Viets, Jr.

7. WATER DEFICITS AND NITROGEN METABOLISM
Aubrey W. Naylor

8. WATER DEFICITS AND HORMONE RELATIONS

Avinoam Livne and Yoash Vaadia

9. PHYSIOLOGICAL BASIS AND PRACTICAL PROBLEMS OF REDUCING TRANSPIRATION

Alexandra Poljakoff-Mayber and J. Gale

10. SOIL WATER CONSERVATION

D. Hillel and E. Rawitz

LIST OF CONTRIBUTORS

Numbers in parentheses indicate the pages on which the authors' contributions begin.

J. GALE (277), Department of Botany, The Hebrew University of Jerusalem, Jerusalem, Israel

D. HILLEL (65, 307), The Hebrew University of Jerusalem, Faculty of Agriculture, Rehovot, Israel

MERRILL R. KAUFMANN (91), Department of Plant Sciences, University of California, Riverside, California

T. T. KOZLOWSKI (1), Department of Forestry, The University of Wisconsin, Madison, Wisconsin

AVINOAM LIVNE (255), The Negev Institute for Arid Zone Research, Beersheva, Israel

AUBREY W. NAYLOR (241), Department of Botany, Duke University, Durham, North Carolina

JOHNSON PARKER (125), Northeastern Forest Experiment Station, Forest Service, U. S. Department of Agriculture, Hamden, Connecticut

ALEXANDRA POLJAKOFF-MAYBER (277), Department of Botany, The Hebrew University of Jerusalem, Jerusalem, Israel

E. RAWITZ (307), The Hebrew University of Jerusalem, Faculty of Agriculture, Rehovot, Israel

GLENN W. TODD (177), Department of Botany and Plant Pathology, Oklahoma State University, Stillwater, Oklahoma

YOASH VAADIA (255), Agricultural Research Organization, Volcani Center, Bet-Dagan, Israel

FRANK G. VIETS, JR. (217), Soil and Water Conservation Research Division, Agricultural Research Service, U. S. Department of Agriculture, Fort Collins, Colorado

PREFACE

The very enthusiastic reception given to the preceding two volumes of this treatise was indicative of a rapidly expanding interest in the importance of water to plants. There clearly is mounting concern throughout the world with diminishing water supplies and the need for water conservation to overcome impending deficiencies of food and fiber at a time when population is increasing at an alarming rate. For such reasons additional research on plant–water relations is needed and is proliferating rapidly along a broad front. These considerations provided the impetus for bringing up-to-date a number of topics involving water deficits in plants which were not covered in adequate depth or omitted in the first two volumes.

This volume includes comprehensive and well-documented chapters on the influence of water deficits on shrinkage of plant tissues, seed germination, reproductive growth, and such internal plant responses as protoplasmic resistance to desiccation, enzymatic activity, nitrogen metabolism, hormonal relations, and mineral nutrition. The final two chapters deal with alleviation and control of water deficits in plants. The terminology of this volume follows that developed and used in Volumes I and II.

The contributors to this volume were chosen for their demonstrated competence and scholarly productivity in the subject areas discussed. I owe each of these eminent scientists a debt of gratitude for his scholarly contribution.

T. T. Kozlowski

CONTENTS OF OTHER VOLUMES

CHAPTER 1

SHRINKING AND SWELLING OF PLANT TISSUES

T. T. Kozlowski

DEPARTMENT OF FORESTRY, THE UNIVERSITY OF WISCONSIN, MADISON, WISCONSIN

I. INTRODUCTION

Variations in size of plants during their development are the result of changes in hydration and temperature as well as the progressive accretion of growth. The size changes caused by recurrent shrinking and swelling, which are superimposed on growth of tissues, sometimes are small but at other times may greatly exceed those resulting from continuous growth of tissues through cell division and enlargement. During the growing season the reversible changes in size of plant tissues that are accounted for by changing levels of hydration are much greater than those caused by direct thermal effects. The latter are discussed by Wiegand (1906), Marvin (1949), Small and Monk (1959), Winget and Kozlowski (1964), and

1

McCracken and Kozlowski (1965). These thermal effects are beyond the scope of this chapter and will not be discussed further. This chapter will consider characteristics, causes, measurement, and significance of shrinking and swelling of vegetative and reproductive tissues of plants as a result of changes in cell turgor.

Shrinkage and swelling of plant tissues reflect changes in the energy status of the water as well as in cell turgor. The turgor changes are controlled by relative rates of absorption of water and transpiration and by internal redistribution of water in plants. During a period of soil drying, cell turgor decreases and plant tissues may shrink more or less progressively during each day if atmospheric conditions are conducive to high transpiration throughout the drought. The rate of shrinkage during a rainless period may be slowed by atmospheric conditions (e.g., cloudy weather or high relative humidity) which decrease transpiration. Small amounts of expansion during each night are superimposed on the trend of net shrinkage of tissues during a drought. However, as plants become severely dehydrated they are less likely to regain turgidity during the night, often resulting in permanent wilting of leaves.

Diurnal contraction and expansion of plant tissues are related to higher transpiration than absorption of water during the day, and the reverse at night. During the day absorption of water through the roots lags behind transpiration because of resistance to water movement through the plant. The internal water deficits in plants which develop and decrease turgor during the day usually are reduced or eliminated during the night, when both absorption of water and transpiration are low, but absorption is somewhat greater of the two (Kozlowski, 1968b; Kramer, 1969).

II. VEGETATIVE TISSUES

Both seasonal and diurnal shrinkage of leaves, stems, and roots have been well documented for many different species of herbaceous and woody plants. A few examples will be given.

A. LEAVES

Several investigators have shown that leaf thickness is related to internal water balance. For example, simultaneous measurements by Meidner (1952) of leaf water content and leaf thickness of detached leaves of *Zizyphus mucronata, Heteromorpha involucrata, Gymnospora buxifolia,* and *Xymalos monospora* showed that these were highly correlated, provided that for each species the changes in thickness did not exceed those measured in the field. A 1% change in leaf moisture content was correlated with a 4% change in leaf thickness in *Zizyphus*, a 4.5% change in *Heteromorpha*, a 7.5% change in *Gymnospora*, and a 7% change in *Xymalos*. However, these correlations ceased when leaf water content changes exceeded 3.3% in *Zizyphus*, 1.8% in *Heteromorpha*, 1.2% in *Gymnospora*, and 1.0% in *Xymalos*.

Gardner and Ehlig (1965) found that changes in leaf thickness of herbaceous plants were related to turgor pressure but the amount of leaf shrinkage varied among species. As leaves of cotton (*Gossypium hirsutum*) were relatively rigid and well supported by veins they exhibited only modest shrinkage and wilting symptoms. By comparison, leaves of pepper (*Capsicum fruitescens*) underwent considerable change in thickness during drying and showed extreme wilting symptoms as turgor pressure approached zero. Brun (1965) related changes in leaf thickness of banana (*Musa acuminata*) to transpiration and stomatal opening. When leaves were excised or the petioles frozen, stomatal opening and transpiration increased and leaf thickness decreased sharply. These changes in leaf thickness suggested rapid development of internal water deficits. Leaf thickness did not decrease appreciably when the air above a leaf was saturated or when a leaf was cut under water.

Raschke (1970a) measured separately the amount of shrinkage of the epidermis and the entire leaf of *Zea mays* in response to internal water stress. The time course of shrinkage of the whole leaf and epidermis was similar but the relative amplitudes of changes in thickness were greater for the epidermis. In one experiment, for example, thickness of the epidermis was reduced by 32% during drying whereas that of the entire leaf decreased by 18%.

Parker (1952) described a number of internal changes in needles of *Pinus strobus* and *P. nigra* var. *austriaca* during progressive drying. As internal water deficits developed, cells of the chlorenchyma, endodermis, and transfusion tissue decreased in size (Fig. 1). In *P. nigra* var. *austriaca* the adaxial needle surface, and in *P. strobus* all three surfaces, were bent inward. The shape of cells inside the hypodermis was altered in a way that varied with location of the cells. Some chlorenchyma cells collapsed in their longest direction; others stretched when attached to the hypodermis at the needle corners. In the latter case the transfusion cells contracted but the corners of the needle did not bend inward. Hence, cells attached between these tissues were stretched. Cells of the endodermis became flattened whereas transfusion tracheids and transfusion parenchyma cells collapsed or were distorted. Xylem and phloem cells were altered little, if at all, during progressive drying of needles.

Chaney and Kozlowski (1969c) showed that leaf shrinkage and expansion in English Morello cherry (*Prunus cerasus* grafted on *Prunus*

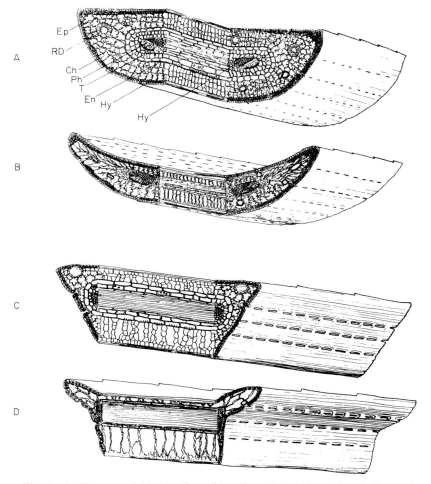

Fig. 1. A. Trans- and longisection of portion of turgid needle of *Pinus nigra* var. *austriaca*. Ep, epidermis; RD, resin duct; Ch, chlorenchyma; Ph, phloem; T, transfusion tissue; En, endodermis; Hy, hypodermis; Xy, xylem. B. Same leaf dried to approximately 55% of original weight. C. Trans- and longisection of portion of turgid needle of *Pinus strobus*. D. Same leaf dried to approximately 55% of original weight. From Parker (1952).

mahaleb rootstock) were closely related to environmental conditions affecting stomatal opening and transpiration. Changes in leaf thickness were negatively correlated with vapor pressure deficit (VPD). As early morning increases in VPD occurred, the leaves shrank rapidly and, when VPD decreased, the leaves expanded (Fig. 2). When environmental conditions

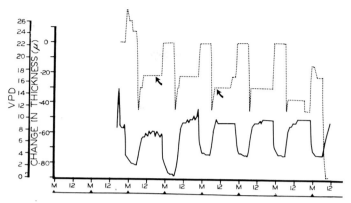

Fig. 2. Diurnal changes in leaf thickness (dotted line) and VPD (solid line) of English Morello cherry. Arrows indicate time of irrigation. From Chaney and Kozlowski (1969c).

were altered so that VPD was high during the night and lower during the day, the correlation of VPD with leaf thickness become positive. Initiation of the light period caused a decrease in leaf thickness whereas darkness triggered leaf expansion. Interruption of the dark period by a 5-min light flash did not influence leaf thickness appreciably, whereas a 90-min period of light caused leaves to shrink, with expansion occurring during a subsequent dark period.

The negative correlation between VPD and leaf shrinkage was much better when soil moisture was high than when it was low. Diurnal patterns of contraction and expansion of two leaves on different trees before and during imposed drought conditions are shown in Fig. 3A. When internal water deficits in the leaves were low early in the week, correlations between leaf shrinkage and VPD changes were typically negative. As soil dried toward the end of the week and internal water deficits in the plants increased greatly, the leaves shrank progressively and their thickness was not correlated with diurnal changes in VPD. Irrigation following the drought caused rapid absorption and translocation of water, restoration of turgor in leaves, and leaf expansion. For a few days following the irrigation, changes in leaf thickness were correlated with variations in VPD (Fig. 3B).

Chaney and Kozlowski (1971) studied diurnal changes in thickness of leaves of potted Calamondin orange trees under conditions of daily irrigation followed by periodic soil-drying cycles. Daily changes in leaf thickness from October 17 to October 21, during which time the tree was irrigated daily, showed high negative correlation with VPD (Fig. 4). Leaf thickness began to decrease near sunrise, at about the time VPD began to increase or slightly later, and the decrease continued until mid- or late afternoon,

T. T. Kozlowski

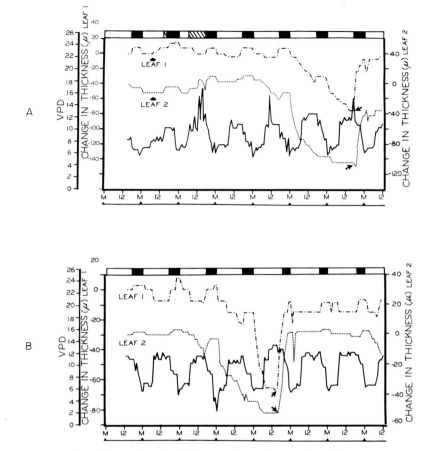

Fig. 3. Diurnal fluctuations in thickness of two leaves of English Morello cherry (broken lines) and VPD (solid lines). Arrows indicate time of irrigation. The photoperiodic regime is shown at top. Cross-hatched area in the photoperiodic regime indicates alternating light and dark due to malfunctioning switch. A. Decline in thickness of leaves as internal water deficits developed followed by rapid increase in hydration and thickness following irrigation. Correlation of changes in leaf thickness and VPD are poor under drought conditions. B. Decrease in thickness of leaves of English Morello cherry under moisture stress and rapid increase in hydration and thickness following irrigation. From Chaney and Kozlowski (1969c).

by which time VPD was decreasing. The tree was not irrigated from the morning of October 25 to the morning of October 29 and small diurnal fluctuations in leaf thickness were recorded during the imposed drought (Fig. 5). Leaves shrank during the day and expanded at night, but the amount and rate of expansion during each successive day of the drought decreased progressively until the night of October 28 when no measurable

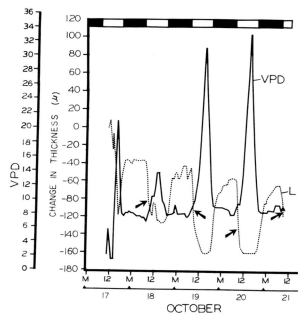

Fig. 4. Changes in leaf thickness (L) of Calamondin orange and VPD from October 17 to October 21. Arrows indicate time of irrigation. Photoperiod is shown at top with darkened portion representing the period from sunset to sunrise. M = midnight; 12 = noon. From Chaney and Kozlowski (1971).

leaf expansion was recorded. During severe drought the leaf began to shrink around sunrise and continued to do so until near sunset, rather than a few hours before sunset, as was the case with leaves of plants that were irrigated daily. Diurnal shrinkage and expansion of leaves from October 29 to November 4 are shown in Fig. 6. Irrigation during the morning of October 29, following a drought, resulted in rapid resumption of diurnal leaf expansion at night and shrinkage during the day. Progressively greater internal water deficits then developed from the morning of October 30 until the morning of November 2. A small amount of leaf expansion was recorded during the night of October 31 and there was practically no expansion during the next night. Irrigation at 8 A.M. on November 2, 3, and 4 was followed by leaf rehydration and resumption of a diurnal pattern of recurrent leaf expansion at night and contraction during the day.

B. STEMS

Many investigations of shoot elongation in herbaceous and woody plants have shown a faster rate of growth during the night, when cells are

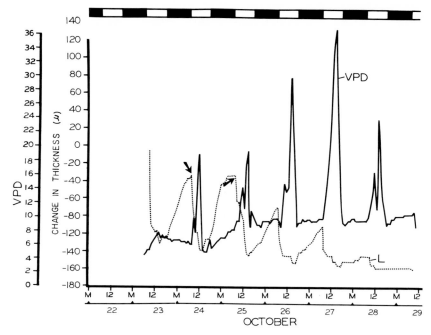

Fig. 5. Continuation of data in Fig. 4 from October 21 to October 28. The tree was not irrigated from the morning of October 25 until the morning of October 29. From Chaney and Kozlowski (1971).

turgid, than during the day. Some investigators have also shown that shoots may decrease in length during the day as a result of dehydration. For example, Brown and Trelease (Miller, 1931) reported that shoots of *Cestrum nocturnum* in the Philippine Islands wilted and decreased in length during the day, regained their original length late in the afternoon, and elongated at night. Wilson (1948) demonstrated a decrease in stem length below the first node of tomato (*Lycopersicon esculentum*) during the day. In Australia, Fielding (1955) recorded shrinkage in the length of the leading shoot of *Pinus radiata* during the daylight hours. Such shoot contraction was most pronounced during the period of rapid spring growth. On one day, a shrinkage of 1 cm was recorded for the apical shoot.

Namken *et al.* (1969) demonstrated a relation between diurnal radial change of stems of cotton (*Gossypium*) plants in wet and dry soils under similar light conditions (Fig. 7). Diurnal changes in stem radius were related to variations in leaf water potential. Contraction of the stem lagged behind leaf water potential in the morning, indicating a lag in establishing a gradient of deficits down the plant to the stem and roots. A 3- to 4-bar

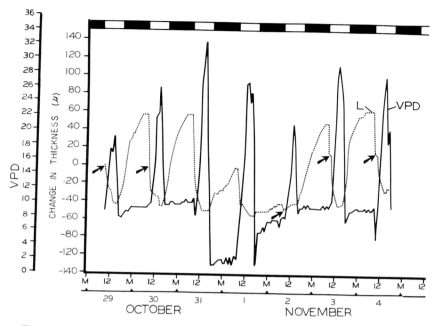

Fig. 6. Continuation of data in Fig. 4 from October 29 to November 4. The tree was not irrigated on October 31 or November 1. From Chaney and Kozlowski (1971).

decrease in leaf water potential from the 6 A.M. condition was noted before the stem began to shrink. A lag response was also observed when the plant began to recover in the afternoon. There was an increase of 2–3 bars in leaf water potential before stems began to expand.

Response of the stem of cotton plants to climatic changes is further shown in Fig. 8. Cloud conditions varied from clear to overcast. A light rain occurred shortly after noon (arrow). Stem contraction and expansion were closely related to changes in solar radiation. Reversal from contraction to expansion occurred readily, depending on increase or decrease in the energy load at the leaf surface.

An interesting result of drought is the progressive shrinkage of tree stems. Continuous stem shrinkage for several days, weeks, or even months during rainless periods has been confirmed for many species of angiosperms and gymnosperms of the Temperate Zone and the tropics (Kozlowski, 1955, 1963, 1969, 1970, 1971a,b; Kozlowski *et al.,* 1962). Some examples will be given.

Bormann and Kozlowski (1962) observed that stems of mature *Pinus strobus* trees contracted for long periods of time during the growing season

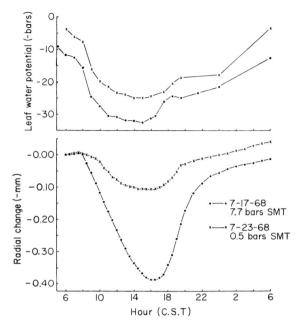

Fig. 7. Diurnal changes in stem radius and leaf water potential of cotton (*Gossypium*) under wet and dry soil moisture conditions. Soil moisture tension (SMT) values are averages for the 0 to 90-cm depth. From Namken *et al.* (1969).

in New Hampshire. Of 272 weekly dendrometer measurements during the growing season (8 trees, 17 wk, 2 oppositely placed dendrometers per tree stem), 41 showed a radial decrease from the previous week's reading. Twenty-six of the 41 observations of stem shrinkage were registered at both dendrometer stations on the same tree; 15 were recorded on one side of a tree while the other side showed radial increase. Thirty-five of the 41 observations of shrinkage were made on four dates: July 11, August 8, August 29, and September 5. The first three of these dates were preceded by 6 to 8 days during which less than 0.1 in. of rainfall was recorded. Only 3 cases of stem shrinkage were recorded before July 4, and 38 cases after that date. Hence, most stem shrinkage occurred during rainless periods in late summer after spring supplies of soil moisture were depleted.

Dendrometer studies of Buell *et al.* (1961) showed that radii of stems of several species of angiosperm trees in New Jersey shrank so much during a severe summer drought in 1957 that all radial increase as a result of xylem increase was cancelled. Only a total of 4.46 in. of rain fell during May, June, and July, whereas a 9-yr average for these months was 9.97 in. Even with the low amount of rainfall early in the season the stems at first

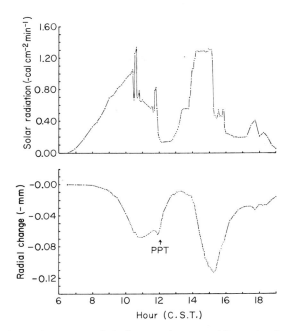

Fig. 8. Relation between radial change of cotton (*Gossypium*) stem and solar radiation on August 2, 1968. From Namken *et al.* (1969).

expanded in a typical sigmoid pattern as a result of cambial growth. By mid-July, however, the trees were severely dehydrated and their stems began to shrink progressively. Thereafter, the amount of stem shrinkage increased until by August 14 the girth of some trees was less than it was before the growing season began. The trees remained in a dehydrated and shrunken condition until soil moisture was recharged by rains in December. The trees then rehydrated rapidly and their stems expanded accordingly.

Leikola (1969a,b) studied daily shrinkage of *Pinus silvestris* stems in relation to environmental conditions over a 4-yr period in Finland. In the spring when the soil was charged with moisture after thawing the amount of diurnal stem shrinkage was small. In late summer during long rainless periods diurnal shrinkage was very marked. As the occurrence of summer rains and droughts was irregular, the times of maximum stem shrinkage varied from year to year.

In Uganda Dawkins (1956) found that stems of the rain forest species, *Lovoa brownii* and *Entandrophragma angolense,* shrank appreciably, often over a 2–3 mo period. Tree stems having a girth of 5 ft could contract more than 0.1 in. in girth in 1 wk and 0.4 in. in 10 wk.

The amount and seasonal duration of stem shrinkage often vary greatly among individual trees. In some trees seasonal shrinkage also varies markedly at different times of the growing season and from year to year. Such variability is especially well known in regions in which the rainfall supply generally is low and variable from year to year. Such conditions prevail in the White Mountains of California where a month of no rainfall may occur at any time. As may be seen in Fig. 9, the amount of seasonal stem shrinkage varied greatly in the same year among six *Pinus aristata* trees on different sites; some tree stems contracted during late July and August whereas others expanded continuously (except for superimposed diurnal shrinkage and expansion). The patterns of seasonal and diurnal radial changes within each tree also varied markedly in three successive years.

In southwestern Colorado, rain in late April and late May caused rapid rehydration and swelling of stems of *Pseudotsuga menziesii* and *Pinus edulis* trees (Fig. 10). As soil moisture decreased during the rainless summer, stems dehydrated and progressively contracted (except for slight daily re-expansion during the night). Some individual trees continued to shrink progressively for more than 3 months in late summer. Following rains in September and October, the trees again rehydrated and expanded rapidly.

Kozlowski and Winget (1964) found that the amount of diurnal

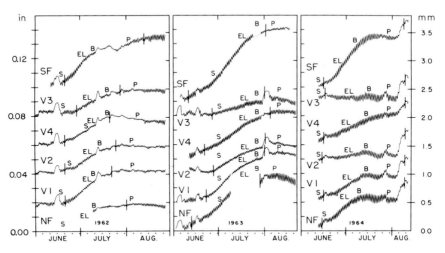

Fig. 9. Variations in stem diameters of 6 young *Pinus aristata* trees during June, July, and August of three successive years, 1962–1964. Tree locations: SF, south-facing site; V1, V2, V3, V4, valley floor; and NF, north-facing site. Phenological stages: S, buds swelling; EL, buds elongating; B, buds opening, needles starting to emerge; and P, pollen shedding. Vertical lines designate time of 5 and 95% of total stem enlargement. From Fritts (1969).

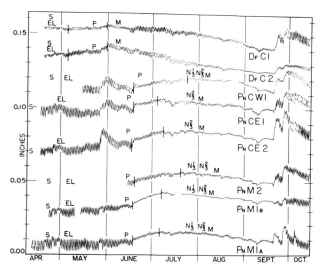

Fig. 10. Diurnal and seasonal changes in stem diameters of trees in south-western Colorado. DfC 1 and DfC 2 are two Douglas fir (*Pseudotsuga menziesii*) trees from Navajo Canyon. PnCW 1 is a pinyon pine (*Pinus edulis*) on a west-facing slope; PnCE 1 and PnCE 2 are pinyon pines on an east-facing slope; PnM 2 is a pinyon pine on the mesa top. PnM 1a and PnM 1b represent a twin pinyon pine on the mesa top with a dendrograph mounted on each stem. Other symbols: S, buds swelling; EL, bud elongation; P, pollen shed; N ⅓, needles ⅓ mature size; M, needles full size. Two vertical bars designate interval of 90% radial increase. From Fritts *et al.* (1965).

shrinkage of stems of trees in Wisconsin varied among species and at differ-ent times in the growing season. Diurnal shrinkage of stems of *Pinus resinosa* was greater than in those of *Populus tremuloides,* and *Quercus ellipsoidalis* stems shrank less than either of these. Early in the growing season, when soil moisture availability was high and leaves were not fully expanded, there was little diurnal stem shrinkage. Later, as leaves expanded and temperature increased, transpiration was accentuated. Reversible diurnal stem shrinkage then became greater. In late summer, when soil moisture was depleted and the trees became severely dehydrated, the amount of diurnal stem shrinkage declined again. Fritts (1958) found that diurnal shrinkage of *Fagus grandifolia* stems in Ohio was low during May and early June. By early July, when temperatures and transpiration were higher, diurnal contraction and expansion of stems increased appreciably.

Using dendrographs, Ogigirigi *et al.* (1970) studied diameter changes in stems of regularly irrigated *Fraxinus americana* and *Pinus resinosa* seedlings, as well as seedlings subjected to soil-drying cycles followed by irrigation. In *Fraxinus,* stem diameters of irrigated seedlings increased gradually as a result of cambial growth whereas stems of unirrigated plants contracted cumulatively. Stems of plants in drying soil began to shrink within 2 days after irrigation was discontinued. Shrinkage began at about noon and ended around 6 P.M. each day. The plants exhibited high rates of transpiration causing leaves to wilt by the fifth day of a soil-drying cycle. Following irrigation of severely droughted plants, most leaves regained turgor within an hour and measurable increases in stem diameter were recorded within 2 hr. Stems expanded to their original diameters within 12 hr after irrigation.

Stem shrinkage in *Pinus resinosa* seedlings during a soil-drying cycle was much more gradual than in *Fraxinus americana* seedlings. The earliest appreciable stem shrinkage was recorded 5 days after irrigation was discontinued. The *Pinus* seedlings required about 4 times as long as those of *Fraxinus* to deplete soil moisture reserves. This reflects much higher transpiration in *Fraxinus*. Following irrigation, *Pinus resinosa* seedlings rehydrated slowly and their stems expanded in about 6 days to their predrought diameters.

Van Laar (1967) found that diurnal fluctuations in stem diameters of *Pinus radiata* trees in South Africa were highly correlated with atmospheric water deficits. Using Thornthwaite's (1948) method to estimate potential evapotranspiration (PET) for 5-day periods, Van Laar (1967) tested a regression equation of the form:

$$y = b_0 + b_1 x_1 + b_2 x_2 + b_3 x_3$$

where $y = $ log diurnal fluctuation (μ); $x_1 = $ log (precipitation — PET) of current 5-day period (cm); $x_2 = $ log (precipitation — PET) of last 5-day period (cm); $x_3 = $ log (precipitation — PET) of last but one 5-day period (cm). The variables x_1, x_2, and x_3 explained 77% of the variation of daily fluctuations in stem diameter.

Development of internal water deficits which induce stem shrinkage often lags behind development of water deficits in leaves. For example, Namken *et al.* (1969) found minimum stem diameters of cotton (*Gossypium*) to lag behind minimum leaf water potential (maximum stress) by about 1 hr. In *Pinus resinosa* trees Turner and Waggoner (1968) noted a delay of about 3 hr between time of maximum water stress in leaves and time of maximum daily stem shrinkage of *Pinus resinosa* trees.

C. Roots

Since the water potential in the root xylem should closely follow the xylem water potential of aerial tissues, recurrent contraction and expansion in roots may be expected. This was confirmed by MacDougal (1936), who

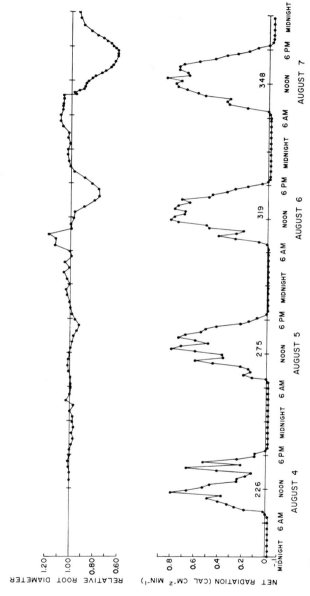

Fig. 11. Variations in root diameter of a cotton plant (*Gossypium hirsutum*) and net radiation during a consecutive 4-day period in Auburn, Alabama. The numbers under the peaks are total net radiation during daylight. From Huck *et al.* (1970).

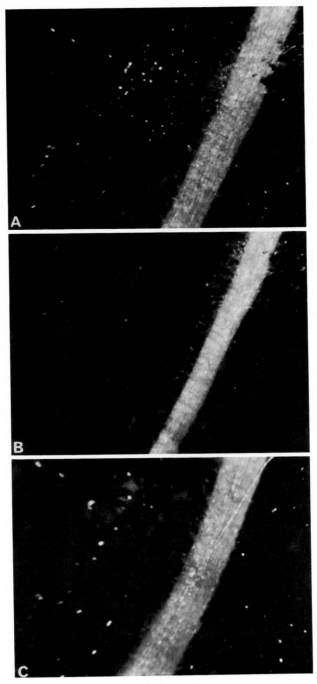

Fig. 12. Diurnal contraction and expansion of roots of *Gossypium hirsutum*.

16

found that woody roots of *Pinus radiata* in California shrank during the day and swelled during the night. In November to March daily contraction of the root began later than in the stem. Shrinkage of the root also continued for a longer period than in the stem. With advance of the growing season, root shrinkage began earlier each day until the time of beginning of shrinkage coincided with that of the stem.

Huck *et al.* (1970) measured extensive shrinkage during the day and expansion at night of roots of a 9-wk-old cotton (*Gossypium hirsutum*) plant in Alabama. Their data were obtained in August in a "rhizotron" where roots of plants growing in the field were observed with time-lapse motion pictures. Figure 11 shows the trend of diurnal changes in diameter of a root over a consecutive 4-day period during which the soil progressively dried. Roots shrank during the day and expanded at night. Diurnal shrinkage was less, and subsequent expansion more rapid, on cloudy days (August 4 and 5) than on sunny days (August 7). On the dry, sunny day minimum root diameters were about 60% of maximum diameter (Fig. 12). There was a lag of about 3 hr between the time of maximum shrinkage and time of peak radiation load.

III. REPRODUCTIVE TISSUES

Reproductive tissues increase in size during their development because of cell division and expansion. Superimposed on such growth changes are reversible changes in size because of hydration changes, with shrinkage occurring during the day and expansion at night. High transpirational losses from leaves during the day lead to extraction of water from reproductive tissues while absorption from the soil is inadequate to supply the leaves. Diurnal expansion and contraction of fruits have been reported for a variety of angiosperm fruits including cotton bolls, acorns, apples, cherries, oranges, lemons, peaches, plums, walnuts, pears, avocados, and cucumbers (MacDougal, 1920, 1924; Bartholomew, 1926; Magness *et al.*, 1935; Anderson and Kerr, 1943; Rokach, 1953; Schroeder and Wieland, 1956; Tukey, 1959, 1960, 1962; Kozlowski, 1961, 1962, 1964, 1965, 1968c; Chaney and Kozlowski, 1969a,c) and cones of gymnosperms (Dickmann and Kozlowski, 1969a; Chaney and Kozlowski, 1969b).

During the afternoons of long, hot summer days in southern California attached lemon fruits decreased in diameter and softened but they recovered turgidity and expanded at night. Fruit shrinkage began at about 6 A.M. and continued until 5 to 6 P.M. During the winter months decrease

---

A. Photograph taken at 12:05 A.M. (August 7); B. Photograph taken 5:05 P.M. (August 7); C. Photograph taken 12:05 A.M. (August 8). In C a soil particle covers part of the root. From Huck *et al.* (1970).

in fruit size began at about 9 A.M. and continued until about 4 to 5 P.M. (Bartholomew, 1926). During a 48-hr period up to 35% more water was lost by lemons attached to the tree than by detached lemons, indicating that attached lemons functioned as a water reservoir for other tissues. The drier the soil became the greater was the quantity of water translocated out of the fruits and the longer the daily period of water deficit. On mornings following a dew the leaves did not begin to extract water from the fruits until most of the moisture evaporated from the leaves and surrounding vegetation. Fog produced similar effects, with daily fruit shrinkage postponed until the fog had dissipated.

Simultaneous measurements of diurnal fluctuations in thickness of a leaf and diameter of a fruit of a Calamondin orange tree under conditions of daily irrigation or periodic drought were made by Chaney and Kozlowski (1971). As may be seen in Fig. 13, while the tree was irrigated daily (October 22–24), the leaf began to increase in thickness in midafternoon and expanded until sunrise of the following day, when VPD began to increase. Subsequent decline in leaf thickness occurred in the morning and early afternoon. Fruit diameter also fluctuated daily, with fruit expansion beginning at sunset or sometimes before sunset and continuing until about 9 A.M. the following morning. At that time the fruit

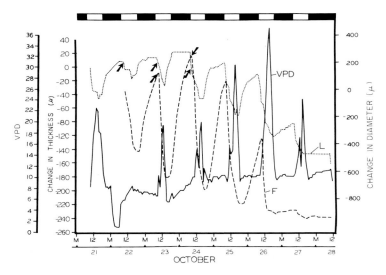

Fig. 13. Changes in fruit diameter (F) and leaf thickness (L) of Calamondin orange and of VPD from October 21 to October 28. Arrows indicate time of irrigation. Photoperiod is shown at top with darkened portion representing the period from sunset to sunrise. M = midnight; 12 = noon. From Chaney and Kozlowski (1971).

began to shrink and continued to do so during most of the day, even when VPD was declining in the afternoon. The beginning of expansion and contraction of the fruit usually occurred a few hours later than in the leaf.

The tree was not irrigated from the morning of October 24 until the morning of October 28. The developing internal water deficits influenced changes in both the leaf and fruit. Leaf thickness was not greatly affected by soil drying until the morning of October 26. The original leaf thickness was not regained by sunrise of October 26 when the leaf began to shrink. Leaf shrinkage then continued until sunset. The leaf then expanded slightly and at a much slower rate than during the period of daily irrigation. Beginning at sunrise on October 27, leaf thickness decreased until sunset and no change in thickness was recorded during the following night. Fruit expansion began 2 hours after sunset on October 25 and continued until about 9 A.M. on October 26, but diameter increase stopped before the fruit attained its original size. Then the fruit shrank during the day and did not expand at night. On the next day (October 27) the fruit shrank very slightly. As may be seen in Fig. 14 drought reduced or prevented expansion of both the leaf and fruit at night following shrinkage of each during the day. During the drought, as during the period of daily irrigation, daily leaf shrinkage preceded that of the fruit.

Following the imposed drought the tree was irrigated on October 28 (Fig. 15). Leaf expansion began shortly thereafter and continued throughout the day and night at least until 1 A.M. In contrast to leaf response, the fruit contracted during the day following irrigation, until 4 P.M., when it began to expand. Fruit expansion then continued until sunrise on October 29, at which time fruit contraction began and continued until or slightly before sunset. The fruit expanded less on the first day than on the second day following the irrigation.

The tree was not irrigated after the morning of October 30 until the morning of November 2. Withholding water did not significantly influence changes in leaf thickness until November 1 when slight diurnal leaf expansion was recorded. After irrigation on November 2, the previously recorded trend of diurnal changes in leaf thickness was recorded. However, as a result of the drought, fruit diameter was not completely regained by sunrise on November 1 and the fruit did not expand at all during the next night. After irrigation at 8 A.M. on November 2, the fruit continued to shrink until sunset and then began to expand. During the subsequent period of daily irrigation, fruit diameter fluctuated daily in limited amounts and original diameter was not regained.

The lag in daily contraction of the fruit over that of the leaf indicates that water was withdrawn from the fruits. While the tree was amply watered, leaf thickness began to decrease around sunrise when stomata

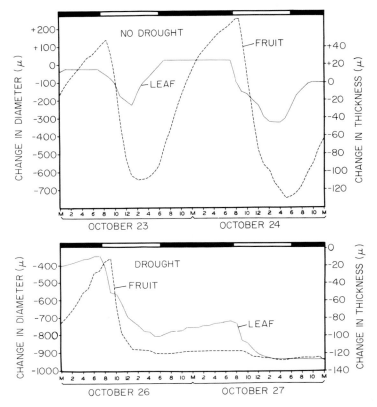

Fig. 14. Changes in fruit diameter and leaf thickness of Calamondin orange during a period of daily irrigation (no drought) and a period during which irrigation was discontinued (drought). Photoperiod is shown at top with darkened portion representing the period from sunset to sunrise. M = midnight; 12 = noon. From Chaney and Kozlowski (1971).

opened and transpiration began. The fruit did not begin to shrink until about 1.5 hr after sunrise. The data indicated that transpiration for about 1.5 hr resulted in a water potential gradient from the fruit to leaves and that water was translocated from the fruit along a free-energy gradient. This conclusion is supported by data of Klepper (1968), who found such a free-energy gradient between fruits and leaves of pear trees. During mid-day the leaves showed a minimum water potential of —15 to —20 bars, whereas the fruits showed a slightly less negative minimum. Kaufmann (1970) reported that in *Citrus sinensis* fruits a diurnal change in exocarp water potential of 5.6 bars corresponded closely to fluctuations in leaf water potential of 5.5 to 6.2 bars.

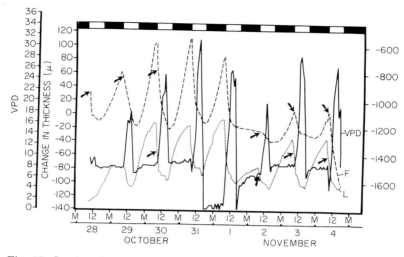

Fig. 15. Continuation of data in Fig. 13 from October 28 to November 4. The tree was not irrigated on October 31 or November 1. From Chaney and Kozlowski (1971).

The best correlations in dimensions of the orange fruit and leaf with VPD showed a lag response of tissue dimensions to changes in VPD. The best negative correlations of leaf thickness and VPD were found when VPD was shifted ahead 1 or 2 hr. The best negative correlation of fruit diameter and VPD occurred when VPD data were shifted ahead 3 or 4 hr, indicating that the VPD influence was exerted indirectly on fruits through leaf transpiration. Evidence that water moved from orange fruits to leaves is also shown in Fig. 16. Percent moisture content of leaves on fruit-bearing branches was consistently higher than it was in leaves on branches without fruits. After 24 hr, leaves on branches without fruits were visibly wilted whereas those on fruit-bearing branches were not.

Kozlowski (1968c) demonstrated that the amount and duration of daily shrinkage of partly developed fruits of Montmorency cherry (*Prunus cerasus*) varied considerably on different days. When appreciable fruit shrinkage occurred, it usually began in the early morning and continued until mid- or late afternoon at which time fruits began to expand. When trees were irrigated frequently, the amount and timing of fruit shrinkage were correlated with atmospheric conditions which influenced transpiration, and these varied greatly from day to day. As may be seen in Fig. 17, fruit shrinkage on Tuesday (6.0 hr of sunlight) was less than on Wednesday (10.7 hr of sunlight). On Thursday (2.1 hr of sunlight) the amount of shrinkage was negligible. The time of beginning of fruit shrinkage also was correlated with amount of sunlight. On Tuesday, fruits began to shrink at

Fig. 16. Turgid Calamondin orange leaves on branch with attached fruit (left) and wilted and curled leaves on branch without fruit (right). From Chaney and Kozlowski (1971).

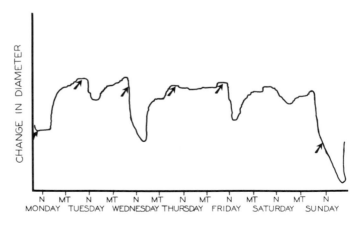

Fig. 17. Diurnal changes in diameters of fruit of Montmorency cherry (*Prunus cerasus*) in a mid-stage of fruit development. Arrows indicate time of irrigation. From Kozlowski (1968c).

about 11 A.M. and on Wednesday by 7:30 A.M. Whereas on Tuesday only 36 min of sunlight were recorded by 11 A.M., on Wednesday 62 min of sunlight were already recorded before 8 A.M. when severe internal moisture deficits developed in leaves. Marked diurnal shrinkage and expansion of cherry fruits occurred even on days with negligible amounts of sunlight, indicating translocation of water out of fruits during the day and into fruits at night. For example, on Saturday the plants were not irrigated and only 0.2 hr of sunlight was recorded by the end of the day. Nevertheless, afternoon shrinkage of fruits was marked and preceded the time of first-recorded sunlight. A combination of dry soil and a sunny day caused maximum withdrawal of water from fruits, resulting in extensive shrinkage, as shown on Sunday. The plants had not been irrigated since Friday morning and 11.4 hr of sunlight were recorded on Sunday.

Expansion and contraction of fruits appeared to be a function of the degree of water deficit in fruits as well as atmospheric conditions influencing transpiration. When trees were unirrigated from Friday A.M. to Sunday A.M. (Fig. 17), water was translocated out of the cherry fruits causing their abrupt and marked shrinkage, beginning at 7 A.M. on Sunday. Within 3 hr the fruits contracted so much that their diameters were smaller than they had been at the beginning of the week. Irrigation at 10:30 A.M. on Sunday caused fruit shrinkage to stop and rapid swelling to follow shortly thereafter. These fruits recovered more than two thirds of their predrought diameters within 2 hr after irrigation. Such rapid swelling of fruits occurred during midday even when atmospheric conditions favored high transpiration rates.

Tukey (1960) also found high correlation between the amount of diurnal shrinkage of apple fruits and atmospheric factors controlling transpiration. Usually, high solar radiation intensities and, to a lesser extent, moderate temperatures and low relative humidity during the morning hours caused increases in transpiration. Water was not absorbed through the roots and translocated to leaves fast enough to meet transpiration requirements. This caused internal water deficits to develop in leaves and water was withdrawn from the fruits. When transpiration was reduced and absorption and translocation of water increased, fruit enlargement occurred. Harley and Masure (1938) reported a strong negative correlation between daily growth of apples and evaporating power of the air.

The amount of diurnal shrinkage and expansion often is less in young fruits than in old ones. Anderson and Kerr (1943) observed that as long as cotton bolls increased in size and primary wall formation occurred in fiber cells, a continuous increase in boll diameter occurred during the day and night. However, when the bolls achieved full size they began to shrink during the day and expand at night. In young cotton plants having many enlarging bolls, the leaves wilted earlier during the day than leaves of old plants with few enlarging bolls. Apparently the bolls on the older plants acted as water reservoirs from which the leaves extracted water. In contrast, the enlarging bolls of the young plants continued to remove water from vegetative tissues, thus promoting early wilting. According to Rokach (1953), fruits of Shamouti orange did not act as water reservoirs before early June, by which time the fruits had not yet reached the size of large olives. Thereafter, water was withdrawn from the orange fruits to other tissues.

The lack of shrinkage of young fruits sometimes is correlated with a low transpiring leaf area. Kozlowski (1965) noted that during early stages of development of Montmorency cherry fruits many leaves were not fully expanded. During early fruit development (stage I) diurnal water deficits in fruits and tree stems were small as shown by lack of shrinkage and intermittent increase in fruit diameter. Young fruits expanded in a series of steps, with diameter increases beginning in early morning, usually around 2 A.M., and continuing for several hours until ceasing at about 6 A.M. However, when plants with young fruits were left unirrigated for several days and the soil became very dry, the fruits shrank appreciably.

Both seasonal and diurnal shrinkage of gymnosperm cones have been reported. Early in their development cones generally show an increase in diameter with little or no superimposed midday shrinkage. In a mid-stage of development, cones show recurrent shrinkage during the day and expansion at night (Fig. 18). During a late period of maturation, cones often show predominant continuous shrinkage (Fig. 19). Chaney and Kozlowski (1969b) observed such a pattern in both *Picea glauca* cones which ripen in 1 yr, and *Pinus banksiana* cones which ripen in 2 yr.

Dendrograph experiments of Dickmann and Kozlowski (1969a) showed three stages related to growth and water relations of *Pinus resinosa* cones during their second year. An early growth stage lasted from resumption of growth in the spring to early July. During this early period, net diameter increment of cones occurred in a stepwise pattern. Diurnal shrinkage was not significant and was followed by rapid expansion. Cone diameter usually increased during the day, but some exceptions occurred.

From mid-July to mid-August, no overall net diameter increase or decrease was recorded. Superimposed on this pattern were marked diurnal fluctuations in diameters of cones. Rapid shrinkage of cones during the late morning and early afternoon was followed by complete and rapid re-expansion during the evening and early night hours. These fluctuations generally took place between 10 A.M. and 2 A.M., but the amount of change varied from day to day.

During the maturation phase the cones dehydrated. Rapid contraction

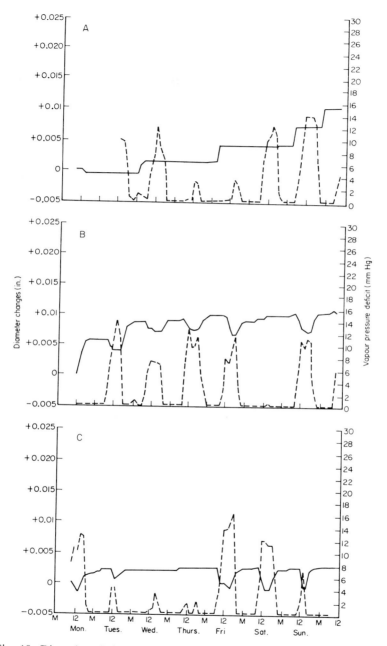

Fig. 18. Diurnal variations in diameters of cones of *Picea glauca* (solid line) and changes in vapor pressure deficit (broken line) at various times during the growing season. A. June 12–19; B. June 19–26; C. June 26–July 3. From Chaney and Kozlowski (1969b).

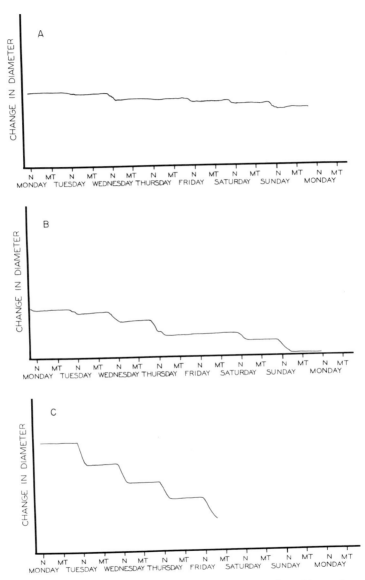

Fig. 19. Progressive late-season shrinkage of maturing *Picea glauca* cones. A. August 22–29; B. August 29–September 5; C. September 5–9. From Chaney and Kozlowski (1969b).

of maturing cones occurred during the 8-hr period from 10 A.M. to 6 P.M. A rapid late-season decline in moisture content was recorded and this reflected both dehydration and continued increase in dry weight. Apparently water in the cells of the cone axis and scales was rapidly displaced by newly forming secondary cell wall materials derived from conversion of starch to cellulose (Dickmann and Kozlowski, 1969b). The stabilization in dry weight plus continued dehydration prior to maturity in *Pinus resinosa* cones was consistent with the decrease in specific gravity of maturing *Pinus ponderosa* and *Picea pungens* cones (Maki, 1940; Cram, 1956).

IV. MEASUREMENT OF SHRINKAGE AND SWELLING

A number of methods have been used to determine changes in expansion and contraction of plant tissues. Various kinds of auxanometers (growth gages) have been used to monitor changes in plant height (Hunter and Rich, 1923; Idle, 1956; Ransom and Harrison, 1955; Klueter et al., 1966). Wilson (1948) showed that the rate of extension of the stem tip of tomato (*Lycopersicon esculentum*) not only varied during the day but there was a pronounced shrinkage in stem length below the first node during the day and elongation during the night. This was determined by attaching two auxanometers to a plant, one to the growing tip and the other to the stem at the first node below the growing tip. As elongation usually occurred above the first node, the auxanometer attached to the growing tip measured both elongation and diurnal fluctuations in stem length. The second auxanometer measured fluctuations in stem length only. The amount of stem tip elongation was determined by subtracting stem fluctuations from total growth measurements (Table I).

For direct measurement of changes in thickness or diameter of soft tissues, such as leaves or stems of herbaceous plants, very sensitive and precise instruments which exert only very slight pressure on the tissues are needed. Several investigators have measured changes in leaf thickness directly. Bachmann (1922) used a lever assembly and Meidner (1952) a system of gear wheels. Brun (1965) recorded changes in thickness of a leaf by the amount of adjustment needed to refocus the leaf under the microscope. He also modified an automatic balance to read in terms of changes in leaf thickness. This was done by calibrating to microns different numbers of microscope cover slips of known thickness. Using a microscope refocusing technique, Raschke (1970a) measured separately the shrinkage of a leaf epidermis as well as of the whole leaf. He mounted a portion of a leaf blade on a plastic slide microscope stage. Particles of safranin were deposited on the leaf surface as markers. One operator focused on the markers, using an immersion objective and oil, and then on the uppermost

TABLE I

DIURNAL FLUCTUATIONS OF STEM TIP AND OF STEM OF TOMATO PLANTS[a]

Time	Stem tip[b]	Stem below first node[c]	Apparent total
8 A.M.	25	−25	0
10 A.M.	87	−51	36
12 M.	102	−27	75
2 P.M.	91	−12	79
4 P.M.	69	−9	60
6 P.M.	60	−12	102
8 P.M.	39	2	41
10 P.M.	50	48	98
6 P.M.–6 A.M.	416	82	498
6 A.M.–6 P.M.	434	−82	352

[a] From Wilson (1948).
[b] Elongation given in 0.001 cm.
[c] Contraction and elongation given in 0.001 cm.

chloroplasts. Another operator recorded the position of the micrometer screw of the vertical adjustment. The difference between the two levels of focus represented the thickness of the epidermis.

Chaney (1970) and Chaney and Kozlowski (1969c, 1971) used a lever mechanism which continuously monitored changes in leaf thickness. A 18-in. long steel rod was counterweighted with a lead wheel and balanced on a razor blade fulcrum. The short arm of the lever, which rested on the upper surface of a leaf, exerted a very light pressure on the leaf. The lower surface of the leaf rested on a stationary screen grid which allowed for diffusion of water vapor, oxygen, and CO_2 into and out of the leaf interior. Changes in leaf thickness were amplified (lever ratio of 40:1). The ink pen attached to the tip of the long arm of the lever traced the amplified changes in leaf thickness to a chart attached to a rotating drum actuated by a clock mechanism. Because of its high sensitivity, the instrument was used only under greenhouse or growth chamber conditions where wind currents did not create problems. Thermal stability of the instrument was shown by horizontal-line traces on days with large temperature extremes when no leaf was placed between the stationary screen grid and the short arm of the lever. In addition to measuring changes in leaf thickness the instrument was used to measure changes in diameters of soft fruits (Chaney and Kozlowski, 1969c, 1971).

A very useful and popular method for measuring water status and shrinkage of leaves is the β-ray technique. This involves placing a source of β-particles on one side of a leaf and a radiation detector on the other side. Absorption of particles by the leaf depends on energy of the β-particles and mass per unit area of the leaf. Over short periods (e.g., up to a few hours) changes in mass per unit area (effective thickness) result primarily from variations in leaf water content. Frequent calibration of β-gages is necessary because the effective thickness of leaves varies with species, age, location in the plant, and environment during leaf growth. There also are differences in effective thickness in different parts of the same leaf. For further discussion of problems with the techniques and methods of β-gage calibration the reader is referred to Mederski (1961, 1964), Nakayama and Ehrler (1964), Gardner and Nieman (1964), Jarvis and Slatyer (1966), Mederski and Alles (1968), and Barrs (1968).

There has been considerable interest in monitoring expansion and contraction in diameters of stems of herbaceous plants. Beeman (1966) and Hawkins (1965) used strain gage measurements but experienced difficulty because of callousing of plants, bonding creep over long periods of time, and thermal gradients across the measuring bridge. By comparison, a stable, trouble-free method involves measurements of microchanges in plant stem radius with a linear variable displacement transducer. Namken et al. (1969) used this method to monitor stem radius of cotton (*Gossypium*) plants. The probe of the instrument was internally spring loaded to maintain light but reliable contact with the plant stem. Using a linear variable differential transformer as a transducer, Splinter (1970) developed a recording micrometer which permitted continuous monitoring of stem diameters of herbaceous plants.

For measuring shrinkage and expansion of hard tissues, such as tree stems or very hard fruits, various kinds of dendrometers, calipers, and tapes have been used (Kozlowski, 1971b). There are two basic kinds of dendrometers: Those that measure changes in circumference by means of a band (Hall, 1944) and those that measure changes of a single radius by gaging the distance between a fixed plane anchored in the wood of a tree and a point on the surface of the bark. In the latter group are dial gage dendrometers (Reineke, 1948; Bormann and Kozlowski, 1962), micrometer caliper dendrometers (Byram and Doolittle, 1950), and recording dendrometers or dendrographs (Fritts and Fritts, 1955; Impens and Schalk, 1965).

Bormann and Kozlowski (1962) found that vernier tree ring bands were much less useful than dial gage dendrometers in detecting small amounts of shrinkage in stems of mature *Pinus strobus* trees. The bands were not precise enough to measure small changes. Also the bands measured circumferential change and gave only an average volume for radial change, whereas the dial gage dendrometer gave direct radial readings at a single point. As a radial increase sometimes occurs at one portion of the stem circumference while another portion undergoes a decrease, shrinkage may be recorded by a dial gage located at a point of stem shrinkage. However, it will not be detected by a band unless there is an average shrinkage for the sum of all radii of the tree. Another limitation of the band in detecting shrinkage of stems is that friction between the overlapping sections may prevent a band from registering small amounts of shrinkage. Some of the recordings of shrinkage by the dial gage dendrometer may also have been errors in measurement. This was shown by the observation that the indicated shrinkage sometimes was only 1 to 5 μ but the dial gage dendrometer had a precision of ± 5.4 μ at the 0.01 level of probability.

Most dendrometers are designed for use with large trees. Some investigators used micrometer calipers to determine periodic changes in stem diameters of tree seedlings (McCully, 1952; Farrar and Zichmanis, 1960). Their observations were widely spaced in time and did not provide a continuous record of diameter changes. Kozlowski (1967a,b, 1968a,c) obtained continuous records of diameter changes in stems of tree seedlings by mounting the heavy Fritts dendrograph on a supporting frame. The crossbar of the dendrograph was suspended on steel balls resting in small holes cut into a projecting shelf of the frame. The stem of the experimental seedling was backed against a removable steel plate. The dendrograph rod, bearing on the opposite side of the stem, registered changes in stem diameter. These changes were transcribed continuously to weekly charts on a rotating drum actuated by a clock mechanism as described by Fritts and Fritts (1955). The movement of the instrument was similar to the movement made when it was mounted on the lag screws driven into the stem of a mature tree. However, because the seedling stem was held between the steel plate and the dendrograph rod, the instrument measured diameter changes rather than radial changes. The adaptation had a high thermal stability as shown by nearly horizontal dendrograph traces obtained when daily diameter changes of a stainless-steel rod (instead of a seedling stem) were measured. Although it was possible to use the adaptation to measure shrinkage and expansion of hard tissues (e.g., stems of woody seedlings, acorns, pine cones) it was not feasible to do so with soft tissues such as soft stems or fruits.

V. BIOLOGICAL IMPLICATIONS OF SHRINKAGE AND SWELLING OF PLANTS

Rhythmic shrinking and swelling of plant tissues and associated changes in turgidity are correlated with variations in growth (Kozlowski, 1964, 1968b; Slatyer, 1967; Crafts, 1968; Kramer, 1969). Internal water deficits influence such processes as seed germination (Chapter 2, Vol. III), absorption of water and transpiration (Crafts, 1968), enzymatic activity (Chapter 5, Vol. III), nutrient availability (Chapter 6, Vol. III), and photosynthesis (Kozlowski and Keller, 1966).

Considerable evidence shows that growth of plants is reduced by internal water deficits through stomatal closure causing reduction in photosynthesis. Cell turgor also influences growth through its effect on cell enlargement. For example, close correlation between turgor pressure and cell enlargement of cotton leaves was shown, with enlargement stopping at approximately zero turgor pressure (Wadleigh and Gauch, 1948). Ordin (1958, 1960) induced water deficits in *Avena* coleoptiles with nonpermeating mannitol and permeating sodium chloride. He was able to create similar water deficits and water potentials but different osmotic potentials and turgor pressures. Cell enlargement was reduced more by mannitol, which caused greatest decrease in turgor but little change in osmotic potential. An increase in osmotic potential did not inhibit metabolism of noncellulosic polysaccharides. However, reduction in turgor influenced cell wall metabolism and cell elongation. These observations were confirmed by Plaut and Ordin (1961) for leaves of sunflower and almond.

This section will discuss some other important biological implications of changes in cell turgor which accompany periodic shrinking and swelling of plant tissues.

A. WILTING

A direct effect of leaf dehydration is decrease in turgor which eventually causes wilting of plants. However, the amount of water that must be transpired to cause visible wilting of leaves varies for different species. Wilting is associated with a rather definite value of leaf water potential in leaves of plants of a given species, age, and cellular solute content (Gardner, 1965). On a given plant the old leaves usually wilt first. Cătský (1965) noted that the relative water content (RWC) of most of the leaves on plants growing in moist soil was rather similar. As the soil dried, however, the RWC dropped more rapidly in old leaves than in young ones. After 5 days the RWC of old leaves was about 45% and that of young leaves 80%.

Wilting usually is classified as incipient, temporary, or permanent. Incipient wilting, which is characterized by slight decrease of turgor, usually does not cause drooping of leaves and occurs whenever conditions favor high transpiration. Incipent wilting grades into temporary wilting, a state characterized by visible drooping of leaves during the day followed by rehydration and recovery from wilting during the night. During sustained periods of soil drying, temporary wilting grades into a state of permanent wilting in which plants do not recover turgidity at night. Permanently wilted plants can recover turgidity only when water is added to the soil. Prolonged permanent wilting usually kills most species of plants (Kramer, 1969).

In the early morning the turgor of leaf cells of plants rooted in wet soil is high. As water is lost during the day by transpiration, water deficits at the root surface increase. Diurnal variations occur and soil water deficits at the root surface decrease at night (leaves transpire negligibly because stomates are closed) when water moves to root surfaces. Water deficits increase during each day of a soil-drying cycle and diurnal changes in leaf water deficits occur, with deficits being lower during the night than during the day. With further soil drying, diurnal recovery in water balance in the soil next to the roots and in leaves occurs less rapidly. Leaf turgor is decreased during each succeeding day of a soil-drying cycle inasmuch as the water deficit in the plant is consistently higher than the deficit in the soil. Finally, when turgor pressure in the leaves drops to some critical threshold value, the leaves reach a state of permanent wilting. According to Slatyer (1957, 1967), this occurs when the water potentials of the leaf, root, and soil around the roots are equal and turgor pressure is zero. However, Gardner and Ehlig (1965) showed that symptoms of visible wilting appeared when turgor pressure dropped to between 2 and 3 bars. The wilting associated with the permanent wilting point was due to a change in elastic properties of the cell when turgor pressure declined below some critical value, rather than reaching a value of zero. Gardner and Ehlig (1965) noted that when turgor pressure was above 2 bars leaf thickness was relatively constant and little flexing occurred with varying turgor pressure. When turgor pressure dropped below the critical value of near 2 bars, the elastic modulus decreased markedly, allowing the leaf to sag.

The relation between turgor pressure and wilting often is complicated by appreciable differences among species in the amount of supporting tissues in their leaves. For example, cotton (*Gossypium*) leaves are well supported by veins and show only modest wilting symptoms. By comparison, pepper (*Capsicum fruitescens*) leaves are rather elastic and show extreme wilting symptoms as turgor pressure approaches zero (Gardner and Ehlig, 1965). Leaves of *Ilex* and *Pinus* are permeated with abundant lignified tissue and do not droop readily even after their parenchyma cells have lost turgidity. When oil palm (*Elaeis guineensis*) trees are severely dehydrated they do not wilt visibly because of the nature of their fibrous leaves, with a thick hypodermis and a well-developed cuticle. When an oil palm leaf is excised an early symptom of water deficit is loss of sheen on the leaflets which then become dull and also exhibit some downward rolling. The leaves change color to brown as they become desiccated, yet they do not droop (Rees, 1961).

B. STOMATAL APERTURE

Stomatal closure occurs when the water content of subsidiary cells is decreased. This creates a water potential gradient between subsidiary cells and guard cells, causing water movement out of the guard cells. As the guard cells lose turgidity the stomatal pore closes; when the guard cells gain turgidity the pore opens. In dicotyledonous plants the guard cell walls contain locally thickened areas on the ventral side (adjacent to the pore). When turgor pressure is increased in a guard cell, the thinner elastic areas of the guard cell wall on the dorsal side become distended while the more rigid, inelastic areas bend so as to tend to open the stomatal pore (Fig. 20). In monocotyledonous plants increase in turgidity of guard cells results in swelling of their bulbous ends, thereby causing a pore to develop (Pallas, 1966; Kramer, 1969) (Fig. 21).

The importance of turgidity in control of stomatal aperture was shown by Heath (1938). He observed that puncturing of a single guard cell induced closure of the stomatal pore on the same side. If a single subsidiary cell was punctured, the stomatal opening was wider on the same side, emphasizing the importance of the difference in turgor between guard cells and subsidiary cells. Changes in turgor of subsidiary cells are responsible for "passive" stomatal opening associated with sudden water deficits as well as passive stomatal closure when water is supplied to a leaf which is under water stress.

In leaves saturated with water, the compression of stomata by turgid epidermal cells occurs commonly. When water is suddenly lost from mesophyll cells of turgid leaves, the turgor of epidermal cells and subsidiary cells is decreased before turgor of guard cells is changed. This causes the stomata to open widely for a short time after which guard cell turgor declines and stomata close. Raschke (1970b) showed, for example, that when accessory cells of *Zea mays* collapsed due to water stress, the pressure on the guard cells was released, causing the stomata to open.

When water is resupplied to the leaf, uptake occurs first in epidermal and subsidiary cells, and the guard cells are pressed together (Heath and Mansfield, 1969).

Stomata usually close during early stages of water stress, often long before visible wilting occurs, and they remain closed during continued drought. According to Iljin (1957), a water loss of 10% (fresh weight basis) was enough to induce stomatal closure in many plants. In some plants (e.g., *Vicia, Chrysanthemum*) stomatal closure occurred when water loss was only 3–5%.

Gardner (1971) has demonstrated a close correlation between stomatal conductance of bean leaves and turgor. Inasmuch as turgor is in-

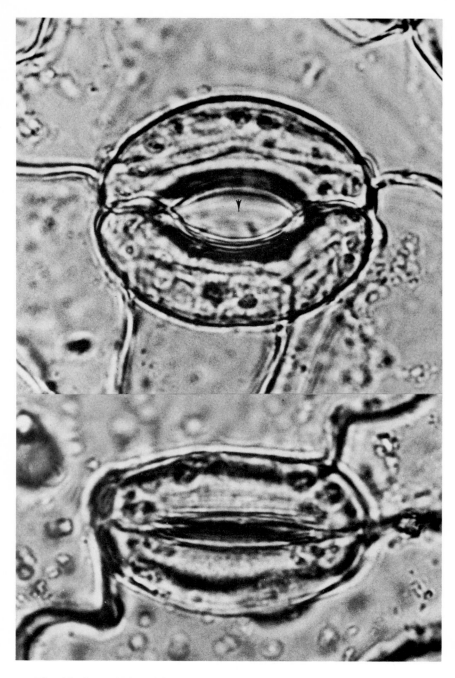

Fig. 20. Bean (*Vicia faba*) stoma which is representative of most evergreen and deciduous plants; in an open (upper) and closed condition (lower). Arrow points to stomatal pore. Magnification: ×1760. Photograph courtesy of J. E. Pallas.

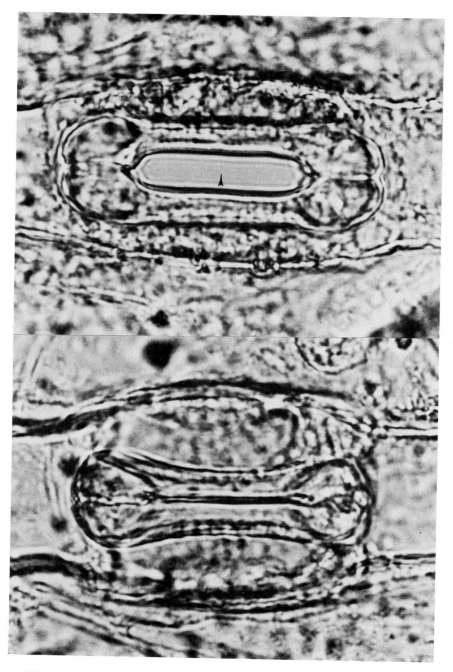

Fig. 21. Typical grass stoma on rye (*Secale cereale*) leaf in a wide-open (upper) and closed condition (lower). Arrow points to stomatal pore. Magnification: ×1600. Photograph courtesy of J. E. Pallas.

fluenced markedly by the rate of transpiration, there is a direct relation between stomatal conductance and transpiration. As transpiration rate increases, stomatal conductance decreases. When the soil water potential was in the range of 0 to −1 bar, stomatal conductance was influenced more by transpiration rate than by soil water potential.

Slatyer (1967) recognized two types of stomatal responses to internal water deficits: (1) a transient change in stomatal aperture because of changes in guard cell turgor relative to turgor of adjacent cells, and (2) a long-term change associated with severe internal water deficits. The latter response occurs at different values of leaf water potential for different species and for plants grown under different environmental conditions (Ehlig and Gardner, 1964; Gardner, 1971). When some critical water deficit is reached in a plant its stomata begin to close and, with continued increase in internal water deficit, the stomata continue to close until closure is almost complete. The long-term response of stomata during progressive drought is also affected by other factors such as internal CO_2 levels. Therefore, a specific degree of stomatal opening does not always occur at the same leaf water potential. Nevertheless, a general relation exists between leaf water potential and stomatal aperture.

Cyclic, short-time changes in stomatal aperture and transpiration often occur in rapidly transpiring plants at intervals of the order of 30 minutes. Such fluctuations occur even in a constant aerial environment. Barrs and Klepper (1968) attributed such persistent cycling behavior to changes in leaf turgor. Stomatal aperture and transpiration rate were least when leaf water potential was high and greatest when water potential was low. Lowest leaf water potential values lagged behind the time of maximum transpiration, and high resistance to water flow occurred in cycling plants. Both appeared to be necessary for the cycling response. Root resistance to water flow was important in initiating cycling by inducing water deficits in the leaves as stomata opened.

Data of Ehlig and Gardner (1964) indicated that in detached leaves of several species of herbaceous plants nearly all of the stomata on a given leaf closed over a narrow range of leaf water content. In pepper (*Capsicum fruitescens*) all stomata appeared to be open at 94% of the turgid weight and closed at 80% of the turgid weight.

Several investigators have shown that water potential of a leaf can vary over a considerable range (above a critical tension) without a significant effect on stomatal aperture. For example, Begg *et al.* (1964) found that at high light intensities leaf resistance of millet did not change when leaf water potential varied between −5 and −15 bars. Stomata closed, however, at a leaf water potential of about −20 bars. Similarly, Turner and Waggoner (1968) did not detect changes in leaf resistance of

Pinus resinosa for leaf water potentials between —5 and —15 bars during the time of day when incoming radiation was above 0.25 cal·cm^{-2}·min^{-1}.

In a given plant the time of stomatal closing during increasing leaf water deficits may vary with leaf age, as well as stomatal size and location. Stomata of shade leaves on a given plant are more sensitive than those of sun leaves to water loss. The stomata of young leaves often close more rapidly than those of old leaves in response to water stress. For example, young leaves of jarrah (*Eucalyptus marginata*) transpired about 20% less (per unit of leaf area) than mature leaves on sunny days because of more effective stomatal closure in the young leaves (Doley, 1967). Similar responses were found in *Eucalyptus stuartiana* (Henrici, 1946). In *Banksia menziesii* and *Stirlingia latifolia* mature leaves transpired more (per unit of fresh weight) than young leaves, but the differences decreased as the leaves developed. The differences were related to less effective stomatal responses of old leaves subsequent to lignification and cutinization (Grieve, 1956).

Kanemasu and Tanner (1969) demonstrated that abaxial and adaxial stomata of snap beans (*Phaseolus vulgaris*) reacted differently to increasing water deficits. Apertures of stomata on the abaxial surface, which were about 7 times as numerous as those on the adaxial surface, were not significantly influenced at leaf water potentials greater than —11 bars, but with further decrease in leaf water potential the resistance increased rapidly. By comparison, resistance of adaxial stomata increased sharply at a leaf water potential of approximately —8 bars and was constant at higher water potentials.

There is some evidence that large and small stomata on the same leaf may react somewhat differently to water deficits. For example, Waisel *et al.* (1969) noted that response of different-sized stomata in leaves of *Betula papyrifera* varied with environmental changes. When light intensity or water status was altered, large stomata tended to open first and close last.

As mentioned, stomatal opening occurs when turgor of guard cells is increased relative to turgor of subsidiary cells. There is no widespread agreement on how the turgor changes are brought about. The often given simple explanation is that stomatal opening in the light is due to enzymatic hydrolysis of starch to sugar in the guard cell plastids. This explanation now seems inadequate. Heath and Mansfield (1969) questioned that this simple mechanism could be primarily responsible for light-operated stomatal movement, but they acknowledged that it probably played a part. They pointed out that the reaction catalyzed by phosphorylase to produce glucose-1-phosphate cannot lead to a change in osmotic pressure, since the total number of solute molecules remains the same. Thus, inorganic phosphate would have to be moved into the guard cells. If this occurs, change of starch to sugar does not itself raise osmotic pressure. On the other hand, additional reactions could cause hydrolysis of glucose-1-phosphate. Energy would be required for reformation of starch to induce stomatal closure, which ought not to occur in the dark in the absence of oxygen. Yet lack of oxygen in the dark does not prevent stomata from closing. Furthermore, stomata of onion (*Allium*) which do not contain starch open and close in response to changes from light to dark, and the reverse. Heath and Mansfield (1969) concluded that starch hydrolysis and reformation did contribute to stomatal responses in the light but were more important in controlling movements due to water strain and in operating endogenous rhythms. The evidence available at present indicates that several internal factors contribute toward controlling stomatal aperture and that guard cell movements often take place independently of changes of starch to sugar, and the reverse.

There is some evidence that turgor changes in guard cells are caused by translocation of substances into them. The evidence for movement of various substances between guard cells and epidermal cells is strong. For example, sugars are translocated from epidermal cells to guard cells (Pallas, 1964). Likewise material sprayed on the leaves can be absorbed by guard cells and translocated to epidermal cells. Heath and Mansfield (1969) believe that the weight of evidence indicates participation of an active transport mechanism using ATP derived from oxidative phosphorylation.

1. Midday Closure of Stomata

Temporary stomatal closure in midday has been reported for a large number of species of plants (Kozlowski, 1964; Slatyer, 1967). Associated with midday stomatal closure are decreases in rates of transpiration and photosynthesis (Kramer and Kozlowski, 1960; Kozlowski and Keller, 1966). Midday stomatal closing is normally followed by reopening and increased transpiration in the late afternoon before the final daily transpiration decline as light intensity decreases.

When soil moisture was readily available, about 70 to 100% of the stomata of apple trees opened in the morning. They usually remained open throughout the day when air humidity was high, even though the soil had dried considerably. However, when the soil was charged with water, the stomata closed before noon when humidity was low and temperature high. With a lowered soil moisture supply, the daily duration of stomatal opening was reduced. When the soil was very dry (at or near wilting percentage), the stomata usually did not open at all unless the air was nearly saturated during the early morning. Under the latter conditions, the stomata opened for a short time but usually closed within an hour (Magness *et al.*, 1935).

Midday stomatal closure can be attributed to several causes. A major factor in its development is an absorption lag behind transpiration. This causes leaf dehydration and reduction in leaf water potential to the critical level associated with stomatal closing. If leaf water deficits are not severe, midday stomatal closure can be brought about by high temperatures causing increase in intercellular CO_2 concentration. When such temporary stomatal closure occurs, transpiration is reduced and turgidity of leaves increases at a time of high evaporative demand (Slatyer, 1967).

Midday stomatal closure in oil palm (*Elaeis guineensis*) trees in Nigeria occurred during the latter half of the dry season (from November to April). Such temporary diurnal closure of stomata disappeared with onset of the rainy season. During the dry season there was a rapid early-morning opening of stomata until about 9 A.M. Stomata then began to close gradually until about 2 P.M., after which time slight opening occurred until 4 P.M. Then late-afternoon stomatal closing began and was completed very soon after sunset. After any significant rain, daily stomatal closure became less pronounced and remained slight or disappeared until the effect of the rain on leaf turgor disappeared and afternoon closure was again evident. The extent of stomatal closure was correlated with total daily evaporation from a tank evaporimeter (Rees, 1958, 1961).

2. Stomatal Closure and Drought Resistance

Water loss from plants is prevented by successively earlier stomatal closing during each day of a developing drought and by temporary stomatal closure during midday. However, stomatal closure during droughts may not prevent killing of plants which have high rates of cuticular transpiration.

Some plants close stomata early during developing droughts and others do not. Gymnosperms usually undergo more leaf dehydration than angiosperms before stomata begin to close. Stocker *et al.* (1943) emphasized that water retention of drought-resistant plants often was a function of rapid stomatal closure during leaf dehydration. The stomata of plants of the Mediterranean maqui close rapidly when soil dries whereas those of *Phillyrea media* do not (Oppenheimer, 1953). During a soil-drying cycle a drought-resistant variety of *Theobroma cacao* closed its stomata at much higher soil moisture contents than did two less drought-resistant types (Nunes, 1967).

Kaul and Kramer (1965) compared the drought tolerance of *Ilex cornuta* var. *Burfordii* and *Rhododendron poukhanensis* by measuring transpiration rates, water deficits and stomatal opening of leaves of plants in drying soil. Transpiration rate declined more rapidly in *Ilex* than in

Rhododendron. Internal water deficits were higher in unwatered *Rhododendron* than in unwatered *Ilex* plants. Stomata closed sooner and at a lower water deficit in *Ilex* than in *Rhododendron* when subjected to water deficits. *Ilex* was considered to be more tolerant of drought than *Rhododendron* because it had more efficient control of transpiration and a higher resistance to cuticular transpiration.

According to Quraishi and Kramer (1970) seedlings of *Eucalyptus rostrata* were injured more during a drought than were seedlings of *E. polyanthemos* or *E. sideroxylon*. The lower leaves of *E. rostrata* were so badly injured by moderate water stress that transpiration rate did not return to normal after rewatering, as did transpiration of *E. polyanthemos*. Transpiration decline curves showed that *E. rostrata* had very poor stomatal control and closed its stomata much later than *E. sideroxylon* whereas *E. polyanthemos* was intermediate. *Eucalyptus rostrata* also had the lowest leaf water deficit before stomata closed.

Rapid wilting of an abnormal diploid potato (*Solanum tuberosum*) mutant which preceded tip scorching and premature leaf fall was traceable to failure of stomatal closure. Water was lost more rapidly from excised leaves of the abnormal mutant than from excised leaves of normal plants. Furthermore, stomata of the abnormal mutant remained open even in wilted leaves (Waggoner and Simmonds, 1966). Excessive wilting of tomato (*Lycopersicon esculentum*) mutants also resulted from high transpirational water loss (Tal, 1966). Whereas mutants and normal plants had similar rates of cuticular transpiration the behavior of their stomata varied. The wilting tendency of the mutants resulted from high stomatal frequency, wide opening of stomata, and resistance to stomatal closing even in the dark. The greatest difference between mutant and normal plants occurred under conditions which caused stomatal closure in the latter group. This is illustrated in Table II, with the mutants having rela-

TABLE II

RATE OF TRANSPIRATIONAL WATER LOSS OF WILTY MUTANT AND NORMAL TOMATO PLANTS[a]

Variety	Water loss/gm leaf dry wt/hr (gm)		Water loss/cm² leaf area/hr (mg)	
	Day	Night	Day	Night
Rheinlands Ruhm (normal)	2.28	0.33	7.75	1.15
flacca (wilty mutant)	5.06	2.15	15.49	6.53
sitiens (wilty mutant)	6.57	3.92	18.47	10.89

[a] From Tal (1966).

tively much higher rates of transpiration during the night than during the day.

3. Aftereffects of Water Deficits on Stomatal Opening

Several investigators have noted that a period of drought, followed by rewatering, continued to inhibit the capacity of stomata for opening despite rapid recovery in leaf turgor. For example, leaf water deficits of 4–6% inhibited stomatal opening in *Rumex acetosa* for 8 hr after rewatering (Stålfelt, 1955). When *Taraxacum officinale* plants were wilted for several hours and then watered, complete stomatal opening did not occur until about 48 hr later (Heath and Mansfield, 1962). Such aftereffects of water deficits have also been reported for stomata of *Pelargonium* (Milthorpe and Spencer, 1957) and *Vicia faba* (Stålfelt, 1963). Following 2 to 4 days of wilting, *Nicotiana tabacum* plants did not show complete recovery of stomatal opening for 2 to 5 days after rewatering. In *Vicia faba* water deficits caused similar but less pronounced effects. In both species, the extent of the aftereffect was proportional to the leaf water deficit attained just before rewatering (Fischer *et al.,* 1970). The major part of the poststress damage was located in the guard cells and the minor part in the mesophyll. The part arising in the mesophyll appeared to be caused by elevated CO_2 concentrations in the stomatal cavity (Fischer, 1970). Although the causes of the aftereffects were not adequately explained, several hypotheses have been proposed for them including: (*a*) increased CO_2 concentration in intercellular spaces, (*b*) interference with energy supply for stomatal opening, (*c*) interference with membrane function, (*d*) shrinkage of cell walls, and (*e*) alteration of metabolic patterns (Fischer, 1970).

Severe drought frequently causes permanent damage, with some stomata opening slowly or not at all when plants are rewatered. For example, Iljin (1957) reported that when plants were rewatered following a prolonged drought the leaves regained turgor and appeared to be perfectly normal, yet up to 45% of the stomata failed to open. Glover (1959) found in East Africa that stomatal responses on irrigation following a drought varied, depending on duration of the drought. Stomata of maize plants subjected to drought for 3 or 4 days opened readily within a day or two of rewatering. However, if a drought lasted for a week or more the plants recovered turgidity and seemed normal after heavy watering but their stomata did not recover full turgor up to 10 days after irrigation. Yet the stomata of leaves which unfolded after the heavy watering responded normally. Thus, on the same maize plant there were two sets of leaves with varying stomatal behavior: those of predrought origin whose stomata were damaged and those of postdrought origin with normal stomata. By contrast, stomata of sorghum recovered readily from severe drought following irrigation and their increase in turgor followed closely the recovery of leaves from wilting. Glover (1959) suggested that the capacity of sorghum stomata to recover from severe drought accounted for its superiority as a grain crop in arid regions.

C. Flow of Oleoresins and Latex

The flow of oleoresins from injured pine trees is related to stem hydration and turgor of epithelial cells that line the resin ducts. When transpiration exceeds absorption of water, the resulting dehydration of trees reduces turgor of epithelial cells and lowers the rate of exudation from the interconnected capillary system.

Oleoresin exudation pressure (OEP) varies both seasonally and diurnally with atmospheric factors which control transpiration and soil moisture availability. Vité (1961) found a close correlation between internal water deficits and OEP of *Pinus ponderosa* trees. Diurnal fluctuations of OEP followed transpiration cycles. The pressure was highest at dawn and decreased as the trees were dehydrated during the day, only to increase again in the late afternoon and during the night. A minimum OEP of 4 to 8 atm usually occurred between 2 and 4 P.M. and a maximum of 8 to 12 atm prevailed between midnight and dawn. However OEP was modified by climatic factors (temperature, light intensity, humidity) which influenced transpiration. A decreasing OEP was reversed within 5 min by sprinkling the crown with water and within 20 min by reducing light intensity. Vité (1961) also found a gradual decline in OEP as the season progressed and soil moisture was depleted. Very low OEP's were found on hot and dry afternoons late in the season, especially in trees growing in shallow soil on exposed sites.

According to J. P. Barrett and Bengston (1964) daily OEP in *Pinus elliottii* var. *elliottii* trees was high when the vapor pressure gradient of air was low. However, as the gradient increased, transpiration increased causing internal water deficits and a decrease in OEP (Fig. 22). Vapor pressure gradients accounted for 83% of the variation in exudation pressure. Lorio and Hodges (1968) found diurnal patterns of OEP of *Pinus taeda* to be related to changes in soil moisture and atmospheric vapor pressure deficits. The relation with vapor pressure deficit was maximal when soil moisture deficit was high and minimal when it was low.

As laticifers of intact rubber-producing plants are under turgor pressure, latex flows when the trees are tapped. The rate of latex flow is influenced by turgor changes and is correlated with changes in environmental factors that control internal water balance of trees. Buttery and Boatman

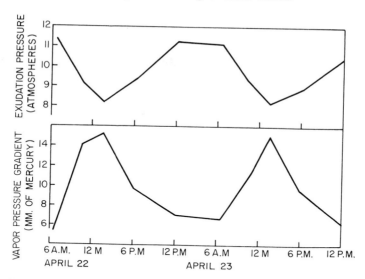

Fig. 22. Variation in oleoresin exudation pressure of *Pinus elliottii* and vapor pressure gradient over a 2-day period. From J. P. Barrett and Bengston (1964).

(1964) recorded diurnal variations in turgor pressure of laticiferous phloem tissues of *Hevea brasiliensis,* with lowest values occurring during the day. As may be seen in Fig. 23, on the first day, which was sunny, the pressure fell to a minimum by 3 P.M. and then rose to remain rather constant during the night. On the second day, the pressure declined until midday but increased again after a heavy rainstorm shortly after noon (12:20 P.M.). Buttery and Boatman (1964) concluded that restriction in turgor of the latex vessel system leading to decrease in latex flow most probably was brought about by loss of water to the xylem.

In another study Buttery and Boatman (1966) showed that early morning turgor pressures in laticiferous phloem tissues of *Hevea brasiliensis* were in the range of 7.9 to 15.0 atm, falling during the day and increasing at night. Such diurnal changes were positively correlated with relative humidity and inversely correlated with changes in temperature, evaporation, leaf water deficits, and stomatal opening (Fig. 24). The changes in turgor did not occur in defoliated trees. These observations indicated that daily loss in turgor resulted from high transpirational losses which caused movement of water out of phloem tissues (Buttery and Boatman, 1966).

The rate of latex flow from *Cryptostegia grandiflora* also has been shown to vary diurnally with changes in internal hydrostatic pressures. The rate of flow was lowest at midday and coincided with the time of greatest atmospheric vapor pressure deficit.

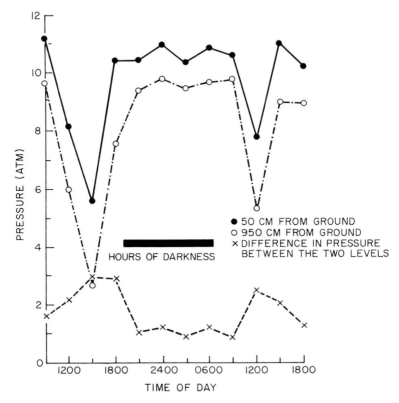

Fig. 23. Diurnal changes in hydrostatic pressure at two stem heights of a *Hevea brasiliensis* tree. From Buttery and Boatman (1964).

When transpiration was high during midday, the leaf blades of *Crypto-stegia* were curled outward and the leaves were partly wilted, even when soil moisture availability was high. This condition of internal water deficit was particularly noticeable when vapor pressure deficits exceeded 20 mm Hg. The rapid dehydration of the trees during midday caused withdrawal of water from the latex vessels, reduction in their turgor, and a decrease in the rate of latex flow (Curtis and Blondeau, 1946).

In British Honduras the stems of sapodilla (*Achras zapota*) trees showed rhythmic daily shrinkage and swelling, reaching a maximum diameter and turgidity at approximately 6:00 A.M. At that time flow of latex was at a daily maximum. However, before tapping and drainage were completed (requiring a period of up to 2 hr), the rate of latex flow declined as transpiration increased and turgor declined because of increased tem-

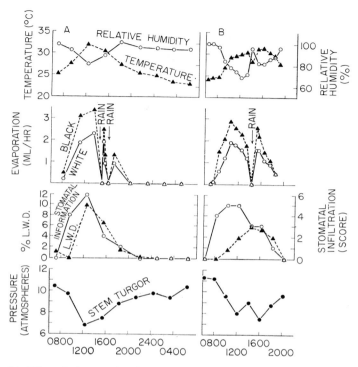

Fig. 24. Diurnal changes in air temperature and relative humidity, evaporation measured by black and white atmometers, stomatal aperture of *Hevea brasiliensis* by infiltration, leaf water deficit, and stem turgor. Observations were made on two different days (A and B). From Buttery and Boatman (1966).

perature and light intensity as well as decreased relative humidity. For these reasons Karling (1934) suggested that tapping of sapodilla trees at night, rather than early morning, would produce maximum yield of latex.

D. DISPERSAL OF SPORES, POLLEN, AND SEEDS

1. Spores

Water relations play an important role in spore discharge. Ingold (1939, 1953) described in detail a number of spore-discharge mechanisms which involved shrinking and swelling. A few types will be mentioned briefly.

Spores of some fungi are discharged by a water-squirting mechanism that depends on bursting by high turgor of a cell in which the wall is stretched to the breaking point. Bursting occurs along specialized regions of weakness in the cell wall. In certain Phycomycetes the spore or sporangium is located on top of the exploding cell. For example, dehiscense in *Pilobulus* results from swelling of mucilage which occurs in the sporangium below the spores. Following rupture of its wall the sporangium is shot away.

In Ascomycetes the elastic wall of an ascus is lined with a cytoplasmic layer which has the properties of a differentially permeable membrane. The thin and elastic cell wall is freely permeable to water and substances in solution. When the ascus becomes turgid it bursts and explosively scatters the spores. The pattern of bursting of asci is definite for various species. For example, in *Sordaria curvula* it occurs by separation of an apical cup; in Pezizales by a hinged lid; and in *Sphaerotheca* by an apical split. In powdery mildews only the spores may be ejected or the asci are first shot away and subsequently burst to scatter the spores (Ingold, 1939, 1953). The asci of some Pyrenomycetes are not explosive. Spore liberation occurs by oozing of a tendril-like thread of spores from the ostiole.

The aecidiospores of rust fungi occur in cups called aecidia. Each of the tightly packed spores has a polyhedral form because of the pressure of adjacent spores. Under damp conditions absorption of water causes turgor increases and a tendency for each spore to become spherical. As some spores in the outer layer of the spore mass suddenly round off, their adhesion to adjacent spores is broken and they are discharged into the air. Discharge of additional spores follows, with the supply of spores in the aecidium renewed from the base.

Characteristic of Basidiomycetes is a drop excretion mechanism, in which the spore excretes a drop of fluid just before it is shot away. Although there is some disagreement about the actual mechanism of spore discharge, Ingold (1939) suggested that the surface-tension energy of the excreted drop might be used to cause discharge.

Ingold (1939) also described an interesting catapult mechanism of spore discharge in *Sphaerobulus* (Gasteromycetes). As may be seen in Fig. 25A, the fruit body consists of two cups fitting one inside the other. Within the inner cup is the spore-containing gleba immersed in a liquid. Under conditions of a moist substratum and high humidity high turgor pressure develops in the palisade layer. The gleba is then projected outward for a considerable distance (up to 6 yd) by a sudden turning inside out of the inner cup (Fig. 25B).

In sphagnum moss the spores are discharged as a result of drying of the spore-containing capsule. As water evaporates from the epidermal cells of the capsule they shrink transversely, causing the rounded capsule to become cylindrical. As the volume of the capsule is reduced, the gas in the capsule is compressed. Eventually the increased air pressure causes an

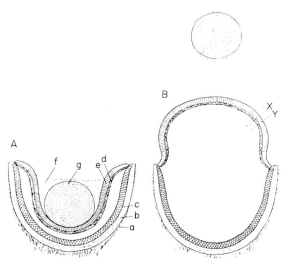

Fig. 25. Mechanism of dispersal of gleba of *Sphaerobolus stellatus:* a, layer of loose hyphae; b, gelatinous layer; c, pseudoparenchymatous layer; d, layer of fine tangential hyphae; e, palisade layer; f, lubricating liquid; g, gleba. A. Immediately before discharge. B. At the moment of discharge. For explanation see text. From Ingold (1939).

explosive dehiscence along the circular line of weakness, the annulus, and the dry powdery spores are discharged.

In ferns, some club mosses, and leafy liverworts a spring-type mechanism is involved in ejection of spores from a sporangium. Ingold (1939) described such a mechanism for the fern *Dryopteris,* in which about two thirds of the spore-containing sporangium is encircled by an annulus and the remaining third by a thin-walled stomium (Fig. 26). The annulus is made up of cells with thick inner and radial walls and thin outer and side walls. As water evaporates from the thin outer walls of the annulus cells, water in turn is imbibed by the walls from the cell interior, causing the cells to decrease in volume. This causes a break to occur between stomium cells and the annulus slowly bends back as water continues to evaporate from it. In each annulus cell there is a strain emanating from the tendency of the inner tangential wall to return to its original shape. Opposing this are forces resulting from cohesion of water in the cell and its adhesion to cell walls, resulting in tension on the water. As the tendency of the tangential wall of the annulus to unbend increases, the tension on the water is increased until the water ruptures and a gas bubble forms. The thick wall of the annulus cell returns to its original form and there is an increase in the volume of the gas phase. As such changes take place more or less

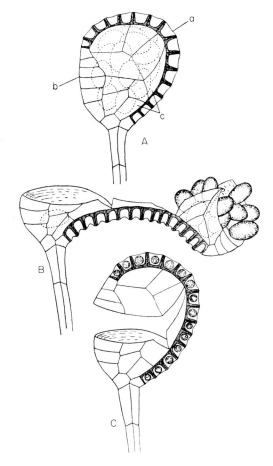

Fig. 26. Dehiscence of sporangium and spore discharge of *Dryopteris*. A. Mature sporangium: a, annulus; b, stomium; c, cells forming side of sporangium; outlines of the spores are shown by broken lines. B. The sporangium after dehiscence. The annulus is bending back and its cells are full of sap. C. The sporangium immediately after spore discharge. For further explanation see text. From Ingold (1939).

simultaneously in all annulus cells, the entire annulus ejects and scatters the spores that are contained in its free end (Ingold, 1939).

The spores of most liverworts are violently ejected from sporangia by elaters. During drying the sporangium splits and bends backward. The exposed elaters, which are reinforced by spiral bands of thickening, react like spiral springs which tighten on drying. When the tensile water inside ruptures, the elaters untwist rather violently and in so doing scatter spores into the air. In some mosses spores are ejected from the capsule by movements of peristome teeth in response to changes in humidity (Ingold, 1939).

2. Pollen

Dehiscence of anthers usually results from hygroscopic shrinkage. When dehiscence is longitudinal, drying causes borders of openings to retract. In poricidal anthers the pores form through localized disintegration of the cell wall and shrinkage of surrounding tissue. Repeated opening and closing of anther sacs often occur with changes in atmospheric humidity (Eames, 1961).

Humidity influences pollen dispersal by controlling evaporation of water from the exothecium cells surrounding the anthers, thereby affecting their opening. During periods of rain the opening of anthers normally stops. According to Sarvas (1962), the daily cycle of pollen dissemination by gymnosperms in Finland showed a maximum dispersal around midday and almost no dispersal at night. These fluctuations reflected changes in both humidity and temperature. During the period of anthesis the days in Finland are warm and the nights are very cool. Also the relative humidity drops below 50% during the day and approaches 100% during the night. On the few unusual days when relative humidity was low at night, considerable pollen was dispersed during the night, emphasizing the importance of dehydration of anthers.

3. Seeds

Various imbibition, cohesion, and turgor mechanisms play a role in seed dispersal and are associated with reversible hydration and dimensional changes in plant tissues.

a. *Imbibition Mechanism.* Shrinkage and swelling mechanisms involve opposing actions of cell walls. The differential shrinkage which plays a role in opening of many dry dehiscent fruits is related to the organization of the cell wall of constituent tissues. Most shrinkage occurs perpendicular to directions of orientation of the cellulose microfibrils which constitute the framework of the cell wall. If differences occur in orientation of microfibrils in different cells, or if the direction of cells varies, bending or torsion results. A few examples of the imbibition mechanism in seed dispersal will be given.

As a follicle dries out it splits along the line of marginal fusion of the carpel. Legume fruits, which split into two valves, vary with respect to relative thickness of pericarp tissues, cell structure, orientation of elements, and fine structure. The two valves of a dried legume usually twist as a result of shrinkage of thick walls of pericarp cells. Shrinkage is greatest at right angles to the direction of microfibril orientation. Tension develops and causes the valves to twist after the forces which account for adhesion in the abscission zone are overcome. The legume then splits and the seeds are ejected. Various modifications and deviations from the above pattern of legume dehiscence are discussed by Fahn (1967).

In capsules splitting usually occurs from the apex downward in one of several patterns; e.g., loculicidal (along the dorsal bundle of each carpel); septicidal (between the carpels); septifragal (outer wall of the fruit separates from the septa which remain attached to the axis); porous (by small pores in the pericarp); circumscissile (by transverse splitting) results in the formation of a lid; and valvate (by outwardly flared teeth).

Dehiscence in some capsules is caused by swelling of cell walls. As epidermal cells are relatively thick-walled they shrink less than the parenchymatous cells below them. The abscission tissue of capsules is developed between the teeth or valves. According to Hofmann (1931), dehiscence of the fruit of *Aspidosperma megalocarpon* is caused by stresses set up in the wall of the fruit where bands of parenchyma alternate with bands of sclerenchyma tissue. The parenchyma, with a higher water content than the sclerenchyma, shrinks more on drying. The resulting tension causes the 3-carpellary fruit to split lengthwise down the sides from apex to base.

In fruits which exhibit circumscissile dehiscence, the lid and the base of the mature fruit are more or less rigid because of production of sclerenchyma cells. A zone of mechanical weakness is formed between the lid and the base because of alignment of the cells in this zone, size of the cells, number of cells, thickness of cell walls, meristematic state of cells, and various combinations of these. The splitting force appears to be due to development of seeds that completely fill the cavity of the ovary and maintain their mature size while the wall of the fruit shrinks on drying (Rethke, 1946).

Swelling mechanisms involving extrafloral organs also are involved in seed dispersal. These include, for example, floral pedicels (*Salvia*), axes of inflorescences (*Plantago*), and whole branches (*Anastatica*). An imbibition mechanism is also involved in hygroscopic movements in bristles, involucre bracts, calyxes, and awns. In some Gramineae changes in moisture content cause the spiral portion of the propagule to twist and untwist (while the upper end is held in place by pressure against the soil) and it is thereby driven into the soil (Fahn, 1967).

Several investigators have demonstrated progressive dehydration and shrinkage of cones of gymnosperms prior to their opening. For example, cones of *Picea mariana, P. glauca,* and *Larix laricina* dehydrated from about 400% of dry weight in midseason to less than 40% in September. Although changes in percent moisture of cones early in the season were influenced by both translocation of water into and out of cones as well as by dry weight increment of cones, the decline in percentage moisture was primarily a dehydration phenomenon (Clausen and Kozlowski, 1965). Beaufait (1960) reported that moisture content of full-sized green cone scales of *Pinus banksiana* varied from 250 to 350%. As the cones matured and turned brown their moisture content decreased rapidly to about 12–15%. Similarly Ching and Ching (1962) recorded marked dehydration of *Pseudotsuga menziesii* cones to a low moisture content of 16% in early September when the cones were opening. Rapid dehydration of maturing *Picea glauca* and *Abies grandis* cones also occurred as they approached maturity (Cram and Worden, 1957; Pfister, 1967).

As shown by Harlow *et al.* (1964) and Allen and Wardrop (1964), the movement of the cone scale away from the cone axis on drying is related to the greater shrinkage of ventral (abaxial) tissues than dorsal (adaxial) tissues within the cone scale (Fig. 27). As expected from the extreme dimensional changes along the axis of the cells which make up the ventral (abaxial) surfaces of cone scales (Table III), electron micrographs showed that the walls of cells were composed of microfibrils oriented at about 90° to the long axis of the cell.

The opening and shedding of female cones of *Pinus radiata* were preceded by severance of the vascular connection between the cone and branch by occlusion of tracheid lumens with resin. Inasmuch as cones opened when the moisture content was low (usually below 20%) there appeared to be a progressive buildup of stress in the scales because of differential shrinkage between adaxial vascular tissue (shrinkage of 1.0–1.5%) and abaxial sclerenchyma tissue (shrinkage of about 15%). Opening of cone scales may occur suddenly, reflecting lateral cohesion of the scales on drying. Actual opening of cones occurs when the stresses in the scales exceed cohesive forces between them. After cones have opened they may close and open again as air humidity changes (Fielding, 1947; Allen and Wardrop, 1964).

b. Cohesion Mechanism. This mechanism necessitates water in cell lumens. Cells of cohesion tissue, which generally are elongated and thin-walled, have their longitudinal axis oriented at right angles to shrinkage tissue. Cell volume decreases because of water evaporation as well as cohesion of water molecules and their adhesion to cell walls. The contracting tissue may draw with it attached organs or may cause breakage at abscission zones. Cohesion tissues are well known in many Compositae and some Umbelliferae. Often cohesion mechanisms operate together with imbibition mechanisms (Zohary and Fahn, 1941).

c. Turgor Mechanism. Living cells are involved in the turgor mechanism. Elastic tissue is stretched by another tissue having high turgor or a tissue

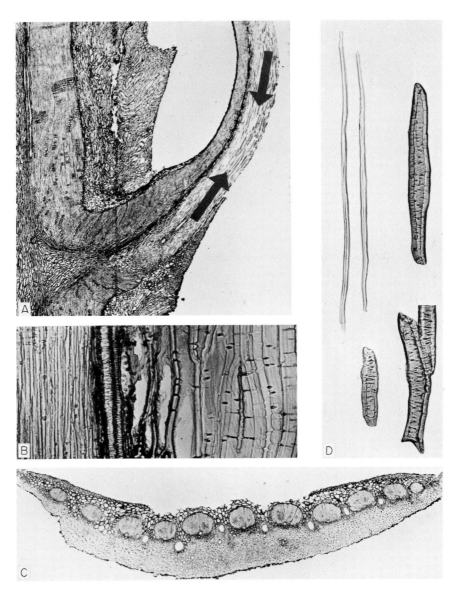

Fig. 27. Structure of cone scales in relation to cone opening on drying. A. Longitudinal section of a cone scale of *Pinus* and part of the cone axis. Magnification: ×18. B. Enlarged section of area between the two arrows in A. Magnification: ×270. C. Cross section of cone scales. Magnification: ×27. D. Two tracheids (left) from one of the woody strands and several of the thick-walled cells (right and below) from the zone of high-dimensional changes. Magnification: ×150. From Harlow *et al.* (1964).

TABLE III

LENGTHWISE SWELLING OF VENTRAL SURFACES OF PINE CONE
SCALES AFTER 24-HR IMMERSION IN WATER[a]

Species	Scale near top of cone (mm)			Scale near base of cone (mm)		
	Dry	Wet	%[b]	Dry	Wet	%[b]
Pinus strobus	1.59	1.79	12.6	1.35	1.50	11.1
Pinus flexilis	1.10	1.50	36.4	1.00	1.32	32.0
Pinus ponderosa	2.00	2.32	16.0	2.04	2.32	13.9
Pinus palustris	3.50	4.06	16.0	3.49	3.85	10.3
Pinus sabiniana	2.30	2.93	27.4	1.91	2.40	25.6

[a] From Harlow *et al.* (1964).
[b] Percent based on air dry length.

having high turgor is stretched when set against a resisting tissue. Since great tensions are involved, the fruit opens along an abscission line when tension exceeds a critical value. As the stretched tissue contracts, the seeds are ejected with appreciable force, often for considerable distances. Such a turgor mechanism is involved in seed dispersal of fleshy fruits of *Ecballium elaterium, Impatiens parviflora, Cardamine impatiens, Lathraea clandestina, Dorstenia contrajerva, Oxalis acetosella,* and *Biophytum.* For excellent discussions of anatomical considerations in dispersal of reproductive tissues by imbibition, cohesion, and turgor mechanisms the reader is referred to Fahn (1967), and Fahn and Werker (1972).

E. ROOT FUNCTIONS

Diurnal variations in root diameters have important implications in mathematical analyses of root functions. As Huck *et al.* (1970) pointed out, models of transport of various materials from the soil to the root interior generally have been based on two important assumptions: (1) that a root has a constant diameter, and (2) that transfer coefficients remain constant with time. Huck *et al.* (1970) concluded that the first of these assumptions is erroneous and the second probably is also. It is likely that transfer coefficients do not remain constant inasmuch as transport properties are related to turgor. When roots shrink the contact between the root and soil solution decreases. Since ion transfer to roots occurs only in contact areas it must also be decreased. At midday, water transport to roots is impeded because a high proportion of it must move across vapor gaps. Furthermore, variations in root diameter alter vertical translocation in roots because of changes in cross-sectional areas available for transport.

F. Errors in Growth Measurements

Because seasonal and diurnal shrinking and swelling of tree stems and branches are superimposed on irreversible diameter changes resulting from cambial growth, care must be taken in estimating cambial growth by measurement of changes in girth over short-time intervals (Kozlowski, 1971b).

Kozlowski (1967a,b) noted that during soil-drying cycles stem diameters of *Acer negundo* seedlings decreased, gradually at first and then more abruptly as soil moisture content approached the wilting percentage. Irrigation following a drought caused rapid rehydration and swelling of stems (Fig. 28). Tree seedlings often shrank so much during a soil-drying cycle that no reliable estimates could be made of diameter increase resulting from cambial growth.

Measurements of radial change of stems usually show much more linear variation due to hydration changes than to increment resulting from cambial growth. For example, the amount of stem shrinkage of *Pinus canariensis* stems in Australia during 1 day approximately equaled the amount of cambial growth increment of 5 days (Holmes and Shim, 1968) (Fig. 29). On clear days shrinkage of *Picea* stems from sunrise to early afternoon approximated a week's cambial growth increment (Kern, 1961). In arid regions the amounts of stem shrinkage in one day may sometimes exceed radial increase from cambial growth over a period of weeks.

Because of large diurnal changes in stem diameters there is danger in estimating cambial growth by taking measurements at different times of days with instruments such as dendrometers which do not provide a continuous record of shrinking and swelling. On the other hand, instruments

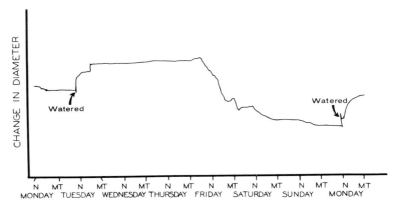

Fig. 28. Shrinkage of stem of potted 4-yr-old *Acer negundo* seedling during a soil-drying cycle, followed by expansion of stem when soil was watered.

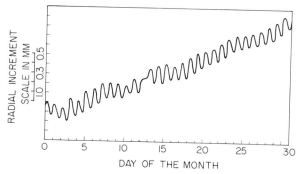

Fig. 29. Diurnal variations superimposed on cumulative radial increment of a *Pinus canariensis* tree. From Holmes and Shim (1968).

such as dendrographs (Fritts and Fritts, 1955) provide some insight, on a day-to-day basis, of the amount of shrinkage and expansion of tree stems. The trend of cambial growth increment can be estimated over a period of several days by connecting daily peaks or valleys of dendrograph traces provided the amplitude of the curve does not change appreciably from day to day. However, during periods of variable weather the ratio of hydration change to total radial change may vary greatly from day to day, making it difficult to estimate cambial growth increment. The hydration component of radial changes is greater over short periods of time than over long ones. For example, whereas total radial change over a 2-yr period may largely represent the increment traceable to cambial growth, radial change for a few-day period may be caused almost entirely by reversible hydration changes.

Many investigators have tried to determine the seasonal duration of cambial growth by measuring radial increase in stems with dendrometers or dendrographs. This is difficult because when cambial growth slows toward the end of the growing season, radial changes often continue as a result of hydration changes. Fielding and Millet (1941) emphasized that diameters of *Pinus radiata* stems were constantly shrinking and swelling and it was therefore difficult to determine with dendrometers exactly when cambial activity began or ended.

G. Cracking and Splitting of Plants

Cracking of plant tissues can sometimes occur when turgor is either very high or very low. For example, when environmental conditions (high availability of soil moisture, high relative humidity) promote unusually high turgor in plants, splitting of fruits may occur. Such injury has been described in fruits of avocado, prunes, oranges, lemons, cherries, grapes, and apples. For a more detailed discussion of splitting and cracking of reproductive tissues as a result of high turgor the reader is referred to Chapter 3 of this volume.

Because various tissues within a plant have different moduli of elasticity, cracking may occur as a result of severe desiccation. For example, tree stems sometimes crack during droughts. Such drought cracks usually are aligned more or less vertically, or they may spiral. They may be short or long and sometimes extend for much of the length of the stem. Drought cracks in tree stems may be superficial or deep. In *Abies* they extended only through the bark, and in *Picea* to the pith (Schädelin, 1942).

Day (1954) described development of stem cracks, during a very dry year, in several species of gymnosperms in England (including *Picea sitchensis, Picea abies, Abies grandis, Abies procera, Tsuga heterophylla, Pseudotsuga menziesii, Larix leptolepis,* and *Larix decidua*). The vertically aligned or spiraling stem cracks, which varied in length from 1 to 10 or more ft, formed consistently within the outer portion of an annual ring and apparently always in the sapwood. The splitting resulted in collapse of unlignified tracheids on each side of the crack. From these collapsed tracheids there sometimes extended other tracheids which did not thicken normally. The tracheids in the zone of failure were abnormally shaped, with curved walls which tended to close together, apparently from severe dehydration. The rays also were distorted. In another study Day (1964) concluded that development of flutes or hollows in tree stems, dying of bark in vertical strips, and longitudinal cracking of bark resulted from internal water deficits of varying degrees. Flutes, which reflected arrested cambial activity, developed as a result of a continuous water deficit which did not become acute. When water deficits were severe, both strip necrosis and splitting of bark followed.

D. K. Barrett (1958) found that over a 48-yr period many cracks had developed in the stem of a 70-yr-old *Abies procera* tree. The cracks were located at or near the extreme parts of annual xylem increments and formed during periods of very dry weather and high temperature.

Lutz (1952) described clefts and distortion of elements in the wood of *Picea glauca* trees in Alaska that were caused by severe internal water deficits. The clefts appeared as diamond-shaped openings within an annual ring. The collapse of wood was attributed to an imbalance between transpiration and absorption of water. Lutz postulated that during late autumn, winter, and early spring, the absorption of water through roots was essentially prevented by frozen soil and low temperatures. As transpirational losses were appreciable, internal water deficits developed rapidly and tremendous tensions developed. The structure of the wood in the damaged trees contributed to the injury. Growth was rapid, earlywood was thin walled, and the latewood zone was poorly developed.

Drought cracks in tree stems may or may not heal. In *Pinus radiata* stems, the space left by drought cracks was filled with cellular tissue. In *Pseudotsuga* it was filled with resin, and in *Picea glauca* the space did not fill (Amos, 1954). Following a prolonged drought in 1952–1953 cracking was observed in stems of plantation-grown *Cupressus lusitanica* trees in the Sudan. The cracks occurred to a depth of 2–3 in. in the wood, usually beginning at the stem base and often extending to a height of 20 ft or more. In some cases the wounds healed; in others the cracks remained open and decay fungi and borers invaded trees through them. The injury occurred because most of the roots of *C. lusitanica* trees ran parallel to the surface in the top 12 in. of soil. When the surface-soil layers dried out, absorption of water was not adequate to meet transpiration demands and severe internal water deficits developed (Anonymous, 1954).

Drought cracks in trees sometimes are mistaken for frost cracks. However, as drought cracks tend to be wider in the middle than at the ends they do not heal as readily as frost cracks (Wartenberg, 1933). Drought cracks may also be mistaken for lightning cracks but the latter generally are longer and may extend the full length of the stem (Parker, 1965).

REFERENCES

Allen, R., and Wardrop, A. B. (1964). The opening and shedding mechanism of the female cones of *Pinus radiata*. *Aust. J. Bot.* **12**, 125.

Amos, G. L. (1954). Radial fissures in the early wood of conifers. *Aust. J. Bot.* **2**, 22.

Anderson, D. B., and Kerr, T. (1943). A note on the growth behavior of cotton bolls. *Plant Physiol.* **18**, 261.

Anonymous. (1954) Cupressus stem crack. *Sudan. Min. Agr. Forest. Dep., Rep.* p. 43.

Bachmann, F. (1922). Studien über die Dickenänderungen von Laubblättern. *Jahrb. Wiss. Bot.* **61**, 372.

Barrett, D. K. (1958). Cracking in the main stem of Noble Fir, Lethan, Nairnshire. *Scot. Forest.* **12**, 187.

Barrett, J. P., and Bengston, G. W. (1964). Oleoresin yields for slash pines from seven seed sources. *Forest Sci.* **10**, 160.

Barrs, H. D. (1968). Determinations of water deficits in plant tissues. *In* "Water Deficits and Plant Growth" (T. T. Kozlowski, ed.), Vol. I, Chapter 8, p. 235. Academic Press, New York.

Barrs, H. D., and Klepper, B. (1968). Cyclic variations in plant properties under constant environmental conditions. *Physiol. Plant.* **21**, 711.

Bartholomew, E. T. (1926). Internal decline of lemons. III. Water deficit in lemon fruits caused by excessive leaf evaporation. *Amer. J. Bot.* **13**, 102.

Beaufait, W. R. (1960). Some effects of high temperatures on the cones and seeds of jack pine. *Forest Sci.* **6**, 194.

Beeman, J. F. (1966). Growth dynamics of small tobacco plants. Ph.D. Thesis. North Carolina State Univ., Raleigh, North Carolina.

Begg, J. E., Bierhuizen, J. F., Lemon, E. R., Misra, D. K., Slatyer, R. O., and Stern,

W. R. (1964). Diurnal energy and water exchange in bulrush millet in an area of high solar radiation. *Agr. Meteorol.* **1**, 294.

Bormann, F. H., and Kozlowski, T. T. (1962). Measurements of tree ring growth with dial gage dendrometers and vernier tree bands. *Ecology* **43**, 289.

Brun, W. A. (1965). Rapid changes in transpiration in banana leaves. *Plant Physiol.* **40**, 797.

Buell, M. F., Buell, H. F., Small, J. A., and Monk, C. D. (1961). Drought effect on radial growth of trees in the William L. Hutcheson Memorial Forest. *Bull. Torrey Bot. Club* **88**, 176.

Buttery, B. R., and Boatman, S. G. (1964). Turgor pressures in phloem: Measurements on *Hevea* latex. *Science* **145**, 285.

Buttery, B. R., and Boatman, S. G. (1966). Manometric measurements of turgor pressures in laticiferous phloem tissues. *J. Exp. Bot.* **17**, 283.

Byram, G. M., and Doolittle, W. T. (1950). A year of growth for a shortleaf pine. *Ecology* **31**, 27.

Cătský, J. (1965). Water saturation deficit and photosynthetic rate as related to leaf age in the wilting plant. *In* "Water Stress in Plants" (B. Slavik, ed.), pp. 203–208. Junk Publ., The Hague.

Chaney, W. R. (1970). A device for continuous monitoring of changes in leaf thickness. *Forest Sci.* **16**, 56.

Chaney, W. R., and Kozlowski, T. T. (1969a). Seasonal and diurnal changes in water balance of fruits, cones, and leaves of forest trees. *Can. J. Bot.* **47**, 1407.

Chaney, W. R., and Kozlowski, T. T. (1969b). Seasonal and diurnal expansion and contraction of *Pinus banksiana* and *Picea glauca* cones. *New Phytol.* **68**, 873.

Chaney, W. R., and Kozlowski, T. T. (1969c). Diurnal expansion and contraction of leaves and fruits of English Morello cherry. *Ann. Bot. (London)* [N. S.] **33**, 991.

Chaney, W. R., and Kozlowski, T. T. (1971). Water transport in relation to expansion and contraction of leaves and fruits of Calamondin orange. *J. Hort. Sci.* **46**, 71.

Ching, T. M., and Ching, K. K. (1962). Physical and physiological changes in maturing Douglas-fir cones and seed. *Forest Sci.* **8**, 21.

Clausen, J. J., and Kozlowski, T. T. (1965). Seasonal changes in moisture content of gymnosperm cones. *Nature (London)* **206**, 112.

Crafts, A. S. (1968). Water deficits and physiological processes. *In* "Water Deficits and Plant Growth" (T. T. Kozlowski, ed.), Vol II, Chapter 3, p. 85. Academic Press, New York.

Cram, W. H. (1956). Maturity of Colorado spruce cones. *Forest Sci.* **2**, 26.

Cram, W. H., and Worden, H. H. (1957). Maturity of white spruce cones and seed. *Forest Sci.* **3**, 263.

Curtis, J., and Blondeau, R. (1946). Influence of time of day on latex flow from *Cryptostegia grandiflora*. *Amer. J. Bot.* **33**, 264.

Dawkins, H. C. (1956). Rapid detection of aberrant girth increment. *Emp. Forest. Rev.* **35**, 3.

Day, W. R. (1954). Drought crack of conifers. *For. Rec. Forest. Comm., London* **26**, 1–40.

Day, W. R. (1964). The development of flutes or hollows on main stems of trees and its relation to bark splitting and strip necrosis. *Forestry* **37**, 145.

Dickmann, D. I., and Kozlowski, T. T. (1969a). Seasonal growth patterns of ovulate strobili of *Pinus resinosa* in central Wisconsin. *Can. J. Bot.* **47**, 839.

Dickmann, D. I., and Kozlowski, T. T. (1969b). Seasonal variations in reserve and structural components of *Pinus resinosa* Ait. cones. *Amer. J. Bot.* **56**, 515.

Doley, D. (1967). Water relations of *Eucalyptus marginata* SM. under natural conditions. *J. Ecol.* **55**, 597.

Eames, A. J. (1961). "Morphology of the Angiosperms." McGraw-Hill, New York.

Ehlig, C. F., and Gardner, W. R. (1964). Relationship between transpiration and the internal water relations of plants. *Agron. J.* **56**, 127.

Fahn, A. (1967). "Plant Anatomy." Pergamon Press, Oxford.

Fahn, A., and Werker, E. (1972). Anatomical mechanisms of seed dispersal. *In* "Seed Biology" (T. T. Kozlowski, ed.), Vol. I, Chapter 4. Academic Press, New York.

Farrar, J. L., and Zichmanis, H. (1960). The accurate measurement of diameter changes in small stems. *Annu. Ring* p. 16.

Fielding, J. M. (1947). The seeding and natural regeneration of Monterey pine in South Australia. *Bull. Comm. For. Timber Bur. Aust.* No. 29.

Fielding, J. M. (1955). The seasonal and daily elongation of the shoots of Monterey pine and the daily elongation of the roots. *Leafl. Forest Bur., Aust.* **75**, 1.

Fielding, J. M., and Millet, M. R. O. (1941). Some studies of the growth of Monterey pine (*Pinus radiata*). I. Diameter growth. *Bull. Forest Bur., Aust.* **27**, 1.

Fischer, R. A. (1970). After-effect of water stress on stomatal opening. II. Possible causes. *J. Exp. Bot.* **21**, 386.

Fischer, R. A., Hsiao, T. C., and Hagan, R. M. (1970). After-effect of water stress on stomatal opening. I. Techniques and magnitudes. *J. Exp. Bot.* **21**, 371.

Fritts, H. C. (1958). An analysis of radial growth of beech in a central Ohio forest during 1954–1955. *Ecology* **39**, 705.

Fritts, H. C. (1969). Bristlecone pine in the White Mountains of California. *Pap. Lab. Tree-Ring Res.* No. 4.

Fritts, H. C., and Fritts, E. C. (1955). A new dendrograph for recording radial changes of a tree. *Forest Sci.* **1**, 271.

Fritts, H. C., Smith, D. G., and Stokes, M. A. (1965). The biological model for paleoclimatic interpretation of Mesa Verde tree-ring series. *Amer. Antiquity* **31**, 101.

Gardner, W. R. (1965). Dynamic aspects of water availability to plants. *Soil Sci.* **89**, 63.

Gardner, W. R. (1971). Internal water status and response in relation to the external water regime. *Proc. UNESCO Symp. Plant Response Climatic Factors, 1970* (in press).

Gardner, W. R., and Ehlig, C. F. (1965). Physical aspects of the internal water relations of plant leaves. *Plant Physiol.* **40**, 705.

Gardner, W. R., and Nieman, R. H. (1964). Lower limit of water availability to plants. *Science* **143**, 1460.

Glover, J. (1959). The apparent behaviour of maize and sorghum stomata during and after drought. *J. Agr. Sci.* **53**, 412.

Grieve, B. J. (1956). Studies in the water relations of plants. I. Transpiration of western Australian (Swan Plain) sclerophylls. *J. Proc. Roy. Soc. West. Aust.* **40**, 15.

Hall, R. C. (1944). A vernier tree growth band. *J. Forest.* **42**, 742.

Harley, C. P., and Masure, M. P. (1938). Relation of atmospheric conditions to enlargement rate and periodicity of Winesap apples. *J. Agr. Res.* **57**, 109.

Harlow, W. M., Coté, W. A., Jr., and Day, A. C. (1964). The opening mechanism of pine cones scales. *J. Forest.* **62**, 538.

Hawkins, G. W., Jr. (1965). Growth chamber and simulation studies of the effect of soil moisture tension on the growth rate of Virginia type tobacco. M. Sc. Thesis, North Carolina State Univ., Raleigh, North Carolina.

Heath, O. V. S. (1938). An experimental investigation of the mechanism of stomatal movement, with some preliminary observations upon the response of the guard cells to 'shock.' *New Phytol.* **37**, 385.

Heath, O. V. S., and Mansfield, T. A. (1962). A recording porometer with detachable cups operating on four separate leaves. *Proc. Roy. Soc., Ser. B* **156**, 1.

Heath, O. V. S., and Mansfield, T. A. (1969). The movements of stomata. *In* "The Physiology of Growth and Development" (M. B. Wilkins, ed.), Chapter 9, p. 303. McGraw-Hill, New York.

Henrici, M. (1946). Transpiration of South African plant associations. Pt. II. *Sci. Bull. Dept. Agr. For. Union S. Afr.* p. 247.

Hofmann, E. (1931). Fruit of *Aspidospermum megalocarpon* and its dehiscence mechanism. *Sitzungsber. Akad. Wiss. Wien, Math.-Naturwiss. Kl., Abt. 1: Min., Biol., Erdk.* **40**(1/2), 83.

Holmes, J. W., and Shim, S. Y. (1968). Diurnal changes in stem diameter of Canary Island pine trees (*Pinus canariensis,* C. Smith) caused by soil water stress and varying microclimate. *J. Exp. Bot.* **19**, 219.

Huck, M. G., Klepper, B., and Taylor, H. M. (1970). Diurnal variations in root diameter. *Plant Physiol.* **45**, 529.

Hunter, C., and Rich, E. M. (1923). An apparatus for the measurement of stem elongation. *New Phytol.* **22**, 44.

Idle, D. B. (1956). Studies in extension growth—a new contact auxanometer. *J. Exp. Bot.* **7**, 347.

Iljin, W. S. (1957). Drought resistance in plants and physiological processes. *Annu. Rev. Plant Physiol.* **8**, 257.

Impens, I. I., and Schalk, J. M. (1965). A very sensitive dendrograph for recording radial changes of a tree. *Ecology* **46**, 183.

Ingold, C. T. (1939). "Spore Discharge in Land Plants." Oxford Univ. Press, London and New York.

Ingold, C. T. (1953). "Dispersal in Fungi." Clarendon, Oxford.

Jarvis, P. G., and Slatyer, R. O. (1966). Calibration of β gauges for determining leaf water status. *Science* **153**, 78.

Kanemasu, E. T., and Tanner, C. B. (1969). Stomatal diffusion resistance of snap beans. I. Influence of leaf-water potential. *Plant Physiol.* **44**, 1547.

Karling, J. S. (1934). Dendrograph studies on *Achras zapota* in relation to the optimum conditions for tapping. *Amer. J. Bot.* **21**, 161.

Kaufmann, M. R. (1970). Water potential components in growing citrus fruits. *Plant Physiol.* **46**, 145.

Kaul, O. N., and Kramer, P. J. (1965). Comparative drought tolerance of two woody species. *Indian Forest.* **91**, 462.

Kern, K. G. (1961). Ein Beitrag zur Zuwachsmessung mit den Arnberg-Mikro-dendrometer. *Allg. Forst-Jagdztg.* **132**, 206.

Klepper, B. (1968). Diurnal pattern of water potential in woody plants. *Plant Physiol.* **43**, 1931.

Klueter, H. H., Downs, R. J., Bailey, W. A., and Krizek, D. T. (1966). Motion detector. *Amer. Soc. Agr. Eng.* **66-836**.

Kozlowski, T. T. (1955). Tree growth, action and interaction of soil and other factors. *J. Forest.* **53**, 508.

Kozlowski, T. T. (1958). Water relations and growth of trees. *J. Forest.* **56**, 498.

Kozlowski, T. T. (1961). The movement of water in trees. *Forest Sci.* **7**, 177.

Kozlowski, T. T., ed. (1962). "Tree Growth." Ronald Press, New York.

Kozlowski, T. T. (1963). Growth characteristics of forest trees. *J. Forest.* **61**, 655.

Kozlowski, T. T. (1964). "Water Metabolism in Plants." Harper, New York.

Kozlowski, T. T. (1965). Expansion and contraction of plants. *Advan. Front. Plant Sci.* **10**, 63.

Kozlowski, T. T. (1967a). Diurnal variation in stem diameters of small trees. *Bot. Gaz.* **128**, 60.

Kozlowski, T. T. (1967b). Continuous recording of diameter changes in seedlings. *Forest Sci.* **13**, 100.

Kozlowski, T. T. (1967c). Water relations of trees. *Midwest. Chapt. Int. Shade Tree Conf. Proc.*, **22**, 34.

Kozlowski, T. T. (1968a). Water balance in shade trees. *Int. Shade Tree Conf., Proc.* **44**, 29.

Kozlowski, T. T. (1968b). Introduction. *In* "Water Deficits and Plant Growth" (T. T. Kozlowski, ed.), Vol. I, Chapter 1, p. 1. Academic Press, New York.

Kozlowski, T. T. (1968c). Diurnal changes in diameters of fruits and tree stems of Montmorency cherry. *J. Hort. Sci.* **43**, 1.

Kozlowski, T. T. (1969). Soil water and tree growth. *In* "The Ecology of Southern Forests" (N. E. Linnartz, ed.), pp. 30–57. Louisiana State Univ. Press, Baton Rouge.

Kozlowski, T. T. (1970). Physiological implications in afforestation. *Proc. World Forest. Congr., 6th, 1966* Vol. 2, p. 1304.

Kozlowski, T. T. (1971a). "Growth and Development of Trees," Vol. I. Academic Press, New York.

Kozlowski, T. T. (1971b). "Growth and Development of Trees," Vol. II. Academic Press, New York.

Kozlowski, T. T., and Keller, T. (1966). Food relations of woody plants. *Bot. Rev.* **32**, 293.

Kozlowski, T. T., and Winget, C. H. (1964). Diurnal and seasonal variation in radii of tree stems. *Ecology* **45**, 149.

Kozlowski, T. T., Winget, C. H., and Torrie, J. H. (1962). Daily radial growth of oak in relation to maximum and minimum temperature. *Bot. Gaz.* **124**, 9.

Kramer, P. J. (1969). "Plant and Soil Water Relationships: A Modern Synthesis." McGraw-Hill, New York.

Kramer, P. J., and Kozlowski, T. T. (1960). "Physiology of Trees." McGraw-Hill, New York.

Leikola, M. (1969a). The effect of some climatic factors on the daily hydrostatic variations in stem thickness of Scots pine. *Ann. Bot. Fenn.* **6**, 173.

Leikola, M. (1969b). The influence of environmental factors on the diameter growth of forest trees. Auxanometric study. *Acta Forest. Fenn.* **92**, 9.

Lorio, P. L., and Hodges, J. D. (1968). Oleoresin exudation pressure and relative water content of inner bark as indicators of moisture stress in loblolly pines. *Forest Sci.* **14**, 392.

Lutz, H. J. (1952). Occurrence of clefts in the wood of living white spruce in Alaska. *J. Forest.* **50**, 99.

McCracken, I. J., and Kozlowski, T. T. (1965). Thermal contraction in twigs. *Nature (London)* **208**, 910.

McCully, W. G. (1952). Measuring diameter changes in small stems. *Ecology* **33**, 300.

MacDougal, D. T. (1920). Hydration and growth. *Carnegie Inst. Wash. Publ.* **297**.

MacDougal, D. T. (1924). Growth in trees and massive organs of plants. *Carnegie Inst. Wash. Publ.* **350**, 50.

MacDougal, D. T. (1936). Studies in tree growth by the dendrographic method. *Carnegie Inst. Wash. Publ.* **462**.

Magness, J. R., Degman, E. S., and Furr, J. R. (1935). Soil moisture and irrigation investigations in eastern apple orchards. *U. S. Dep. Agr., Tech. Bull.* **491**.

Maki, T. E. (1940). Significance and applicability of seed maturity indices for ponderosa pine. *J. Forest.* **38**, 55.

Marvin, J. W. (1949). Changes in bark thickness during sap flow in sugar maple. *Science* **109**, 231.

Mederski, H. J. (1961). Determination of internal water status of plants by beta ray gauging. *Soil Sci.* **92**, 143.

Mederski, H. J. (1964). Plant water balance determination by beta gauging technique. *Agron. Abstr.* **56**, 126.

Mederski, H. J., and Alles, W. (1968). Beta gauging leaf water status: Influence of changing leaf characteristics. *Plant Physiol.* **43**, 470.

Meidner, H. (1952). An instrument for the continuous determination of leaf thickness changes in the field. *J. Exp. Bot.* **3**, 319.

Miller, E. C. (1931). "Plant Physiology." McGraw-Hill, New York.

Milthorpe, F. L., and Spencer, E. J. (1957). Experimental studies of the factors controlling transpiration. III. The interrelationships between transpiration rate, stomatal movement, and leaf water content. *J. Exp. Bot.* **8**, 414.

Nakayama, F. S., and Ehrler, W. L. (1964). Beta ray gauging technique for measuring leaf water content changes and moisture status of plants. *Plant Physiol.* **39**, 95.

Namken, L. N., Bartholic, J. F., and Runkles, J. R. (1969). Monitoring cotton plant stem radius as an indication of water stress. *Agron. J.* **61**, 891.

Nunes, M. A. (1967). A comparative study of drought resistance in cacao plants. *Ann. Bot. (London)* [N. S.] **31**, 189.

Ogigirigi, M. A., Kozlowski, T. T., and Sasaki, S. (1970). Effect of soil drying on stem shrinkage and photosynthesis of tree seedlings. *Plant Soil* **32**, 33.

Oppenheimer, H. R. (1953). An experimental study on ecological relationships and water expenses of Mediterranean forest vegetation. *Palestine J. Bot., Rehovot Ser.* **8**, 103.

Ordin, L. (1958). The effect of water stress on the cell wall metabolism of plant tissue. *Radioisotop. Sci. Res., Proc. Int. Conf., 1957* Vol. 4, pp. 553–564.

Ordin, L. (1960). Effect of water stress on cell wall metabolism of *Avena* coleoptile tissue. *Plant Physiol.* **35**, 443.

Pallas, J. E. (1964). Guard cell starch retention and accumulation in the dark. *Bot. Gaz.* **125**, 102.

Pallas, J. E. (1966). Mechanisms of guard cell action. *Quart. Rev. Biol.* **41**, 365.

Parker, J. (1952). Desiccation in conifer leaves: Anatomical changes and determination of the lethal level. *Bot. Gaz.* **114**, 189.

Parker, J. (1965). Physiological diseases of trees and shrubs. *Advan. Front. Plant Sci.* **12**, 97.

Pfister, R. D. (1967). Maturity indices for grand fir cones. *U. S. Forest Ser., Res. Note* **INT-58.**

Plaut, Z., and Ordin, L. (1961). Effect of soil moisture content on the cell wall metabolism of sunflower and almond leaves. *Physiol. Plant.* **14,** 646.

Quraishi, M. A., and Kramer, P. J. (1970). Water stress in three species of *Eucalyptus. Forest Sci.* **16,** 74.

Ransom, S. L., and Harrison, A. (1955). Experiments on growth in length of plant organs. I. A recording auxanometer. *J. Exp. Bot.* **6,** 75.

Raschke, K. (1970a). Leaf hydraulic system: Rapid epidermal and stomatal responses to changes in water supply. *Science* **167,** 189.

Raschke, K. (1970b). Stomatal responses to pressure changes and interruptions in the water supply of detached leaves of *Zea mays* L. *Plant Physiol.* **45,** 415.

Rees, A. R. (1958). Field observations of midday closure of stomata in the oil palm, *Elaeis guineensis,* Jacq. *Nature* (*London*) **182,** 735–736.

Rees, A. R. (1961). Midday closure of stomata in the oil palm, *Elaeis guineensis* Jacq. *J. Exp. Bot.* **12,** 129.

Reineke, L. H. (1948). Dial gage dendrometers. *Ecology* **29,** 208.

Rethke, R. (1946). The anatomy of circumscissile dehiscence. *Amer. J. Bot.* **33,** 677.

Rokach, A. (1953). Water transfer from fruits to leaves in the Shamouti orange tree and related topics. *Palestine J. Bot.* **8,** 146.

Sarvas, R. (1962). Investigations on the flowering and seed crop of *Pinus silvestris. Commun. Inst. Forst. Fenni.* **53,** 1.

Schädelin, W. (1942). "Die Auslesedurchforstung als Erziehungsbetreib hochster Wertleistung." Paul Haupt, Bern-Leipzig.

Schroeder, C. A., and Wieland, P. A. (1956). Diurnal fluctuation in size of various parts of the avocado tree and fruit. *Proc. Amer. Soc. Hort. Sci.* **68,** 253.

Slatyer, R. O. (1957). Significance of the permanent wilting percentage in studies of plant and soil water relations. *Bot. Rev.* **23,** 585.

Slatyer, R. O. (1967). "Plant-Water Relationships." Academic Press, New York.

Small, J. A., and Monk, C. D. (1959). Winter changes in tree radii and temperature. *Forest Sci.* **5,** 229.

Splinter, W. E. (1970). An electronic micrometer for monitoring plant growth. *Amer. Soc. Agr. Eng., Pap.* **67-111.**

Stålfelt, M. G. (1955). The stomata as a hydrophotic regulator of the water deficit of the plant. *Physiol. Plant.* **8,** 572.

Stålfelt, M. G. (1963). Diurnal dark reactions in the stomatal movements. *Physiol. Plant.* **16,** 756.

Stocker, O., Rehm, S., and Schmidt, H. (1943). Der Wasser-und Assimilationshaushalt dürresistenter und dürreempfindlicher Sorten landwirtschaftlicher Kulturpflanzen II. Zuckerrüben. *Jahrb. Wiss. Bot.* **91,** 278.

Tal, M. (1966). Abnormal stomatal behavior in wilty mutants of tomato. *Plant Physiol.* **41,** 1387.

Thornthwaite, C. W. (1948). An approach toward a rational classification of climate. *Geograph. Rev.* **38,** 55.

Tukey, L. D. (1959). Periodicity in growth of fruits of apples, peaches, and sour cherries with some factors influencing this development. *Pa., Agr. Exp. Sta., Bull.* **661,** 1.

Tukey, L. D. (1960). How apples grow—some recent findings. *Pa. Fruit News* **39,** 12, 14, 16, and 18.

Tukey, L. D. (1962). Factors affecting rhythmic diurnal enlargement and contraction in fruits of the apple (*Malus domestica* Bork.). *Proc. Int. Hort. Congr., 16th, 1962* Vol. 3, p. 328.

Turner, N. C., and Waggoner, P. E. (1968). Effects of changing stomatal width in a red pine forest on soil water content, leaf water potential, bole diameter, and growth. *Plant Physiol.* 43, 973.

Van Laar, A. (1967). The influence of environmental factors on the radial growth of *Pinus radiata*. *S. Afr. Forest. J.,* June, p. 1.

Vité, J. P. (1961). The influence of water supply on oleoresin exudation pressure and resistance to bark beetle attack in *Pinus ponderosa*. *Contrib. Boyce Thompson Inst.* 21, 37.

Wadleigh, C. H., and Gauch, H. G. (1948). Rate of leaf elongation as affected by the intensity of the total soil moisture stress. *Plant Physiol.* 23, 485.

Waggoner, P. E., and Simmonds, N. W. (1966). Stomata and transpiration of droopy potatoes. *Plant Physiol.* 41, 1268.

Waisel, Y., Borger, G. A., and Kozlowski, T. T. (1969). Effects of phenylmercuric acetate on stomatal movement and transpiration of excised *Betula papyrifera* Marsh. leaves. *Plant Physiol.* 44, 685.

Wartenberg, H. (1933). Kälte und Hitze als Todesursache der Pflanze und als Ursache von Pflanzenkrankheiten. *In* "Pflanzenkrankheiten" (P. Sorauer, ed.), pp. 475–592. Parey, Berlin.

Wiegand, K. M. (1906). Some studies regarding the biology of buds and twigs in winter. *Bot. Gaz.* 41, 373.

Wilson, C. C. (1948). Diurnal fluctuations of growth in length of tomato stem. *Plant Physiol.* 23, 156.

Winget, C. H., and Kozlowski, T. T. (1964). Winter shrinkage in stems of forest trees. *J. Forest.* 62, 335.

Zohary, M., and Fahn, A. (1941). Anatomical-carpological observations in some hygrochastic plants of the oriental flora. *Palestine J. Bot., Jerusalem Ser.* 2, 125.

CHAPTER 2

SOIL MOISTURE AND SEED GERMINATION

D. Hillel

THE HEBREW UNIVERSITY OF JERUSALEM, FACULTY OF AGRICULTURE,
REHOVOT, ISRAEL

I. INTRODUCTION

The problem of ensuring proper germination of planted seeds and subsequent establishment of seedlings is of general importance in agriculture. The germination, emergence, and establishment phase is critical in the growth cycle of plants, as it determines the density of the stand obtained, influences the degree of weed infestation, and limits the eventual yield. Unless the success of this early phase is assured, an entire planting may be doomed from the outset.

The problem of germination and establishment is especially acute under arid zone conditions, where the soil surface is wetted only infrequently and irregularly and the rate of evaporation is high. In such circumstances a seedling must compete with the process of atmospheric drying for the rapidly diminishing moisture of the surface layers. Often these layers dry out too rapidly for a seed to germinate, or for a germinated

65

seedling to extend its roots downward into the deeper layers where available moisture can be found. Consequently seedlings may fail to survive, even though overall ecological conditions may be favorable for the mature plants. Still another hindrance to germination and seedling emergence is the tendency of certain soils to slake when wetted and to form a hard crust upon drying. Tender seedlings, unable to emerge through the crust, may lie smothered only a few millimeters below the surface. Where the crust is kept moist and prevented from hardening, it may nevertheless hinder germination or seedling growth by limiting aeration. The extreme temperature fluctuations which take place at the bare soil surface may further limit the chances of a seed to germinate and of a seedling to survive. Arid zone soils may also contain soluble salts which may tend to accumulate and form excessive concentrations at the soil surface as a consequence of evaporation.

Especially sensitive and susceptible to adverse soil conditions are range plants with small seeds and tender seedlings. Even where overall ecological conditions are favorable for a given species, the difficulty of obtaining an adequate stand may decide at the outset against successful reseeding operations. The author has observed repeated failures of range seeding attempts carried out without sufficient regard of the basic problems involved. Such practical factors as depth and pattern of seed placement, seedbed preparation, compaction, etc., must be based on knowledge of the physiological and physical relationships involved, of the specific requirements of the species to be sown, and of the soil physical and climatic conditions which prevail.

For the most part, prevalent methods of seedbed preparation and of seeding are still based on the accumulated experience of generations and on trial and error methods of experimentation. Until relatively recently there has been little fundamental research on the basic interactions of physiological and environmental factors in germination and establishment. Such factors as soil moisture, temperature, salinity, aeration, and mechanical strength which prevail at the surface are of known importance, yet only sketchy data are available on their separate and combined effects. This may be due to the fact that many of the studies heretofore lacked a sufficiently sound physical basis, theoretical as well as experimental, and are therefore difficult to interpret and to apply under different circumstances.

In particular, it is unclear how seeds and seedlings of different species are affected by soil moisture potential (matric and/or osmotic) per se, or by the dynamics of water supply from the soil via the contact zone and the seed coat into the seed. What is the threshold, or critical, suction for germination? Is there, similarly, a critical supply rate, or flux, below which germination will fail? What are the relative magnitudes of the hydraulic resistances of the soil, the contact zone, the seed coat, and the seed itself? What are the possible effects of geometric factors such as depth of seed placement, as well as size, shape, orientation, sowing pattern, and density of the seeds? How does the rate of evaporation or redistribution of soil water affect germination? How does soil moisture interact with other pertinent factors such as temperature, aeration, and mechanical strength, all of which affect germination? Finally, how can the seedbed environment be managed and optimized to ensure rapid and uniform germination and seedling establishment in actual field practice, particularly under arid conditions?

II. SOME BASIC ASPECTS OF GERMINATION

Seed germination is the renewal of growth of the dormant embryo generally caused by a change in the environmental conditions. A comprehensive review and analysis of the physiology of germination was published by Mayer and Poljakoff-Mayber (1963). Physiological and environmental aspects of seed dormancy and the onset of germination have been investigated by Evenari (1949), Pollock and Toole (1961), Wood and Buckland (1966), Went (1961), and Abdul-Baki (1969a,b); and have most recently been reviewed by Koller (1971). Biochemical (enzymatic and hormonal) processes involved in the termination of seed dormancy and the onset of germination were described by van Overbeek (1966), as well as by Whalley and McKell (1967), Woodstock and Grabe (1967), and Wilson et al. (1970). It must be recognized that germination actually consists of a series of sequential processes leading to the resumption of embryo growth. Drought influences on germination and seedling emergence were recently reviewed by Wright (1971).

For germination to be triggered, the environment must provide a specific set of physical and chemical conditions optimal for the particular species. Among the conditions required are an adequate supply of water, a suitable temperature and composition of gases, illumination for certain seeds, and the absence of toxic or inhibitory substances.

Under natural conditions, the seeds of even a particular species vary greatly in germinability. This property often depends on the age of the seed. In many instances germinability is lowest soon after ripening and improves with time (though excessive aging can eventually result in loss of viability). As pointed out by Koller (1971), the dormancy that is involved in preventing germination from taking place readily even under "normal" conditions is an evolutionary safeguard against the uncertainty of the natural environment. Thus, catastrophes which may occur will involve only a fraction of the seeds which had been produced, and their effect on the population of the species as a whole will be only temporary.

In agriculture, on the other hand, it is desirable to have seed lots characterized by the ability to germinate rapidly and uniformly in response to favorable conditions. Cultivated plants, in fact, have in most cases been selected for high, rapid, and simultaneous germination.

The need to evaluate and control germinability has led to the establishment of standard procedures for testing seeds, designed to determine seed quality or "viability." However, these procedures are rather arbitrary, and in many cases do not simulate the actual environment so that they might often fail to predict field performance (Isely, 1958). Evaluations of seed germinability are often carried out in petri dishes, on a substrate of filter paper or agar, or in lint rolls moistened with some standard amount of distilled water. As pointed out by Negbi et al. (1966), such conditions of moisture and aeration are rarely, if ever, encountered by seeds in their natural environments.

The two main components of seed, embryo and endosperm, provide the basic potential of determining response during germination and initial growth (Wright, 1971). The endosperm provides nutrients for growth and development of the embryo (Stanley and Butler, 1961), and the quantity of the food reserve available during germination is related to seed size (Cooper and MacDonald, 1970). In a given seed lot, heavier seeds usually possess greater growth potential than light seeds (McDaniel, 1969), as they are better able to furnish the energy needed for germination and initial growth of seedlings until adequate leaf area is developed for photosynthesis.

In general, the physiological effects of water stress on plants are most pronounced in rapidly growing tissues, a fact which is particularly exemplified by the developmental phases of germination, emergence, and initial seedling growth (Slatyer, 1969).

The process of water uptake in germination consists of at least two distinct stages: (1) *an imbibition stage,* during which water sorption by seeds is largely passive and in which nonviable seeds can be expected to absorb water at the same rate as viable ones; and (2) *a growth stage,* which begins with the appearance of a radicle (or rootlet), and during which water uptake becomes increasingly due to this rootlet as it seeks and penetrates into moister layers in the soil. Between these two stages, it is often possible to recognize an apparent pause, a time period which might actually constitute a distinctly separate stage, during which internal physiological processes (respiratory, enzymatic, etc.) set the stage for the onset of actual growth. During this stage, growth inhibitors, such as there might be, are dissipated either by chemical transformation or by outward diffusion.

As a first step in any investigation of germination in the soil, it is necessary to establish the physiological behavior and basic environmental requirements of the particular species of interest. In this connection, the following characteristics, among others, have been found pertinent:

1. *The sorption isotherm,* or "seed moisture characteristic": the relation of the seed's water content to its water potential.
2. *The critical potential,* or "threshold potential": the lowest value of water potential at which the seed can germinate (Hunter and Erickson, 1952).
3. *The possible presence of germination inhibitors,* and the mode and rate of their dissipation.
4. *The time rate* of imbibition, the time required for germination (radicle emergence), and time rate of rootlet elongation at different ambient temperature and water potential values.
5. *Critical hydration levels of the seed:* the minimal water content at which the seed begins to germinate, and the hydration level at which seed water uptake becomes biologically irreversible.
6. *Critical depth of emergence:* the maximal depth from which the seedling, once germinated, can successfully emerge.

III. SEVERAL ENVIRONMENTAL FACTORS AFFECTING GERMINATION

Prior to elucidating the moisture relations of germination, which are the main topic of this chapter, it is well to reemphasize the importance of a number of other environmental factors that can exercise a crucial effect on germination. Included among these are temperature, oxygen supply, light, pH, solutes, microorganisms, and the mechanical strength (or impedance) of the soil (Toole *et al.,* 1956). Most of these factors interact very strongly, and at times inextricably, with the moisture factor, so that at least a brief mention of some of their effects seems necessary.

A. TEMPERATURE EFFECTS

In the field, the most important factor interacting with both matric potential and osmotic suction is temperature. This has been studied by McGinnies (1960) and Evans and Stickler (1961), who found that the adverse effect of moisture stress is increased at extreme temperatures.

Soil temperature is extremely variable, both seasonally and diurnally. In semiarid regions, diurnal temperature fluctuations in the seedbed zone of shallow-seeded plants normally reach and often exceed an amplitude of about 20°C on bright dry days and about 10°C during cloudy and rainy periods (Tadmor *et al.,* 1964). Soil temperature depends (in addition to latitude and altitude) on soil color, moisture, evaporation, bulk density, roughness, and the presence of mulches. Moreover, steep temperature gradients with depth usually exist in the soil profile.

Each species of seeds appears to have definite temperature requirements for germination (minimal, optimal, and maximal temperatures). Departures from the required range, even temporary, can reduce the metabolic activity leading to germination (Laude *et al.,* 1952; Laude, 1957; and Sosebee and Herbel, 1969). Extreme and prolonged departures from the optimum can affect the very survival potential of the seed species in its particular environment (Ellern and Tadmor, 1966, 1967). On the other hand, some species apparently respond favorably and germinate better at alternating temperatures (Harrington, 1923; Morinaga, 1926; Lehmann and Eichele, 1931; Stotzky and Cox, 1962; Cuddy, 1963). The germination of some seeds is enhanced by moist pretreatment at near-freezing temperature, a practice which came to be known as "stratification" (Koller, 1971).

Studies on the effect of temperature on germination and emergence have been reported by S. J. Richards *et al.* (1952), Koller (1955), and Mayer and Poljakoff-Mayber (1963). McGinnies (1960) studied the interaction of temperature and moisture stress as affecting germination of grasses. Studies on the interactions between temperature and other environmental factors have also been reported by Evans and Stickler (1961) and by Read and Beaton (1963), among others. The effects of both salinity and soil matric suction are more pronounced whenever the temperature is not optimal for germination.

In regions where rainfall occurs primarily during the cool season, low temperatures may limit germination and seedling growth at the very time that moisture conditions are favorable, even as the unavailability of moisture limits growth during the warm season. This may have been the cause of frequent past failures of range seeding or reseeding operations (Tadmor and Hillel, 1956; Negbi, 1957; Arnon, 1958).

The effect of soil temperature persists beyond germination into the seedling growth stage. The response of root development to temperature apparently differs greatly among species (Sprague, 1943; Lovvorn, 1945). Unfavorable soil temperatures reduce the rate of root penetration (Troughton, 1957). According to Kleindorst and Brouwer (1965), the shoot/root ratio increases with temperature.

Since, in an arid environment, the soil surface zone tends to dry rapidly during the dry spells which generally follow the infrequent rains, the time rate of germination and of initial seedling growth can be a critical factor in the success or failure of seeding operations. In particular, the rate of seedling root elongation and its penetration into the deeper layers of the soil, where moisture is retained longer against evaporation, often determines whether the seedling will win or lose its race against drought, and thus whether it will thrive or die.

B. Aeration

As pointed out by Mayer and Poljakoff-Mayber (1963), germination is a process requiring an expenditure of energy by the cells of the seed. Energy-requiring processes in living cells are usually sustained by processes of oxidation in the presence or absence of oxygen. These processes, respiration and fermentation, involve an exchange of gases, an output of carbon dioxide in both cases, and the uptake of oxygen in the case of respiration. Consequently, seed germination is markedly affected by the composition of the ambient air and its exchange with the external atmosphere.

The beginning of germination is characterized by a marked increase in the respiration rate of the seeds (Toole et al., 1956). Respiration during germination was described by Yemm (1965). As soon as the seeds start to take up water there is a steep increase in their respiration rate. After several hours, however, the increased respiration rate levels out to become constant. At this stage there may be an increase in the respiration quotient, signifying the possible occurrence of anaerobic respiration. Rupturing of the seed coat, or increasing the oxygen content of the environment, tends to shorten this period. At the third stage there is again an increase in the respiration rate, until a maximum is attained. These three stages tend to parallel the stages of water uptake by the seeds (Oota, 1957; Stanley, 1958; Stiles, 1960).

It has been suggested that the enzyme system for aerobic respiration develops only gradually during the process of germination (Wareing, 1963). Some seeds, like rice, have a well-developed system for alcoholic fermentation, using the energy output of this process for germination (Vlamis and Davis, 1943). Most seeds, however, require an adequate supply of oxygen during the process of germination. They show a decrease in the rate of germination and in the final germination percentage when the partial pressure of oxygen in the environment is lowered (Siegel and Rosen, 1962).

The evidence available on the effects of carbon dioxide on germination is rather conflicting. According to some authors CO_2 may inhibit respiration and root growth (Norris et al., 1959), but other authors show that low concentrations of CO_2 have a growth-promoting effect provided that oxygen tensions are not limiting (Thornton, 1944; Unger and Danielson, 1965). Recently it was found that CO_2 can have either a toxic or a tonic effect on radicle growth, depending on the time and duration of application and other factors (Grable and Danielson, 1965).

Gingrich and Russell (1956) found that corn radicle growth was limited at partial pressures of oxygen below 5.25%. Gill and Miller (1956) devised a method for varying oxygen pressure and mechanical impedance separately. They found that as oxygen pressure was reduced, the effect of mechanical pressures was more pronounced in limiting corn radicle elongation. A study of the effects of oxygen and carbon dioxide concentrations on germination of seeds of range grasses was reported by Dasberg et al. (1966a). They found that the rate of seed water uptake, seed respiration, and germination were lowered, and also the final germination percentage attained was lowered at oxygen concentrations below atmospheric. On the other hand, the germination of certain legumes can be enhanced with CO_2 concentrations in the range of 0.5–2.5% (Koller, 1971).

The oxygen supply in the air-filled pores of the soil is renewed continuously by the process of diffusion, but the O_2 can be taken up by the seeds only after passing through the moisture film covering them. The diffusion coefficient of oxygen in water is smaller by a factor of 10^4 than the diffusion coefficient in air, and it is possible, therefore, that the rate of oxygen supply to the seeds is governed by diffusion through water films. In principle, the oxygen availability to the germinating seeds can be dependent on several processes:

1. Gas diffusion in the soil air phase; low soil porosity or crusts at the soil surface may limit the supply of oxygen and the disposal of carbon dioxide.
2. Oxygen diffusion through the water films covering the seeds; excess water in the soil causes a thickening of the water films and may affect the rate of oxygen supply to the seeds.
3. Oxygen dissolved in the water and taken up by the seeds during germination may answer part of the oxygen demand of the seeds (though in most cases this amount is probably negligible).

Sedgley (1963) reported that submerged seeds failed to germinate unless provided with small amounts of hydrogen peroxide (optimal concentration, 0.04 vol strength).

The requirement of a germinating seed for ample water to hydrate its embryonic tissues may conflict with its requirement for efficient gaseous exchange. According to Negbi et al. (1966), this is apparently the cause of water sensitivity, where germination is inhibited by an overabundant supply of moisture (Cavazza, 1953; Kirsop and Pollock, 1957; Jansson,

1959). Negbi *et al.* (1966) suggested that germination in some seeds is inhibited by excessive hydration owing to the added resistance to diffusion of oxygen into the seed through the enclosing mucilaginous seed coat. Small differences in water potential can cause appreciable differences in the amount of water held in the mucilage and thus affect the length of the diffusion pathway for respiratory gases.

C. MECHANICAL IMPEDANCE

Many soils, and particularly arid-zone soils that are low in organic matter content and are often bare of vegetative cover, tend to form a dense crust under the beating and slaking action of raindrops (Hillel, 1961). Soil crusting has long been known to restrict germination and seedling emergence, either by forming a mechanical obstruction or by limiting aeration, or by a combination of these effects. L. A. Richards (1965) reported that bean seedling emergence in a fine sandy loam was decreased from 100 to 0% when the crust strength (as measured by the modulus of rupture) increased from 108 to 273 millibars. Hanks and Thorp (1957) found that seedling emergence of wheat was apparently not related to crust thickness or seed spacing, but was highly correlated with crust strength and moisture. Arndt (1965) also showed that the mechanical impedance of the soil to the penetration and emergence of seedling shoots increases with decreasing soil moisture.

Like crusting, seedbed compaction often has an adverse effect on germination as well as on subsequent plant growth (Parker and Taylor, 1965; Philips and Kirkham, 1962). On the other hand, it has been found that under marginal soil moisture conditions a certain degree of compaction may improve germination and establishment (Hudspeth and Taylor, 1961; Hyder and Sneva, 1956; McGinnies, 1962; Parker and Taylor, 1965; Triplett and Tesar, 1960; Dasberg *et al.,* 1966b). In fact, limited compaction over the seeding row is a common agronomic practice and most seeding implements have attachments for compacting the excessively loose soil that falls over the seeds.

IV. MOISTURE RELATIONS OF GERMINATION

A. SOIL MOISTURE POTENTIAL

Surface soil moisture can vary from saturation, as during a rain, to near zero, as after a prolonged dry spell. Some soils are permanently water-logged (e.g., in swamps), whereas others are nearly always dry (e.g., in deserts). In a given soil, the water content depends on climatic conditions and season, as well as on texture and plant cover.

The energy state of soil water, expressed in terms of the potential, depends upon soil water content, nature of the soil matrix (capillary and adsorptive phenomena), and concentration of solutes. Soil water potential is often expressed in terms of suction, which is the equivalent "negative" pressure imparted to soil water by its affinity to the soil matrix and by the solutes (salts) present in the soil solution. This suction can be given in pressure units (dynes/cm², bars, or atmospheres), or in head units (i.e., the height of a vertical column of water or mercury, as specified, equivalent to the given pressure). A fundamental discussion of soil water potential and related phenomena was given by Hillel (1971).*

It has long been established that the rate of germination decreases with increasing soil water suction (Doneen and McGillivray, 1943). The final germination percentage similarly diminished at low soil water contents. The ability of seeds to germinate at low soil moisture contents depends upon the species; each species of seed appears to require the absorption of a certain minimum amount of moisture in order to germinate and appears to have its own threshold suction value for germination (Hunter and Erickson, 1952). The values reported are: 12.5 bars for corn, 7.9 bars for rice, 6.6 bars for soybean, 3.5 bars for sugar beets. These are mean values for the soil bulk, while real values must be determined for the seed-soil interface.

Suction in dry seeds is frequently in excess of 500 bars; thus dry seeds will absorb water from soils below the permanent wilting percentage (Black, 1968). This fact was noticed very early. Thus, Whitney and Cameron (1904), mixed 50 gm of cowpea seeds with 50 gm of soil containing 7.5 gm of water and found that in 12 hr the cowpea seeds had absorbed all but 0.65 gm of the soil water, leaving the soil in essentially air-dry condition. Doneen and McGillivray (1943) found that seed size did not affect germination, and that seeds of some species germinated fairly well even in soil below the permanent wilting percentage (presumably, at a suction greater than 15 bars).

Under favorable conditions of water supply, temperature and seed-coat

---

* The total potential ϕ of soil water can be considered as having three components:

$$\phi = \psi + \pi + z$$

where ψ is the matric potential (consequence of capillary and adsorptive attraction of the soil matrix for water); π is osmotic potential (caused by the dissolved solutes in soil water); and z is gravitational potential (determined by the vertical position, or elevation, of the point with reference to some specified datum).

permeability, some seeds may absorb enough water to double their weight within a period of two days (Black, 1968). Normally, the volume of soil required to contribute this amount of water is considerably greater than the volume of the seed that absorbs it. (Hence, the "distance of travel" may be several millimeters or even centimeters.) Large seeds may not be able to absorb enough water from the soil (especially sands) to permit germination unless the soil were rewetted by at least one rainfall or irrigation during the period required for germination and seedling establishment.

Seeds that absorb an amount of water too small to permit germination may subsequently dehydrate without apparent loss of viability. Repeated cycles of partial wetting may sometimes help to dissipate inhibitors, but in other cases may result eventually in the reduction of germinability. Fungal hyphae can develop at values of matric suction well above those at which plants will grow (Griffin, 1963). In time the fungi can damage the seed so that germination will not take place even if more water is added.

Collis-George and Melville (1969) reported a system in which wheat grains were situated 1 cm above a water surface, with no soil present, such that around the seed the diffusion coefficient was that of water vapor through air at atmospheric pressure. The seeds then germinated in 60 hr on the average.

The germination behavior of a seed in soil can be affected by soil permeability as well as suction. Each species of seed may possess its own water requirements in terms of the energy state and the rate of supply of water.

B. Matric versus Osmotic Effects

The question arises as to whether plants respond equally to matric and to osmotic potentials. L. A. Richards and Wadleigh (1952) argued that, as osmotic potential causes reductions of vapor pressure (and changes in the other colligative properties) identical with those caused by the numerically equivalent matric potential, the two potentials should affect plant water uptake and growth in the same degree. Ayers (1952) also held that the effects of soil matric potential and osmotic potential on seed germination are equivalent.

Gingrich and Russell (1957), on the other hand, found that, more often than not, a decrease in matric potential (i.e., an increase in matric suction) had a greater effect on reducing growth than a corresponding decrease of osmotic potential (i.e., an increase in solute concentration). They attributed this result to the fact that an increase in matric suction entails a decrease in soil moisture content and hence a decrease in conductivity (or permeability). This effect might restrict the soil moisture supply rate. Furthermore, biological systems do not have perfectly semipermeable membranes so that solutes are in fact not totally excluded from the living cells and tissues. Plants can thus adjust to osmotic "stress." This adjustment, however, is strictly limited, since internal concentrations of salt may soon become toxic (Ayers, 1952).

Uhvits (1946) attempted to simulate the effect of soil water suction on germination by the use of osmotic solutions. Some investigators (Evans and Stickler, 1961; Helmerick and Pfeifer, 1954; Knipe and Herbel, 1960; Younis *et al.,* 1963) have subsequently tried to characterize the drought resistance of species and varieties by their ability to germinate in osmotic solutions. Seeds react in specific ways to solutions of different chemicals at equivalent osmotic concentrations: NaCl causes a toxic effect as the seeds absorb Cl (Uhvits, 1946; Wiggans and Gardner, 1959); $CaSO_4$ also has a toxic effect when compared with mannitol (Collis-George and Sands, 1962).

Collis-George and Sands (1962) found that germination rates of seeds on a suction plate at a low matric potential were much slower than those of seeds exposed to similar osmotic potential. No effect of degree of contact between seed coat and matrix on germination rate was detected, and it was concluded that permeability was not a factor in the experiments. They further concluded that in germination of seeds, osmotic potential does not manifest any restriction on germination until large concentrations (probably of a toxic nature) are reached.

Thus, germination behavior does not support the hypothesis that matric and osmotic potentials should have similar biological consequences merely because their free energy measurements are identical. Collis-George and Sands (1962) found that, for some osmotic systems, -100 cm of matric potential is as effective as $-10,000$ cm of osmotic potential in retarding seed germination rates. When moisture potential is controlled either by osmotic potential (Uhvits, 1946; Ayers and Hayward, 1948; Rodgers *et al.,* 1957; Wiggans and Gardner, 1959; McGinnies, 1960) or directly by vapor pressure (Owen, 1952) retardation of germination rate is associated with large reductions (-597 to -785 J/kg) in the moisture potential of the seed environment. However, when moisture potential is controlled by matric potential (Collis-George and Sands, 1959, 1962; Collis-George and Hector; 1966) detectable retardation of germination is associated with very small (-23.8 J/kg) reductions in moisture potential.

Potential energy considerations are apparently not enough to explain seed and seedling response, and the rate of water movement in the soil and through the soil-to-seed contact zone can be important. This may be especially so during the early germination stage, when the seeds must absorb large amounts of water.

Sedgley (1963) criticized the assumption of Collis-George and Sands

(1962) that the matric suction of water being supplied to seed sited on a matrix can be varied independently of the permeability factor. The effect of retreat of water menisci as the matric suction is increased is to reduce the area of seed coat in contact with a readily available supply of water. Improving the seed-to-soil moisture contact increases the rate of water uptake and hastens germination. When permeability is nonlimiting, small differences in suction do not affect water uptake, and hence germination.

Collis-George and Hector (1966), Collis-George and Williams (1968), and Collis-George and Melville (1969) later suggested that matric suction influences seed germination through its positive contribution to the effective stress within the solid framework of the soil surrounding the seed.[*] These investigators attempted to separate the effects of matric suction into those that can be attributed to the free energy of the soil water and those that can be attributed to the effective stress in the soil system. They suggested that the influence of matric suction on seed germination in the range 0–400 cm of water (0–0.4 bar) can be "wholly attributed" to the mechanical control this stress exerts on the solid framework (which, in turn, affects the shearing strength of the soil). Seed germination was apparently impaired by an effective stress of the order of 10^5–10^6 dynes/cm^2. Presumably, this stress restricts seed swelling.

The effects of osmotic and of matric suction appear to be additive even though they are not simply equivalent. The salinity effect is more severe at low soil water contents (Ayers, 1952; Chapin and Smith, 1960; Read and Beaton, 1963). Bernstein and Hayward (1958) found that, contrary to common opinion, plants in the germination stage are not necessarily more susceptible to salinity than at later growth stages. However, the germination zone, or seedbed, being the uppermost soil horizon, is naturally subject to much greater extremes of salinity, as well as to water content and suction, than the root zone as a whole.

C. Dynamics of Soil Water Uptake during Germination

According to Mayer and Poljakoff-Mayber (1963), the extent (and rate) at which a seed imbibes water from soil during the initial stage of germination is determined by composition of the seed, permeability of the seed coat or fruit to water, and availability and mobility of water in the environment (i.e., the soil).

Imbibition is a physical process that is related to the properties of seed

---

[*] They defined the effective stress σ' of a soil matrix as:

$$\sigma' = \sigma - \psi$$

where σ is the externally applied stress (pressure) and ψ is the soil matric potential. It seems more correct to write $\sigma' = \sigma - \alpha\psi$, where $\alpha < 1$.

colloids. It is in no way related to viability of seeds and occurs equally in nonviable as in live seeds. The strong sorption of water by hydrophilic colloidal micelles results in swelling and considerable pressures, called imbibition pressures, which may reach hundreds of atmospheres. This swelling may lead to rupturing of the seed coat. The chief component that imbibes water within a seed is the protein, whereas the starch component apparently does not add to the total swelling of the seed.

The rate of water entry into a seed is influenced by permeability of the seed coat. Impermeable seed coats are frequently found in seeds of the Leguminosae, as well as in other groups. Morris et al. (1968) measured seed coat permeability of 11 varieties of snap beans (*Phaseolus vulgaris* L.) and found that the permeability was of the order of 0.8 mg of water/mm²/hr. Phillips (1968), who studied the germination dynamics of corn, soybean, and cotton, did not find seed coat permeability to be a rate-limiting factor in water uptake by seeds. A method for testing seed coat permeability by determining changes in concentration of sucrose during imbibition was reported by Morris et al. (1968).

It is likely that seeds do not absorb water uniformly over their entire seed coat. Permeability of the seed coat is apparently greatest at the micropyle, where the seed coat is generally thinner than elsewhere over the seed. Thus, Manohar and Heydecker (1964) reported that at equal levels of water stress (up to 10 bars) and seed-to-water contact area, seeds with Vaseline-covered micropylae germinated more slowly than otherwise.

Phillips (1968) presented a procedure for determining the "average" diffusion coefficient (diffusivity) for internal water sorption by seeds, based on immersion of the seeds in free water and measurement of the imbibition rate. His procedure makes use of a well-known equation derived from the analysis of diffusion systems (Crank, 1956) applicable to spherical geometry with water moving only in the radial direction:

$$\frac{\partial \theta}{\partial t} = \left(\frac{D}{r}\right) \frac{\partial^2 (r\theta)}{\partial r^2} \tag{1}$$

where θ is the water content of the seed (on the volume basis); D is the seed's diffusivity to water (assumed constant, regardless of water content); r is the radial distance from the center of the seed; and t is time. Phillips' procedure is based on solution of this equation for the following initial and boundary conditions:

$$\begin{aligned} \theta &= \theta_0 \text{ at } t = 0, \, 0 < r < a \\ \theta &= \theta_1 \quad t > 0, \quad r = a \end{aligned} \tag{2}$$

where a is the radius of the seed (which is taken to be spherical). That is to say, the initial internal moisture content of the seed is assumed to be uniform, and the water content of the seed surface is considered to be constant throughout the germination period.

The resulting D values varied from 0.2 to 33 $\times$ 10^{-4} cm^2/hr for different seed types. Diffusivity of soybean seed in aerated, distilled water was approximately 4 times greater than that of corn seed and 18 times greater than that of cotton seed. In a silt loam soil at "field capacity," the apparent diffusivity of soybean seed decreased to 11 $\times$ 10^{-4} cm^2/hr, as compared to 32 $\times$ 10^{-4} cm^2/hr in water. In the soil, moreover, diffusivity values of seeds tended to increase with increasing soil moisture content. Thus, the imbibition rate in soil appears to be affected by the seed-to-soil moisture contact area.

Doering (1965) reported moisture diffusivities for a silt loam to vary from 4 cm^2/hr at a water content of 22% to about 2 $\times$ 10^{-2} cm^2/hr at a low water content of 6%, values which exceeded diffusivities of all three seed species tested by Phillips (1968) in aerated, distilled water. It therefore does not seem likely that diffusivity of the soil to water could limit absorption of water by seed unless the soil immediately surrounding the seed were depleted of moisture to considerably below 4%. It is possible, however, that a seed germinating in relatively dry soil does not obtain a continuous film of water over its surface, so that the rate of imbibition might well be limited by the contact zone. The importance of the wetted area of contact between seed and soil is shown in the work of Collis-George and Hector (1966).

A mathematical description of the movement and uptake of soil water by germinating seeds was first attempted by Shull (1920), and later by Stout et al. (1960), by Hadas (1969), and by Ehrler and Gardner (1971).

When a seed takes up water from the soil in contact with it, water begins to move within the soil toward the seed in response to the matric suction gradients which are created. Soil water movement under such conditions can be described by a flow equation, which in one-dimension has the form:

$$\frac{\partial \theta}{\partial t} = \frac{\partial}{\partial x}\left[K(\psi)\frac{\partial \psi}{\partial x}\right] \tag{3}$$

where θ = water content, K = hydraulic conductivity, ψ = water potential, t = time, and x = distance. This can also be expressed, with certain limitations, in the form of the diffusion equation:

$$\frac{\partial \theta}{\partial t} = \frac{\partial}{\partial x}\left[D(\theta)\frac{\partial \theta}{\partial x}\right] \tag{4}$$

where D is the diffusivity. When a weighted mean diffusivity, $\overline{D}$, can be assumed, Eq. (4) becomes:

$$\frac{\partial \theta}{\partial t} = \bar{D}\, \frac{\partial^2 \theta}{\partial x^2} \tag{5}$$

In three dimensions, using spherical coordinates, this equation has the form:

$$\frac{\partial \theta}{\partial t} = \bar{D}\left(\frac{\partial^2 \theta}{\partial r^2} + \frac{2}{r}\,\frac{\partial \theta}{\partial r}\right) \tag{6}$$

To solve Eq. (6), it is necessary to know the geometry of flow (i.e., the shape and size of the seed, depth of seed placement, interseed spacing, etc.). It is convenient (though not necessarily realistic) to assume the soil medium to be homogeneous and to extend infinitely in all directions. It is also necessary to know the initial potentials of the seed and the soil, and their hydraulic properties.

One interesting problem in this connection is to determine the hydraulic resistances of the various segments of the flow path, and to determine which of these might be rate-limiting.

Hadas (1969) attempted to calculate the time rate of imbibition by the following equation (taken from Crank, 1956):

$$\frac{M}{M_\infty} = 1 - \frac{6}{\pi^2} \sum_{n=1}^{\infty} \left(\frac{1}{n}\right)^2 \exp\left\{-\frac{n^2\pi^2\bar{D}t}{r^2}\right\} \tag{7}$$

where M is the cumulative water uptake at time t, M_∞ is the final value of cumulative water uptake; r is radial distance from the seed center; and $\bar{D}$ is mean soil moisture diffusivity. The calculations based on the use of this equation were criticized by Collis-George and Melville (1969), and by Phillips (1969), who gave an alternative equation. In particular, these authors pointed out that use of a single diffusion coefficient for soil and seed which appears to be three orders of magnitude larger than that appropriate for seed would give rise to an underestimation of the time required for imbibition.

Ehrler and Gardner (1971) also attempted to calculate the rate at which a seed could extract water from a soil if water uptake were limited only by the soil and not at all by the contact zone or by the seed. An appropriate solution of the flow Eq. (6), obtainable from Carslaw and Jaeger (1959), gives the seed water content at large times, θ_t:

$$\theta_t = \theta_\infty(1 - a^3/2\, k(Dt)^{3/2}\pi^{1/2}) \tag{8}$$

where a = radius of the seed; k = the ratio of the water content of the soil to the final equilibrium water content of the seed, θ_∞; θ_t = seed water content at large times; D = soil diffusivity, assumed constant; and t = time.

With $a = 0.1$ cm, $D = 10$ cm²/day (the soil diffusivity value at a matric potential of about -15 bars), and $k = 0.05$, the calculation indicates that about 7 hours are needed to achieve 99% of the equilibrium water uptake. The experimental data reported by these investigators for wheat seeds germinated in three soils fit the relation:

$$\theta_t = \theta_\infty(1 - 0.88 \exp - 0.77\, t) \qquad (9)$$

indicating that under actual conditions equilibration is slower than the computed value. The discrepancy can be attributed to contact, seed coat, or internal seed resistance. Ehrler and Gardner conclude that water transport within the soil cannot ordinarily limit seed hydration significantly. Accordingly, the limitation of germination in seedbeds with a water potential higher than -15 bars is probably due to the effect of the potential itself, or to the rate of dissipation of an inhibitor, rather than to the rate of transport of soil water toward the seed.

These conclusions may not hold where soil moisture is a decreasing function of time (as in a seedbed subject to evaporation), since under such conditions the potential may not stay high enough, long enough. The problem may be especially acute when the seeds are large.

D. Effects of Evaporation and Redistribution of Soil Moisture

Germination and seedling establishment under arid conditions can be pictured as a competition between the processes of water uptake by the seed and of root elongation of the seedling, on the one hand, versus the simultaneously occurring process of atmospheric evaporation of soil moisture, on the other hand. Where the soil surface is left bare, the rate of evaporation (particularly in an arid zone) can be such as to desiccate the soil surface within hours. Theoretical solutions of the flow equation, verified by experimental data, indicate that after this (when the drying rate is no longer limited by external evaporativity) the drying effect propagates into the soil according to the square root of time. The cumulative evaporation is proportional to $t^{1/2}$, and the time required for soil moisture to fall to a given water content is proportional to the depth squared. It was shown by Gardner (1959), that a plot of $(\theta - \theta_s)/(\theta_0 - \theta_s)$ (where θ_0, θ_s, θ are, respectively, the initial wetness of the soil, the "final" surface wetness, and the wetness of any depth at any time) versus the composite variable $x/2\sqrt{D_s t}$ gave a relative water content distribution dependent only on the ratio of D_0/D_s (where D_0, D_s are the diffusivities at θ_0, θ_s). From such a curve one can calculate the value of θ at any given depth, x, and time, t.

The presence of a mulch reduces the initial evaporation rate and prolongs the initial, constant-rate stage of the evaporation process (Hillel, 1971). A description of this stage of evaporation was given by Covey (1963). Where the drying rate is slow, the soil profile dries nearly uniformly with depth with a relatively flat profile of water content. Thus, while the drying process is discernible to greater depth, the upper layer of the soil loses moisture at a slower rate and may remain for a much longer time above the critical potential for germination.

Gardner *et al.* (1970) found that where the processes of redistribution and evaporation occur simultaneously, the rate of decrease of water content can be significantly greater than where evaporation occurs by itself. Under such conditions the water content, θ, at any depth decreases with time, t according to the relation

$$\theta = \frac{\theta_i}{(t + c)^b} \tag{10}$$

where θ_i is the initial water content, and b, c are constants. Accordingly, the length of time a given depth will remain above the critical moisture content, θ_c, will obey the relation:

$$t = \left(\frac{\theta_i}{\theta_c}\right)^{1/b} - c \tag{11}$$

While these relationships are based on solution of the flow equation, the constants involved should be evaluated empirically for a particular soil and set of conditions.

In this connection, it is noteworthy that structurally stratified soil profiles do not behave as do uniform profiles of even the same texture. In particular, tilled or aggregated top layers tend to dry more rapidly than corresponding depths of a uniform profile, though such aggregated layers can very effectively conserve the moisture of underlying layers (Hillel, 1968) provided they are finely pulverized. If the surface aggregates are too coarse, however, convective air currents can penetrate into the profile and thus transport a greater flux of vapor through the interaggregate voids.

V. CONTROL MEASURES

A wide variety of control measures has been proposed by various workers. However, this variety is of no avail unless one has some criteria by which to choose the measure or measures most likely to be effective under a given set of conditions.

When it is established that physiological rather than environmental factors are limiting germination, such prior treatments as soaking in water or solutions for removal of inhibitors or for partial imbibition may be effective (Bleak and Keller, 1969). If seed coat resistance is shown to be a problem, such treatment as abrasion or scouring may be of some help.

Once it is assured the seeds can germinate easily, the environmental conditions remain to be optimized. Mulching over the soil surface, either with plant residues or by means of sprayable chemicals or plastic membranes, can, at least for a time, reduce the drying rate of the seedbed. A similar effect can be obtained by increasing the depth of placement, provided the critical depth is not exceeded and the mechanical impedance of the soil to seedling emergence and root penetration is kept low.

The problem of poor seed-to-soil contact can be helped by proper compaction. It must be remembered, however, that seedbed compaction can result in two opposite effects: On the one hand, firmer contact and greater capillary conductivity of the soil can enhance water transport to the seed; on the other hand, greater mechanical impedance to seedling emergence and root penetration, as well as restricted aeration, may be undesirable consequences of compaction. One possible compromise between these opposing effects is to compact the layer below the seeds (improving soil-to-seed moisture supply) and at the same time to loosen the layer above the seeds (thus decreasing mechanical impedance to emergence, and improving aeration). The optimal compaction, in any case, should depend upon the soil type and structure, as well as on seeding depth, seed species, initial soil moisture content, and rate of drying (i.e., the external and surface conditions governing atmospheric evaporation).

Among the main controllable factors in seeding practice are the density, spacing, and pattern of seed placement. Little is known about the influence of intraspecific competition at an early stage of seedling growth and its interaction with habitat factors such as soil moisture, depth of seeding, root elongation, emergence, and stand density. Several instances of "negative competition" have been reported, where high seed densities apparently enhanced early seedling growth. Dense seeding, row seeding, or cluster seeding have been observed to enhance emergence, especially where crusting phenomena were prevalent (Ferguson, 1962; McGinnies, 1960). On the other hand, these same practices can result in more intense competition at a somewhat later stage in plant development. Altogether too little is known about the stages and magnitudes of these contradictory effects, particularly with regard to the species under investigation. Moreover, most of the data available in the literature do not allow separation of direct physiological effects from environmental effects (i.e., biochemical interactions of neighboring seeds versus physical competition for moisture, oxygen, etc.).

A particularly promising method for both increasing temperature and conserving moisture content of seedbeds is to coat the seeding rows with a thin film of plastic material or of oil or wax. Work reported by Tadmor and Hillel (1968) indicated a number of advantageous effects of sprayable asphaltic mulches. Blackening the soil surface can reduce reflectivity from 30 to 45% down to below 10%, and thus increase absorption of radiant energy by the soil. This warming effect can allow planting to be carried out in the fall and spring, the transition seasons which are often too cool for normal planting to succeed. In particular, the possibility of earlier planting in spring makes it possible for plants to utilize soil water of upper soil layers that would otherwise be lost to the atmosphere. The mulch protects the soil against slaking and can help to prevent formation of a hard surface crust which might impede emergence mechanically. The mulch itself can be very thin (of the order of 0.25 mm) and rather soft, and does not appear to form a barrier against the seedlings. The mulch can help to conserve soil moisture in the seedbed—a critical factor, especially in coarse-textured soil where the amount of moisture retained after an irrigation is low.

As against these advantages, the petroleum mulch treatment can present certain disadvantages. The asphalt film over the seedbed, though it is soft and penetrable by most seedlings of ordinary crops and grasses, may nevertheless obstruct the emergence of extremely delicate and weak seedlings. The mulch, just as it encourages the growth of desirable plants, may also stimulate the early development of weeds along the seeding row. The petroleum mulch is not equally effective in all soil types. It is most efficient over light-colored (and hence cool) sandy soils, with a smooth surface which does not crack. The heating effect of the dark mulch, so beneficial in winter, may be excessive in late spring and in summer. Finally, the chief drawback to the use of such a mulch on a large scale is that of cost. The extra outlay of materials and labor required make the economics of this treatment questionable at present. However, it is to be hoped that in time such techniques will become more economical and feasible on a large scale.

REFERENCES

Abdul-Baki, A. A. (1969a). Metabolism of barley seed during germination. *Plant Physiol.* **44**, 773–778.

Abdul-Baki, A. A. (1969b). Relationship of glucose metabolism to germinability and vigor in barley and wheat seeds. *Crop Sci.* **9**, 732–737.

Arndt, W. (1965). The impedance of soil seals and the forces of emerging seedlings. *Aust. J. Soil Res.* **3**, 55–68.

Arnon, I. (1958). The improvement of natural pasture in the Mediterranean region. *Herb. Abstr.* **28**, 225–231.

Ayers, A. D. (1952). Seed germination as affected by soil moisture and salinity. *Agron. J.* **44**, 82–84.

Ayers, A. D., and Hayward, H. E. (1948). A method for measuring the effects of soil salinity on seed germination with observations on several crop plants. *Soil Sci. Soc. Amer., Proc.* **13**, 224–226.

Bernstein, L., and Hayward, H. E. (1958). The physiology of salt tolerance. *Annu. Rev. Plant Physiol.* 9, 25–46.

Black, C. A. (1968). "Soil-Plant Relationships." Wiley, New York.

Bleak, A. T., and Keller, W. (1969). Effects of seed age and pre-planting seed treatment on seedling response in crested wheatgrass. *Crop Sci.* 9, 296–299.

Carslaw, H. S., and Jaeger, J. E. (1959). "Conduction of Heat in Solids." Oxford Univ. Press, London and New York.

Cavazza, L. (1953). L'influenza di basse tensioni dell'acqua sulla germinazione di alcuni semi. *Nuovo G. Bot. Ital.* 60, 759–762.

Chapin, J. S., and Smith, F. W. (1960). Germination of wheat at various levels of soil moisture as affected by applications of ammonium nitrate and muriate of potash. *Soil Sci.* 89, 322–327.

Collis-George, N., and Hector, J. B. (1966). Germination of seeds as influenced by matric potential and by area of contact between seed and soil water. *Aust. J. Soil Res.* 4, 145–164.

Collis-George, N., and Melville, M. D. (1969). Letter to the editor. *Agron. J.* 61, 971–972.

Collis-George, N., and Sands, J. E. (1959). The control of seed germination by moisture as a soil physical property. *Aust. J. Agr. Res.* 10, 628–637.

Collis-George, N., and Sands, J. E. (1961). Moisture conditions for testing germination. *Nature (London)* 190, 367.

Collis-George, N., and Sands, J. E. (1962). Comparison of the effects of physical and chemical components of soil water energy on seed germination. *Aust. J. Agr. Res.* 13, 575–585.

Collis-George, N., and Williams, J. (1968). Comparison of the effects of soil matric potential and isotropic effective stress on the germination of *Lactuca sativa*. *Aust. J. Soil Res.* 6, 179–192.

Cooper, C. S., and MacDonald, P. W. (1970). Energetics for early seedling growth in corn (*Zea mays* L.). *Crop Sci.* 10, 136–139.

Covey, W. (1963). Mathematical study of the first stage of drying of a moist soil. *Soil Sci. Soc. Amer., Proc.* 27, 130–134.

Crank, J. (1956). "The Mathematics of Diffusion." Oxford Univ. Press, London and New York.

Cuddy, T. F. (1963). Germination of the bluegrasses. *Proc. Ass. Off. Seed Anal.* 53, 85–90.

Dasberg, S., Enoch, H., and Hillel, D. (1966a). Effect of oxygen and carbon dioxide concentration on the germination of range grasses. *Agron. J.* 58, 206–209.

Dasberg, S., Hillel, D., and Arnon, I. (1966b). Response of a grain sorghum to seedbed compaction. *Agron. J.* 58, 199–201.

Doering, E. J. (1965). Soil-water diffusivity by the one-step method. *Soil Sci.* 99, 322–326.

Doneen, L. D., and McGillivray, J. H. (1943). Germination of vegetable seeds as affected by different soil moisture conditions. *Plant Physiol.* 18, 524–529.

Ehrler, W. L., and Gardner, W. R. (1971). Dynamics of seed hydration. *Soil Sci. Soc. Amer., Proc.* (in press).

Ellern, S. J., and Tadmor, N. H. (1966). Germination of range plant seeds at fixed temperatures. *J. Range Manage.* 19, 341–345.

Ellern, S. J., and Tadmor, N. H. (1967). Germination of range plant seeds at alternating temperatures. *J. Range Manage.* 20, 72–77.

Evans, W. F., and Stickler, F. C. (1961). Grain sorghum seed germination under moisture and temperature stresses. *Agron. J.* **53**, 369–372.

Evenari, M. (1949). Germination inhibitors. *Bot. Rev.* **15**, 53–194.

Ferguson, R. B. (1962). Growth of single bitterbrush plants versus multiple groups established by direct seeding. *Res. Note, Intermountain Forest Range Exp. Sta.* No. 90.

Gardner, W. R. (1959). Solutions of the flow equation for the drying of soils and other porous media. *Soil Sci. Soc. Amer., Proc.* **23**, 183–187.

Gardner, W. R., Hillel, D., and Benyamini, Y. (1970). Post-irrigation movement of soil water. II. Simultaneous redistribution and evaporation. *Water Resour. Res.* **6**, 1148–1153.

Gill, W. R., and Miller, R. D. (1956). A method for study of influence of mechanical impedance and aeration on the growth of seedling roots. *Soil Sci. Soc. Amer., Proc.* **20**, 154–157.

Gingrich, J. R., and Russell, M. B. (1956). Effect of soil moisture tension and oxygen concentration on the growth of corn roots. *Agron. J.* **48**, 517–521.

Gingrich, J. R., and Russell, M. B. (1957). A comparison of effects of soil moisture tension and osmotic stress on root growth. *Soil Sci.* **84**, 185–194.

Grable, R. A., and Danielson, R. E. (1965). Effects of CO_2, O_2 and soil moisture suction on germination of corn and soybeans. *Soil Sci. Soc. Amer., Proc.* **29**, 12–18.

Griffin, D. M. (1963). Soil moisture and the ecology of soil fungi. *Biol. Rev.* **38**, 141–166.

Hadas, A. (1969). Effects of soil moisture stress on seed germination. *Agron. J.* **61**, 325–327.

Hanks, R. J., and Thorp, F. C. (1957). Seedling emergence of wheat, grain, sorghum and soybeans as influenced by soil crust strength and moisture content. *Soil Sci. Soc. Amer., Proc.* **21**, 357–360.

Harrington, G. T. (1923). Use of alternating temperatures in the germination of seeds. *J. Agr. Res.* **23**, 295–332.

Helmerick, R. H., and Pfeifer, R. P. (1954). Differential varietal response of winter wheat germination and early growth to controlled limited moisture conditions. *Agron. J.* **46**, 560–565.

Hillel, D. (1961). Crust formation in loessial soil. *Trans. Int. Congr. Soil Sci., 7th, 1960* Vol. 1, pp. 330–340.

Hillel, D. (1968). Soil water evaporation and means of minimizing it. *Res. Rep. Fac. Agr., Hebrew Univ.* pp. 98.

Hillel, D. (1971). "Soil and Water: Physical Principles and Processes." Academic Press, New York.

Hudspeth, E. B., and Taylor, H. M. (1961). Factors affecting seedling emergence of Blackwell switchgrass. *Agron. J.* **53**, 331–335.

Hunter, J. R., and Erickson, A. E. (1952). Relation of seed germination to soil moisture tension. *Agron. J.* **44**, 107–110.

Hyder, D. N., and Sneva, F. A. (1956). Seed and plant soil relations as affected by seedbed firmness on sandy loam range land soil. *Soil Sci. Soc. Amer., Proc.* **20**, 416–418.

Isely, D. (1958). Preliminary report on moisture level control in seed testing. *Proc. Ass. Off. Seed Anal.* **48**, 125–131.

Jansson, G. (1959). Germination experiments with water-sensitive barley. *Ark. Kemi* **14**, 161–169.

Kirsop, B. H., and Pollock, I. R. A. (1957). Studies in barley and malt. *J. Inst. Brew., London* **63**, 383–385.

Kleindorst, A., and Brouwer, R. (1965). The effect of temperature on two different clones of perennial ryegrass. *Jaarb. I.B.S.* pp. 29–39.

Knipe, D., and Herbel, C. H. (1960). The effects of limited moisture on germination and initial growth of six grass species. *J. Range Manage.* **13**, 297–302.

Koller, D. (1955). The regulation of germination in seeds. (Review.) *Bull. Res. Counc. Isr., Sect. D* **5**, 85–108.

Koller, D. (1971). Environmental control of seed germination. *In* "Seed Biology" (T. T. Kozlowski, ed.), Vol. 2, Chapter 1. Academic Press, New York.

Laude, H. M. (1957). Comparative pre-emergence heat tolerance of some seeded grasses and of weeds. *Bot. Gaz.* **119**, 44–46.

Laude, H. M., Shrum, J. E., and Biechler, W. E. (1952). The effect of high soil temperatures on the seedling emergence of perennial grasses. *Agron. J.* **44**, 110–112.

Lehmann, E., and Eichele, F. (1931). "Keimungsphysiologie der Gräser (Gramineen)." Enke, Stuttgart.

Lovvorn, R. L. (1945). The effect of defoliation, soil fertility, temperature and length of day on the growth of some perennial grasses. *J. Amer. Soc. Agron.* **37**, 570–582.

McDaniel, R. A. (1969). Relationships of seed weight, seedling vigor, and mitochondrial metabolism in barley. *Crop Sci.* **9**, 823–827.

McGinnies, W. J. (1960). Effects of moisture stress and temperature on germination of six range grasses. *Agron. J.* **52**, 159–163.

McGinnies, W. J. (1962). Effect of seedbed firming on establishment of crested wheatgrass seedlings. *J. Range Manage.* **15**, 230–234.

Manohar, M. S., and Heydecker, W. (1964). Effects of water potential on germination of pea seeds. *Nature (London)* **202**, 22–24.

Mayer, A. M., and Poljakoff-Mayber, A. (1963). "The Germination of Seeds." Pergamon Press, Oxford.

Morinaga, T. (1926). Effect of alternating temperatures upon the germination of seeds. *Amer. J. Bot.* **13**, 141–158.

Morris, I. J., Campbell, W. F., and Wiebe, H. H. (1968). A refractometric method for testing seed coat permeability. *Agron. J.* **60**, 79–80.

Negbi, M. (1957). "Germination of Desert Pasture Plants." FAO Working Party on Mediterranean Pasture and Fodder Development, Tel Aviv.

Negbi, M., Rushkin, E., and Koller, D. (1966). Dynamic aspects of water-relations in germination of *Hirschfeldia Incana* seeds. *Plant Cell Physiol.* **7**, 363–376.

Norris, W. E., Wiegand, C. L., and Johanson, L. (1959). Effect of CO_2 on respiration of excised onion root tips in high O_2 atmospheres. *Soil Sci.* **88**, 144–149.

Oota, Y. (1957). Carbohydrate change in water-absorbing bean germ axes. *Physiol. Plant.* **10**, 910–921.

Parker, J. J., and Taylor, H. M. (1965). Soil strength and seedling emergence relations. I. Soil type, moisture tension, temperature and planting depths effects. *Agron. J.* **57**, 289–291.

Phillips, R. E. (1968). Water diffusivity of germinating soybean, corn, and cotton seed. *Agron. J.* **60**, 568–571.

Phillips, R. E. (1969). Letter to the editor. *Agron. J.* **61**, 971.

Phillips, R. E., and Kirkham, D. (1962). Mechanical impedance and corn seedling root growth. *Soil Sci. Soc. Amer., Proc.* **26**, 319–322.

Pollock, B. M., and Toole, V. K. (1961). After ripening, rest period, and dormancy. *In* "Seeds," pp. 106–112, Yearbook Agr., U. S. Dep. Agr., U. S. Government Printing Office, Washington.

Read, D. W. L., and Beaton, J. D. (1963). Effect of fertilizer, temperature and moisture on germination of wheat. *Agron. J.* **55**, 287–290.

Richards, L. A. (1965). Physical condition of water in soil. *In* "Methods of Soil Analysis," Agron. Ser. No. 9, Vol. 1, pp. 128–137. Amer. Soc. Agron., Madison, Wisconsin.

Richards, L. A., and Wadleigh, C. H. (1952). Soil water and plant growth. *In* "Soil Physical Conditions and Plant Growth" (B. T. Shaw, ed.), pp. 74–253. Academic Press, New York.

Richards, S. J., Hagan, R. M., and McCalla, T. M. (1952). Soil temperature and plant growth. *In* "Soil Physical Conditions and Plant Growth" (B. T. Shaw, ed.), 303–480. Academic Press, New York.

Rodgers, J. B. A., Williams, G. G., and Davis, R. L. (1957). A rapid method for determining winter hardiness in alfalfa. *Agron. J.* **49**, 85–92.

Sedgley, R. H. (1963). The importance of liquid seed contact during the germination of *Medicago tribuloides. Aust. J. Agr. Res.* **14**, 646–654.

Shull, C. A. (1920). Temperature and rate of moisture intake in seeds. *Bot. Gaz.* **69**, 361–390.

Siegel, S. M., and Rosen, L. A. (1962). Effects of reduced oxygen tension on germination and seedling growth. *Physiol. Plant.* **15**, 437–444.

Slatyer, R. O. (1969). Physiological significance of internal water relations to crop yield. *In* "Physiological Aspects of Crop Yield" (J. D. Eastin, ed.). Amer. Soc. Agron. and Crop Sci. Soc. Amer., Madison, Wisconsin.

Sosebee, R. E., and Herbel, C. H. (1969). Effects of high temperatures on emergence and initial growth of range plants. *Agron. J.* **61**, 621–624.

Sprague, V. G. (1943). The effects of temperature and day length on seedling emergence and early growth of several pasture species. *Soil Sci. Soc. Amer., Proc.* **8**, 287–294.

Stanley, R. G. (1958). Gross respiratory and water uptake patterns in germinating sugar pine seeds. *Physiol. Plant.* **11**, 503–515.

Stanley, R. G., and Butler, W. L. (1961). Life processes of the living seed. *In* "Seeds," pp. 88–94, Yearbook Agr., U. S. Dep. Agr., U. S. Government Printing Office, Washington.

Stiles, W. (1960). Respiration. *In* "Handbuch der Pflanzenphysiologie" (W. Ruhland, ed.), Vol. 12, Part 2, pp. 465–492, Springer-Verlag, Berlin and New York.

Stotzky, G., and Cox, E. A. (1962). Seed germination studies in *Musa.* II. Alternating temperature requirement for the germination of *Musa babisiana. Amer. J. Bot.* **49**, 763–770.

Stout, B. A., Synder, F. W., and Buchele,W. F. (1960). The effect of soil compaction on moisture absorption by sugar beet seeds. *Mich., Agr. Exp. Sta., Quart. Bull.* **42**, 528–557.

Tadmor, N. H., and Hillel, D. (1956). Survey of the possibilities for agricultural development of the southern Negev. *Bull. Agr. Res. Sta., Rehovot, Israel* No. 61.

Tadmor, N. H., and Hillel, D. (1968). Establishment and maintenance of seeded dryland range under semi-arid conditions. *Spec. Res. Rep. Volcani Inst., Israel.*

Tadmor, N. H., Dasberg, S., Ellern, S. J., and Harpaz, Y. (1964). Establishment and maintenance of seeded dryland range under semi-arid conditions. *Res. Rep.* for the period April 1963–March 1964, submitted to U. S. Department of Agriculture.

Thornton, N. C. (1944). Carbon dioxide storage. XII. Germination of seeds in the presence of carbon dioxide. *Contrib. Boyce Thompson Inst. Pl. Res.* **13**, 355–360.

Toole, E. H., Hendricks, S. B., Borthwick, H. A., and Toole, V. K. (1956). Physiology of seed germination. *Annu. Rev. Plant Physiol.* **7**, 299–325.

Triplett, G. B., and Tesar, M. B. (1960). Effects of compaction, depth of planting and soil moisture tension on seedling emergence of alfalfa. *Agron. J.* **52**, 681–685.

Troughton, A. (1957). The underground organs of herbage grasses. Distribution of roots in the soil. *Bull. Commonw. Bur. Pastures Field Crops, Hurley Berks* **44**, 41–46.

Uhvits, R. (1946). Effect of osmotic pressure on water absorption and germination of alfalfa seeds. *Amer. J. Bot.* **33**, 278–285.

Unger, P. W., and Danielson, R. E. (1965). Influence of oxygen and carbon dioxide on germination and seedling of corn. *Agron. J.* **57**, 56–58.

van Overbeek, J. (1966). Plant hormones and regulators. *Science* **152**, 721–731.

Vlamis, J., and Davis, A. R. (1943). Germination, growth and respiration of rice and barley seedlings at low O_2 pressures. *Plant Physiol.* **18**, 685–692.

Wareing, P. F. (1963). The germination of seeds. *Vistas Bot.* **3**, 195–227.

Went, F. W. (1961). Problems in seed viability and germination. *Proc. Int. Seed Testing Ass.* **26**, 674–685.

Whalley, R. D. B., and McKell, C. M. (1967). Interrelation of carbohydrate metabolism, seedling development, and seedling growth rate of several species of *Phalaris. Agron. J.* **59**, 223–226.

Whitney, M., and Cameron, F. K. (1904). Investigation in soil fertility. *U. S., Dep. Agr., Bur. Soils Bull.* **23**.

Wiggans, S. C., and Gardner, E. B. (1959). Effectiveness of various solutions for simulating drought conditions as measured by germination and seedling growth. *Agron. J.* **51**, 315–318.

Wilson, A. M., Nelson, J. R., and Goebel, C. J. (1970). Effects of environment on the metabolism and germination of crested wheatgrass seeds. *J. Range Manage.* **23**, 283–288.

Wood, G. M., and Buckland, H. E. (1966). Survival of turfgrass seedlings subjected to induced drouth stress. *Agron. J.* **58**, 19–23.

Woodstock, L. W., and Grabe, D. F. (1967). Relationships between seed respiration during imbition and subsequent seedling growth in *Zea mays* L. *Plant Physiol.* **42**, 1071–1076.

Wright, L. N. (1971). Drought influence on germination and seedling emergence. *In* "Drought Injury and Resistance in Crops" (K. L. Larson and J. D. Eastin, eds.), pp. 19–44, CSSA Spec. Publ. No. 2, Crop Sci. Soc. Amer., Madison, Wisconsin.

Yemm, E. W. (1965). Respiration of germinating seeds. *In* "Plant Physiology" (F. C. Steward, ed.), Vol. 4A, Chapter 3. Academic Press, New York.

Younis, M. A., Stickler, F. C., and Sorenson, E. L. (1963). Reactions of seven alfalfa varieties under simulated moisture stresses in the seedling stage. *Agron. J.* **55**, 177–182.

WATER DEFICITS AND REPRODUCTIVE GROWTH

Merrill R. Kaufmann

DEPARTMENT OF PLANT SCIENCES, UNIVERSITY OF CALIFORNIA, RIVERSIDE, CALIFORNIA

I. INTRODUCTION

During the reproductive cycle, the physiology of many plants is dominated by internal processes and conditions associated with the development of fruits. This chapter deals with the interrelationships among plant water relations, fruit physiology, fruit structure, and growth. Production of viable seeds is an important step in the life cycle of most species of higher plants. Agriculturists, taking advantage of plant requirements to regenerate, have harvested fruits for many centuries because of their food or fiber content.

Because of the great agricultural interest in production of fruits or fruit parts, much research has been devoted to the effects of water availability on reproductive growth. Most of the work has centered on determining the effects on yield of limited water supply at various stages of growth within the reproductive cycle. The type of yield studied has been determined largely by the commercial value of fruit parts. For example, cereals are grown for their seed, and the remainder of the reproductive organ is normally discarded. In contrast, the seeds of many plants such as apples, peaches, and strawberries are of less value than the fleshy portions of the reproductive structure. Cotton is grown both for its seed (oils) and for the surrounding cellulose fibers. Agricultural aspects of water deficits and reproductive growth have been reviewed thoroughly by Salter and Goode (1967) and will be noted here only occasionally. However, much of the agriculturally oriented research provides information useful for understanding the water relations and physiology of the reproductive cycle. No attempt is made here to cite all the literature bearing useful information. Where literature is abundant, representative examples have been selected to illustrate certain phenomena. The usual case, however, is that too little physiological information is available, and some citations are only peripherally related to the subject being discussed.

Growing fruits have much in common with shoot meristematic regions, particularly in the early stages of growth. Cell division and tissue differentiation begin with initiation of flower bud primordia and continue well beyond the time of fertilization. Growth of the flower and fruit involves rapid accumulation of dry matter and water. The growing fruit is a strong sink for carbohydrates, generally resulting in reduced vegetative growth. Particularly in large, fleshy fruits the vascular system must be well-developed to facilitate distribution of large quantities of carbohydrate to the growing tissues. Water is the major component of fruits. Water content varies from less than 50 to more than 90%, depending upon the type of the fruit and stage of development (Table I). Water content as a percent of fresh weight generally decreases as the fruit enlarges and ripens. This decrease in percent water content is caused both by dry matter accumulation and by actual loss of water.

The reproductive cycle can be divided into three stages for consideration of the effects of water stress on growth. First is the phase of flowering, beginning with the initiation of flowers and culminating in fruit set. This phase largely determines the number of ripe fruits which will be produced, but not necessarily their individual size. Fruit size depends primarily upon conditions during the second phase, the period of fruit enlargement, when considerable amounts of carbohydrates and water are transported into the fruit. The third phase is the period of fruit ripening. Ripening connotes different morphological features which depend upon the type of fruit and frequently upon taste or appearance. In cereals such as wheat, for example, ripening involves conversion of sugars into starch and dehydration until

TABLE I
WATER CONTENT OF FRUITS[a]

Fruit	Water content (%)		Reference
	Immature	Mature	
Valencia orange			
Pericarp	—	70–77	Turrell *et al.* (1964)
Pulp	—	86–88	Turrell *et al.* (1964)
Marsh grapefruit			
Pericarp	—	75–84	Turrell *et al.* (1964)
Pulp	—	87–91	Turrell *et al.* (1964)
Avocado	84–87	—	Haas (1936)
Red oak			
Nut	95	52	Chaney and Kozlowski (1969a)
Cup	67	48	Chaney and Kozlowski (1969a)
Red maple			
Pericarp	84	72	Chaney and Kozlowski (1969a)
Seed	83	66	Chaney and Kozlowski (1969a)
Black cherry			
Embryo	63	37	Chaney and Kozlowski (1969a)
Pericarp and seed coat	80	73	Chaney and Kozlowski (1969a)
Jack pine cone (2 yr.)	—	43	Chaney and Kozlowski (1969a)

[a] All values are expressed as percent of fresh weight.

the moisture content becomes quite low. In other small fruits such as maple seeds, water content is high at ripening, and dehydration similar to that in cereals would result in death. Ripening in large, fleshy fruits such as cherries, citrus, and peaches is generally defined in terms of food quality rather than maturation of viable seeds.

II. WATER DEFICITS AND GROWTH OF FRUITS AT DIFFERENT STAGES OF DEVELOPMENT

A. FLOWERING AND FRUIT SET

The effects of water deficits on flowering are remarkably complex because the flowering process can be influenced in many ways. The importance of water stress during the early stages of reproductive growth was examined by Salter and Goode (1967), who summarized 132 reports on the effects of water deficits on yield of cereals. Over one hundred of these reports demonstrated that yield was affected most adversely when water stress occurred during the periods of internode elongation just before flowering, ear emergence from the leaf sheath, and flower opening. Similar effects occur in other plants. For example, Dubetz and Mahalle (1969) found that a soil water stress of 8 bars at flowering reduced the weight of green bush bean pods by 71%, while a similar stress before flowering reduced yield by 53% and after flowering by 35%. Inhibition of flower initiation after the first flowers had formed was greater in plants subjected to water stress at the time early flowers formed than in plants stressed before or after early flowering.

No specific effects of water stress on flowering appear to be common in all plants, however. It is apparent that environmental conditions interact with internal physiological features such that the limiting factor varies among plants, geographical location, season, etc. The importance of rather small environmental differences is shown by Campbell et al. (1969a), who found that seed set in wheat was higher at 60% relative humidity than at 80% when water availability in the soil was not limiting.

Little is known about the effects of water stress on floral initiation, but available evidence generally suggests that water deficits reduce the number of flowers produced. For example, Hartmann and Panetsos (1961) found that wilted olive plants had fewer flowers per inflorescence than did plants receiving adequate moisture (Table II). The higher percentage of perfect flowers resulting when wilting occurred early in the floral development period probably resulted in a greater availability of nutrients to the reduced number of flowers. Even though the number of fruits produced per 100 inflorescences was greater in plants subjected to an early water stress, the total number of fruits was considerably lower than that of well-watered plants because the total production of flowers per tree was sharply reduced by water stress. Water stress also reduced the number of flower buds and delayed the time of differentiation in apricots (Brown, 1953). Fruits from

TABLE II

EFFECT OF WATER STRESS ON BAROUNI OLIVES[a,b]

Treatment	No. of flowers per inflorescence	Percent perfect flowers	No. of fruits per 100 inflorescences
No soil water deficiency	15.7	27.4	3.3
Water stress from March 3 to 11	4.9	65.4	4.3
Water stress from March 7 to 21	8.7	4.0	0.1
Water stress from March 18 to April 4	8.3	9.3	0.6
Water stress from March 1 to April 4	6.7	0.6	0.3

[a] From Hartmann and Panetsos (1961).

[b] Water stress was imposed by allowing container-grown plants to become wilted; small amounts of water were provided to prevent excessive drying during the treatment period.

late-developing flowers had longer stems and smaller stones than normal; the fruits were smaller and matured several weeks later. The decrease in number of flowers in apricots subjected to water stress resulted from fewer nodes bearing flowers (Jackson, 1969). The number of flower buds per node was not affected. Modlibowska (1961) found that early and prolonged water stress in apple trees caused smaller flowers and slower floral development. Flowers developing from buds having a reduced moisture content 2 to 4 weeks before opening failed to open properly, and petals abscised while stuck together ("parachute" blossoms).

Similar effects of water stress on flower production have been observed in annual crops. Vittum *et al.* (1963) observed that irrigation of tomatoes, snap beans, and lima beans increased the percent of flowers that set fruit and decreased blossom-end rot and cracking in tomatoes. Pollination failure occurred in corn subjected to water stress during periods of tasseling, silking, and pollination (Robins and Domingo, 1953). Yield was reduced because grain formed on only part of the ear. Lack of fertilization apparently was related to unreceptiveness of silks, since viable pollen was available from adjacent moist plots. Schreiber *et al.* (1962) found that the number of rows of seeds in corn ears was not affected by water availability at any stage of development, while early application of nitrogen had an effect.

Flower and early fruit abscission is generally increased and seed or fruit set is often decreased by plant water stress, but the effect of water stress is not universal. Erickson and Brannaman (1960) observed that the highest rate of flower bud and leaf abscission of Valencia and Washington navel oranges occurred 3 days after a hot dry spring day. Soil and root temperatures were still low, and presumably high resistance of the root system to water absorption resulted in an abnormally high plant water stress. A similar hot day in the summer, when root temperatures were warmer, had no effect on the rate of abscission, but the stage of development also had changed. Grimes *et al.* (1970) showed that water stress resulting in severe leaf wilting of cotton during the period of early flowering caused shedding of new flower buds (squares) but had no effect on current flowering or on boll retention. A similar stress during peak flowering caused square shedding and also reduced boll retention, while a late stress reduced current flowering and almost completely prevented boll retention. In contrast to the increased flower abscission apparently related to water stress in oranges and cotton, Dubetz and Mahalle (1969) found no change in flower abscission of bush beans when soil water stress reached 8 bars before, during, or after flowering.

Occasionally increased fruit set has been observed when plants are subjected to water stress. Kaufmann (1968) found that reducing the water potential of pepper plants in nutrient solutions from -2.9 to -6.2 bars during the first week of flowering increased fruit set from 5 out of 217 flowers to 22 out of 210 flowers. Campbell *et al.* (1969a) found that seed set in primary and secondary florets of two wheat varieties was greater when soil water stress reached 15 bars than at a soil water stress of 1.4 bars. A period of water stress may be necessary in certain tropical plants to induce flowering. Alvim (1960, 1964) observed that regularly irrigated coffee plants and bamboos growing in moist soil did not flower, but irrigation or precipitation after a drought resulted in flowering. Nakata and Suehisa (1969) showed that flowering of *Litchi chinensis* occurred on only 50% of the branches when soil water was not limiting, but soil water stress caused flowering on 75 to 95% of the branches, and flowering occurred earlier. The amount of flowering was negatively correlated with growth in circumference of the tree trunk.

Environmental conditions and internal plant water stress may have direct effects on the fertilization process. When relative humidity is high, pollen may not be dispersed from the anther. Dry conditions, on the other hand, may result in excessive dehydration of pollen or the stigma. Coyne (1968) also observed that the relative position of anther and stigma in certain tomato varieties was altered by water stress. Such changes in flower structure could reduce the likelihood of viable pollen reaching the stigma.

Abscission of young fruits is common in oranges and some other species. June drop of oranges is generally related to environmental conditions when the new fruit is beginning to expand. For example, Jones and Cree (1965) examined data collected over more than three decades and concluded that yield was significantly related to the maximum temperature reached in June. Coit and Hodgson (1918) concluded that June drop of navel oranges was related to water stress. It is difficult to separate the effects on early fruit abscission of temperature and other factors from those of water stress, because high temperatures are often accompanied by high plant water stresses. Undoubtedly fruit abscission results from interactions of these environmental effects with internal physiological processes such as respiration and carbohydrate supply to the fruit.

The high requirement for carbohydrates during reproductive growth may contribute indirectly to drought sensitivity at flowering by affecting the root system. Salter and Drew (1965) observed that root growth of a pea variety decreased sharply after flower initiation and some older roots died. The reduced effectiveness of the root system in absorbing water could result in a pronounced plant water stress when only a moderate stress exists in the soil. This study and others point to the critical need for measuring plant water stress for careful evaluation of the role of water deficits in the flowering process.

B. Fruit Enlargement

Not surprisingly, most research on water availability and fruit enlargement has been conducted on large fruits such as cherries, oranges, and cotton rather than on grains and other small fruits. Perhaps the most detailed evidence of the effects of climate and soil water availability on enlargement has been accumulated for citrus fruits. The climatic factors having the greatest effect on fruit enlargement are also likely to have the most effect on internal plant water relations. Hales *et al.* (1968) found that the daily volume increase of Valencia oranges was significantly affected by soil water tension and the maximum and average daily relative humidities. Lombard *et al.* (1965) concluded that when the soil was allowed to reach a water tension of 1.0 bar before irrigation, evaporative power of the atmosphere affected the rate of volume increase of navel oranges, but rewatering at 0.2 bar nullified environmental effects on fruit enlargement. Erickson and Richards (1955) found that irrigation at a soil water tension of 0.3 bar resulted in larger Valencia oranges than irrigation at 0.7 bar (Fig. 1). The size class having the most fruit was 6.35 cm

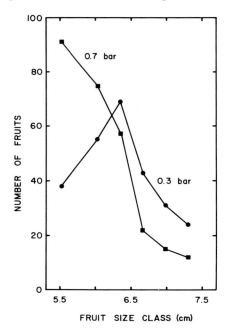

Fig. 1. Influence of soil water stress on distribution of fruit sizes in 17-yr-old Valencia orange trees. Trees were irrigated when tensiometers indicated 0.3 bar at a depth of 30 cm or 0.7 bar at a depth of 60 cm. From Erickson and Richards (1955).

diameter when water stress was low compared with 5.52 cm at the higher stress.

Similar effects of water stress on fruit enlargement have been observed in other plants. Feldstein and Childers (1965) increased the size of peaches grown in Pennsylvania by supplementing normal rainfall each year with irrigation, even though total rainfall was below average only one of the six years of study. Anderson and Kerr (1943) noted that the size of cotton bolls was reduced when enlargement occurred during dry weather. Water stress reduced the period of boll enlargement to 15 days compared with a normal period of 16–18 days. Water deficits during boll enlargement also reduced the length of fibers produced (Sturkie, 1934). Reuther and Crawford (1945) found that irrigating date palms when the soil water stress reached about 0.8 bar reduced fruit size 10–15% and leaf elongation 15–20% from those of control plants watered at a soil water stress of 0.1–0.2 bar. Zimmerman (1958) noted that seeds of castor bean may not fill out properly if leaves are allowed to wilt during morning hours.

Water deficits also have reversible diurnal effects on fruit size, but diurnal changes in size may occur only at certain stages of development. Diurnal changes in fruit volume are also discussed in Chapter 1. The diurnal growth pattern of developing cotton bolls is shown in Fig. 2 (Anderson and Kerr, 1943). During the stage of cell enlargement (until day 15) only moderate fluctuations were observed in the growth rate. After diameter growth ceased, however, the boll diameter decreased sharply during the day and fully recovered at night. The diameter of young cotton bolls increased even when leaves were visibly wilted by 10 A.M. During boll

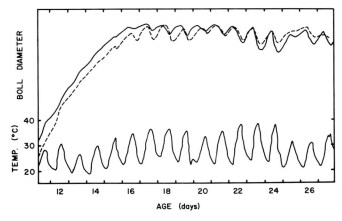

Fig. 2. Growth and diurnal changes in diameter of two cotton bolls for a 17-day period beginning 11 days after flowering. Air temperature is also given. Rain prevented shrinkage on day 25. From Anderson and Kerr (1943).

enlargement, fiber growth involves primary wall formation. Secondary wall growth begins after the fibers and boll have reached their full size, and it is at this time that the water content and diameter of the boll become responsive to diurnal plant water deficits. A reason for this change in response is discussed below. Other fruits such as oranges exhibit diurnal changes in volume by the time they reach 10% of their mature volume (Elfving and Kaufmann, 1971).

Kozlowski (1968) also found an insensitivity of young Montmorency cherries to diurnal water deficits. As fruits matured, similar environmental conditions resulted in diurnal decreases in fruit diameter, but during ripening diurnal changes became less pronounced. Chaney and Kozlowski (1969b) reported that diameter changes of forest tree fruits were related to atmospheric vapor pressure deficits. As the vapor pressure deficit increased during the day, fruit diameter generally decreased; evening decreases in vapor pressure deficits were accompanied by diameter increases. Curiously, however, Chaney and Kozlowski (1969c) also found that the diameter of nearly ripe English Morello cherries increased during the day when the vapor pressure deficit increased, even though leaf thickness decreased. The authors concluded that thermal effects were responsible for this behavior.

Most of the environmental conditions and also soil water availability have their effects on changes in fruit size by altering the internal plant water status. Long-term, irreversible effects of water deficits on fruit enlargement probably result from decreased turgor pressure in the fruit tissues and, during severe stress, from reduced production of photosynthate for growth. The role of water stress in fruit enlargement was clearly demonstrated in olives treated with a film-forming antitranspirant (Fig. 3). Fruits on treated plants were 3% larger in diameter than fruits from untreated plants 16 days after treatment (14 days after rain). Leaf diffusion resistance and water potential were higher in treated plants than in control plants (Davenport et al., 1969).

Perhaps night conditions are more important than day conditions in some fruits. While the maximum relative humidity reached during the night had a significant effect on daily growth of Valencia oranges, the daytime minimum exerted no significant control (Hales et al., 1968). In other cases warm night temperatures are important for rapid fruit growth (Cooper et al., 1963). Many plants are known to grow more at night than during the day. More research on night fruit water relations and temperature may reveal considerable information about environmental effects on fruit growth, particularly in citrus fruits where strength of the pericarp is greatest at low temperatures and high water potentials (Kaufmann, 1970a).

The continuous enlargement of some fruits such as cotton bolls even

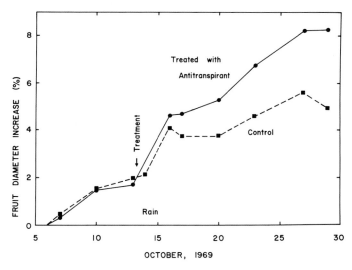

Fig. 3. Effect of a film-forming antitranspirant on the increase in diameter of olive fruits. Natural rain occurred on October 14 and 15 (Davenport *et al.,* 1969).

during periods of water stress suggests that the water relations of young fruits may differ appreciably from those of larger fruits. Shoot meristematic regions are known to compete favorably with the rest of the plant for water (Slatyer, 1957). In cotton bolls, when fiber elongation and boll enlargement cease, the osmotic potential of both fibers and seeds becomes less negative (Kerr and Anderson, 1944). Therefore, at a given low water potential young, expanding bolls may retain turgor because of a more negative osmotic potential whereas older bolls lose their turgor because the osmotic potential is too high.

Diurnal changes in fruit diameter can be explained, with rare exceptions, on the basis of fluctuations in plant water stress. Various reports have related midday decreases in fruit diameter or volume to environmental conditions favorable for transpiration. Direct evidence that changes in diameter of orange fruits are correlated with internal water deficits has been collected (Elfving and Kaufmann, 1971). During the day, fruit diameter decreased when stomata were open and water potential of the leaves was low (Fig. 4). However, closure of stomata in the evening and reduction of the vapor pressure deficit allowed recovery from water stress (water potential increased), and fruit diameter increased. Opening of stomata the following morning caused these changes to be repeated the next day. Similar changes have also been observed during the second day of a 48-hr irrigation. Water potential of orange fruit pericarp tissue has been observed to decrease 5.6 bars from −5.1 bars at 7:30 A.M. to −10.7

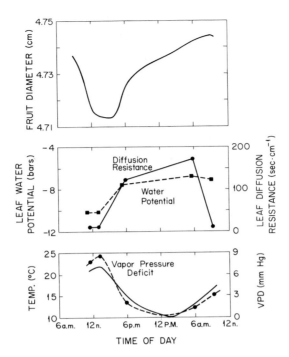

Fig. 4. Diurnal changes in fruit diameter, leaf water potential, and leaf diffusion resistance of Valencia oranges, with air temperature and vapor pressure deficits for the same period. Soil moisture tension at a depth of 46 cm was 0.2 bar (Elfving and Kaufmann, 1971).

bars at 1:30 P.M. (Kaufmann, 1970b). Klepper (1968) also found that diurnal changes in pear fruit diameter were directly related to changes in fruit water potential.

Recovery from fruit shrinkage caused by water stress often begins quickly after water is supplied to the plant. A rapid decrease in size of Montmorency cherries approaching maturity was stopped by watering, and within 2 hr more than half of the diameter decrease was recovered (Kozlowski, 1968). Younger fruits responded more slowly to irrigation, however. Tukey (1964) found that cucumbers began to increase in diameter 15 min after watering and recovered completely in the next 40 min. While the onset, rate, and duration of recovery after watering depend upon severity of the stress and upon the resistances within the plant to movement of water to the fruit, it is clear that fruit shrinkage caused by water stress is readily reversible. Hales et al. (1968) found that the daily volume increase of Valencia oranges was 2 to 4 times greater on days of irrigation than the average volume increase between irrigations.

C. Fruit Ripening

While the biochemical physiology of fruit ripening and its control are receiving considerable attention, little is known about the role of water availability in the ripening process. Extensive changes occur during fruit ripening, often including massive conversions of chemical components, alterations in cell membranes, changes in cell wall structure, and formation of abscission zones. A feature common to most of these processes is high enzyme activity. While researchers often acknowledge that water deficits can influence ripening, little is known about the specific effects of water stress on the physiology of ripening, particularly on the enzymatic processes involved.

Simons (1963) studied the anatomy of apple fruit abscission in irrigated and unirrigated orchards in Illinois. Irrigation increased fruit size and retarded development of the abscission zone between the fruit pedicel and cluster base in the three varieties examined. However, as late as September 30 water and nutrients apparently could still be transported into the fruits of stressed trees even though the abscission zone was well developed. In contrast, observations on navel orange fruits suggest that water is not translocated into ripe fruits as readily as immature fruits (Kaufmann, 1970c). Reuther and Crawford (1945) observed that date fruits on plants irrigated at a soil water tension of 0.8 bar ripened earlier than those on plants irrigated at 0.1–0.2 bar. Aljibury and May (1970) found that less frequent irrigation of tomatoes resulted in a greater percentage of ripe fruit at harvesting but the yield was reduced (Table III).

While ripening may be hastened by water stress in some species and under some conditions, it may be delayed in other situations. For example, when most of the available soil water was depleted before the first irriga-

TABLE III

Effects of Irrigation Frequency on Production of Processing
Tomatoes in San Joaquin Valley, California[a,b]

Irrigation treatments[c]	Percent green fruit at harvest	Yield (tons/acre)	Percent solids	Percent sunburned fruits
10 day (24)	21.90	33.0	5.50	1.07
10 day (12)	20.15	31.9	5.85	1.07
15 day (24)	17.22	27.2	5.90	1.55
15 day (12)	17.32	26.5	6.50	1.45
20 day (24)	13.87	29.4	6.25	1.42
20 day (12)	10.72	27.4	6.45	1.47

[a] From Aljibury and May (1970).
[b] Plants were irrigated at 10-, 15-, and 20-day intervals.
[c] Numbers in parenthesis indicate hours of irrigation.

tion of cotton, the crop was less mature at harvest than it was when soil water depletion was not as great (Table IV). Fruit ripening also is delayed in Elberta peaches (Feldstein and Childers, 1965). Irrigated peaches ripened uniformly, but unirrigated peaches ripened slightly later and more irregularly, requiring two harvests instead of one. Fruit color and splitting of pits were not affected by irrigation, however. Thus the effects of water deficits can include both hastening and retarding of ripening. If severe enough, water stress can accelerate the abscission process, leading in some cases to premature dropping of fruits. If water stress results in only partial development of the abscission zone, the supply of water, minerals, or carbohydrates to the fruit may be decreased, resulting in an extension of the period required for the fruit to reach maturity. The irregularity of ripening of peaches subjected to water stress observed by Feldstein and Childers (1965) was also reported for bush bean pods (Dubetz and Mahalle, 1969).

A change in water content of certain fruit parts, particularly the seed, often accompanies fruit ripening. Grains, for example, become dehydrated as they ripen and are harvested when the plant and seeds have dried thoroughly. Every farmer knows that humid, cool weather postpones dehydration and therefore delays harvesting. The sensitivity of the process of seed dehydration to environmental conditions depends largely on the type of fruit structure. Wheat and barley seeds are directly exposed to the atmosphere, whereas corn seeds are protected by a husk and cucumber or tomato seeds are enclosed in a fleshy placenta. Prokofév and Kholodova (1968) observed that the moisture content of poppy seeds enclosed in a capsule decreased at a steady rate for the first 18 days after flowering regardless of environmental conditions. During the next 20 days, however, high temperature and low relative humidity accelerated seed dehydration and cool, moist conditions delayed it. Thus the time required to produce

TABLE IV

INFLUENCE OF DATE OF FIRST IRRIGATION ON GROWTH, MATURITY, AND YIELD OF COTTON IN SAN JOAQUIN VALLEY, CALIFORNIA[a,b]

Date of first irrigation	Available soil water (%)[c]	Plant height (cm)		Crop maturity (Oct. 2)[d]	Seed cotton yield (lb/acre)
		June 24	Nov. 4		
May 15	80	58	86	92	2250
June 3	52	56	89	94	2610
June 24	20	41	94	84	2920
LSD (5%)	—	2.5	5	3	290

[a] From Stockton et al. (1967).
[b] Soil was preirrigated in early March; emergence occurred in mid-April.
[c] Available soil water at the time of irrigation.
[d] Based on percent of total crop harvested on Oct. 2.

mature seeds was affected by environmental factors only during the last half of the period after flowering.

Dehydration of seeds is a common feature in ripening, and it can proceed even though the environment immediately surrounding the seed may contain considerable water. Abrol and McIlrath (1966) observed that the percent and absolute water contents of tomato seeds reached a minimum 40 to 50 days after flowering. When fruits were detached for 5 days the pattern of dehydration was similar to that of attached fruits. Thus the change in seed water content was caused by internal fruit processes, perhaps involving compositional changes in the seed. Water content of the fruit placenta and pericarp remained above 90% throughout the ripening period and apparently had no influence on water content of the seeds (McIlrath et al., 1963). Prokofév and Kholodova (1968) observed that even though the husk leaves of maize and the fleshy pericarp of pepper kept the relative humidity of the air surrounding the growing seeds quite high, the seeds lost water during the period of 15 to 40 days after flowering.

D. TOTAL FRUIT PRODUCTION

Numerous reports of the effects of water availability on yield of fruits or seeds can be found in the literature. Virtually all reports, except those dealing with saturated soil and poor aeration, indicate that yield decreases as water availability decreases. Salter and Goode (1967) have reviewed the effects of water deficits on yield of crop plants. Therefore only a few recent reports will be emphasized here.

The effects of water availability on fruit or seed yield have generally been studied by analysis of precipitation and yield records or by irrigation studies based on soil water stress measurements. The latter experiments are more pertinent in this chapter because soil and plant water stresses are generally well correlated. Maurer et al. (1969) examined the effects of three levels of soil water stress on growth and yield of bush beans (Table V). Fresh weights of the plants and pods decreased when soil water stress was allowed to reach 0.3 or 0.5 bar. The percent dry matter significantly increased in response to water stress in the pods but not in the rest of the plant. Fresh weight of the pods and plants was more severely decreased when a water stress of 0.5 bar occurred after first flowering than before first flowering. Size and number of the pods was also decreased by water stress. While the amount of water used by the plants decreased sharply as the water stress regime became more severe, the water use efficiency (weight of dry matter produced per volume of water) was not affected significantly. Dubetz and Mahalle (1969) also observed that the yield of bush bean pods was affected by water stress but the most sensitive period

TABLE V

EFFECTS OF SOIL WATER REGIMES ON YIELD AND WATER USE OF
BUSH BEANS (VAR. HARVESTER)[a,b]

Soil water stress treatment	Fresh wt. of plants less pods (gm)	Fresh wt. of pods (gm)	Dry matter in pods (%)	Total water used (l)
Rewatered at −0.2 bar	1061	806	8.75	116.5
Rewatered at −0.3 bar	760	660	9.66	86.8
Rewatered at −0.5 bar	612	461	11.52	68.6
−0.2 bar until first flowers, then −0.5 bar	701	484	12.91	80.5
−0.5 bar until first flowers, then −0.2 bar	765	674	8.82	85.3

[a] From Maurer et al. (1969).

[b] Experiments were performed in weighing lysimeters (12 plants each), with irrigation needs determined on the basis of decrease in available water.

was not the same. A soil water stress of 8 bars during flowering reduced yield from 334 gm (control) to 96 gm, whereas similar stresses before and after flowering reduced the yield to 158 and 217 gm.

Yield of processing tomatoes was reduced when the interval between irrigations was extended from 10 to 20 days, and irrigation for a 24-hr period resulted in higher yields than irrigation for a 12-hr period (Table II). Higher yields from frequently irrigated plants were attributable to better fruit set and to a lower percentage of sunburned fruits. Flocker and Lingle (1961) found that yield of tomatoes was highest when soil was irrigated at a tension of 2.0 bars. A soil water stress of 7.0 bars reduced yield, and irrigation at 0.7 bar was excessive, causing decreased yield and delayed maturation. Stockton et al. (1967) observed that utilization of most of the available soil water before the first irrigation of cotton delayed vegetative growth and maturation but increased total growth and yield (Table IV). In an analysis of irrigation and nitrogen requirements for cotton, Grimes et al. (1967) found that increments of irrigation water resulted in proportional increases of cotton lint yield, but the amount of response to water depended upon the availability of nitrogen. Maximum yield of alfalfa seed occurred when the mean annual integrated soil water potential was between −2 and −8 bars (Taylor et al., 1959). Adequate water was needed until blossoming, but water stress could reach 15 bars by the time of harvest without affecting yield.

The effects of soil water stress on corn yield were examined by Fulton (1970). When water stress at a depth of 40 cm exceeded 5 bars yield was reduced. Grain yield was affected severely by water stress after the time of pollination. Closer spacing of plants increased yield only when water was not limiting. Comparisons at two plant densities were made of yield and rate of water use of unirrigated plants and plants irrigated when soil water stress reached 5 bars (Table VI). Since the year of study was dry, irrigation increased the yield and amount of water used. The water use efficiency also was increased by irrigation, but the increase was more pronounced when plant density was high. The higher plant density resulted in greater evapotranspiration but lower yield in unirrigated plants, and irrigation caused a proportionately greater increase in yield than in evapotranspiration.

Campbell *et al.* (1969b) also showed that the interaction of the plant canopy with the atmosphere is important in wheat. In dry years when soil moisture was limiting, shading of plants increased yield, presumably by protecting the plants from extreme desiccation. In moist years light was limiting, and shading reduced yield. Soil water stress reduced the yield of two wheat varieties by reducing the number of heads per plant (Campbell *et al.,* 1969a). Derera *et al.* (1969) found that the grain yield of wheat was greatest in varieties which produced ears early in the growing season because these varieties avoided the late-season drought. The water use was also less for early flowering varieties.

Similar effects of water stress on fruit production are observed in tree crops. For example, Erickson and Richards (1955) found that the average fruit weight of Valencia oranges decreased significantly from 156 gm for trees irrigated at a soil water stress of 0.3 bar to 144 gm at a stress of 0.7 bar. Fruit size distribution was also affected (see Fig. 1), but the number of fruits per tree was not significantly changed. Richards *et al.* (1960) found that yield of avocados was sharply reduced when soil was irrigated at a water stress of 10 bars rather than at stresses of 0.5 or 1.0

TABLE VI

Effects of Irrigation and Spacing of Corn on Yield,
Evapotranspiration, and Water Use Efficiency[a,b]

Plants per hectare	Irrigation treatment	Yield (metric tons)	Evapotranspiration (mm)	Water use eff. (kg dry matter/mm)
39,536	Control[c]	4.12	181.4	22.8
	Irrigated[d]	6.82	251.6	27.2
59,304	Control[c]	3.76	204.4	18.4
	Irrigated[d]	9.06	246.2	40.4

[a] From Fulton (1970).

[b] Plants were grown in floating lysimeters in the center of large plots.

[c] Control plants receiving no supplemental water experienced severe soil moisture stress.

[d] Irrigated plants received water when soil water tension increased to 5 bars.

bar (Table VII). Trunk diameter after 7 treatment years was also reduced by water stress. However, Richards and Warneke (1969) found no differences in yield or tree growth of lemons in a coastal area of California when trees were irrigated at soil water stresses of 0.6 and 1.5 bars. Irrigation of three varieties of peaches increased 5-year yields by 1.4 to 1.9 bushels per tree (Feldstein and Childers, 1965). When water was deficient, irrigation during the second week before harvesting swelled fruits and significantly increased yield. Irrigation of apples in Michigan tended to reduce the alternate year bearing of fruits (Gamble, 1962).

Most of the effects of water deficits on final yield of reproductive organs probably are related to specific effects during relatively early stages of the reproductive cycle. For example, when water deficits have reduced the number of fruits, the effect is most likely the result of a water stress during flowering. Reductions in fruit size are caused by water deficits after flowering when the fruit is enlarging. Better understanding of the relationship between internal plant water relations and reproductive yield will come only as a result of more detailed studies at each of the stages in the reproductive cycle.

E. Fruit Composition and Quality

The quality of fruits as food and the concentration of fruit constituents can be influenced by water deficits. Maurer *et al.* (1969) found that the fiber content of bush bean pods was increased by about 20% when water stress occurred after flowering. Water stress apparently made the pods more woody and resulted in a less marketable product. Water stress in broccoli decreased succulence (Massey *et al.,* 1962). The dry matter content of the central inflorescence from irrigated plants was 10.9% compared with

TABLE VII
EFFECT OF SOIL WATER STRESS ON YIELD OF AVOCADOS[a,b]

Maximum soil water stress (bars)	Mean No. of irrigations	Mean yield (kg/tree)[c]	Trunk diameter after 7 years (cm)[d]
0.5	38	178.5	19.3
1.0	25	153.1	17.3
10.0	11	61.7	13.2

[a] From Richards *et al.* (1960).

[b] Plants were irrigated when tensiometers or soil resistance blocks indicated that the maximum stress for each treatment had been reached.

[c] Yield values are the means for the sixth and seventh years after planting.

[d] Trunk diameter shows the cumulative effect of water stress treatments for the 7-yr period.

14.4% for inflorescences from unirrigated plants. Ascorbic acid and carotene concentrations and flavor were not affected by water stress. When irrigation frequency in tomatoes was extended from 10 to 20 days, the percent solids in fruits increased (Table III). Soil water stress increased the percent dry matter in bush bean pods, particularly when the maximum stress occurred after flowering (Table V). Uriu *et al.* (1964) observed that as soil water stress increased from 0.4 or 0.8 bar to 5 bars, the soluble solids in peach fruits increased and water content decreased. Although rag content of mature Valencia oranges was increased by water stress, juice content, soluble solids, and acid concentration were unaffected (Erickson and Richards, 1955). Studying the irrigation of grapes, Hendrickson and Veihmeyer (1951) found that the Tokay variety of table grapes had better color after storage when water stress occurred for relatively long periods during growth. The intensity of wine color from Carignane and Barbera varieties was slightly greater for dry plots, probably because the ratio of skin (which contains most of the pigments) to volume was greater in the stressed plants.

Feldstein and Childers (1965) measured mineral concentrations in three varieties of peaches. Potassium concentration (as a percent of dry weight) was increased as a result of water stress by about 0.2% in two varieties, while phosphorus, calcium, magnesium, and nitrogen remained the same. In a lysimeter experiment, Campbell *et al.* (1969b) collected data supporting the field observation that protein content of wheat was increased by water deficits. Protsenko *et al.* (1968) found that severe water deficits increased the concentrations of free amino acids in the ears of three varieties of wheat by 2.8 to 3.7 times on a percent dry weight basis. Similar increases were observed in stems and leaves. The concentration of certain amino acids, notably proline, increased as Valencia orange fruits aged on the tree (R. L. Clements and Leland, 1962). Proline accumulates in a number of plants subjected to water stress. Perhaps increased proline concentration in oranges is related to the lower water potentials which occur after the fruits ripen (Kaufmann, 1970c).

III. FACTORS AFFECTING THE WATER RELATIONS OF FRUITS

A. WATER AVAILABILITY

It is clear that water availability to plants and environmental factors which affect transpiration influence the water content of fruits. The relationship between plant water stress and diurnal fluctuations of fruit diameter has already been considered (Section II,B). Hilgeman (1966) observed that irrigation or precipitation caused hydration of grapefruits and resulted in high internal hydrostatic pressures. Kaufmann (1969) also concluded that irrigation increased the turgor pressure in oranges. Water potential of orange pericarp tissue measured after an irrigation in autumn decreased from about —5 bars to —8 bars during a 29-day period (Table VIII). Greater decreases would be expected during the summer when transpiration is normally higher. Hales et al. (1968) concluded that nighttime relative humidities lower than 37% prevented recovery of Valencia oranges to full volume.

While high water potential is considered desirable for fruit growth because it is accompanied by high turgor, another factor also deserves consideration. In Valencia oranges and probably in all citrus fruits, high water potentials cause high stretch moduli in the pericarp tissue (Fig. 5). Therefore the sphere of peel surrounding the vesicles is the strongest (least extensible) at high water potentials. Fruit growth, including irreversible extension of the pericarp tissue, must require high turgor pressures to overcome the high tissue strengths which occur when water deficits are minimal. More research is needed on the relationship among tissue strength, water stress, and growth because pericarp or epidermis strength may be a factor in growth of many fruits, including cherries, apples, plums, and tomatoes.

For more than half a century, beginning with Hodgson (1917) and Bartholomew (1926), it has been stated that under certain conditions fruits may act as a reservoir of water for the rest of the plant. Since the water content of large fruits is high (Table I), a significant portion of the

TABLE VIII

WATER POTENTIAL OF NAVEL ORANGE PERICARP TISSUE FOLLOWING IRRIGATION[a,b]

Days after irrigation	Water potential (bars)[c]
2	—5.5 ± 0.2
7	—5.1 ± 0.3
9	—6.2 ± 0.2
16	—6.1 ± 0.2
21	—6.8 ± 0.6
23	—6.9 ± 0.6
27	—8.2 ± 0.3
29	—7.9 ± 0.4

[a] From Kaufmann (1970b).

[b] Trees were irrigated 9/30/69. During the next month samples were collected between 7:30 and 8:00 A.M. for measurement of water potential with a thermocouple psychrometer.

[c] Standard error of the mean water potential is indicated.

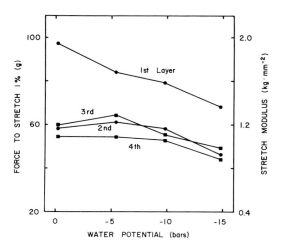

Fig. 5. Effect of water potential on the tensile force and stretch modulus of Valencia orange pericarp strips 5 mm wide and 1 mm thick taken from the fruit equator. Measurements were made on successive layers of tissue beginning at the exocarp (first layer) and ending at the inner mesocarp. Tensile force was measured as the load required to stretch the excised strip 1%. Strips were incubated in quarter strength Hoagland's solution containing polyethylene glycol 6000 for 1 hr prior to and during the measurement. From Kaufmann (1970a).

water in some plants occurs in the fruit. The classic experiments have involved measurements of water loss by cut branches and fruits. When fruits are left on detached branches, the branches dehydrate more slowly than those without fruits, and the fruits lose more water than do completely detached fruits. More recently, Klepper (1968) observed that diurnal changes in water potential of Vaseline-coated and uncoated pear fruits attached to the tree were nearly similar and concluded that Vaseline-coated fruits lost water to the leaves. In these experiments it is unquestionable that fruits are losing water to the rest of the plant. Some fruits, however, seem to be capable of substantial transpiration, and it is not yet clear whether these fruits act as a reservoir of water. Even if diurnal water deficits and recovery of fruits lag behind those of leaves, suggesting the development of water potential gradients from fruit to leaves, the possibility remains that changes in water content of some fruits are governed largely by fruit transpiration. Additional experiments are needed to determine whether water moves from fruits to leaves in species where fruit transpiration is significant.

B. Water Vapor Exchange in Fruits

Gas movement between the fruit and surrounding air may significantly influence the water balance of the fruit and also the effects of oxygen, carbon dioxide, ethylene, and perhaps other gases on fruit metabolism. Differences exist in permeability of the fruit surface. Some fruits such as citrus allow gas exchange whereas others are highly impermeable to gases. Tomatoes, for example, have a very high resistance to gas movement. When the pedicel scar on detached fruits is coated, essentially all exchange of gases ceases. Haas (1936) concluded that mature avocado fruits require gas exchange to prevent internal fruit decomposition. When resistance to gas flow is prominent, it normally occurs at the pericarp or epidermal regions of the fruit. Hoff and Dostal (1968) measured the resistance to flow of air from the core to the exterior of apple fruits. Removing the peel reduced the resistance to only about half of the total, indicating that the thin layer of peel tissue had as much resistance as all the rest of the tissue. Ben-Yehoshua (1969) found that drying of the pericarp of uncoated mature orange fruits apparently increased peel resistance to diffusion of carbon dioxide during storage. Dehydration also decreased the commercial value of the stored fruit.

Development and composition of the fruit cuticle have direct effects on permeability of fruits to water vapor and other gases. During growth of apple fruits, cutinous substances on the epidermal cells expand inward, although the cuticle structure changes little (de Vries, 1968). In full-grown fruits the epidermal cells become completely surrounded by cuticle. In orange fruits, however, cell division in the epidermis and hypodermis is continuous until the fruit matures, and the cuticle is thin over recently divided epidermal cells (Bain, 1958). Several experiments provide information on the contribution of specific components of the cuticle structure to reduction of water vapor movement. Hall (1966) showed that under uniform conditions the removal of surface wax from the cuticle of apples increased water loss from 0.34% of fresh weight to at least 0.50%, depending on the method of wax removal. Possingham et al. (1967) found that chemical disorganization of the platelet structure of grape berry wax increased the rate of cuticular transpiration. Using cuticle models having grape wax components as coatings, Grncarevic and Radler (1967) concluded that oleanolic acid, the main component of grape cuticle wax, was not effective in reducing evaporation. Evaporation was controlled largely by hydrocarbons, long chain alcohols and aldehydes, and probably esters found in the wax.

While stomata are important pathways for gas exchange in leaves, their significance in fruits seems variable. Data are meager, but stomate densities range from 0/mm^2 in many fruits to at least 80/mm^2 in some varieties of snap bean (Table IX). Orange and banana fruits have only about 20 and 3%, respectively, as many stomata per unit area as do leaves. Other fruits may have higher stomatal densities, but it is likely that

TABLE IX
DENSITY OF STOMATA IN FRUITS

Fruit	Age	Fruit part	Density (No./mm²)	Reference
Snap bean (49 varieties)	Mature	—	20–80	Hoffman (1967)
Banana	Immature	Top	11	Johnson and Brun (1966)
		Bottom	3	Johnson and Brun (1966)
	Mature	Top	6	Johnson and Brun (1966)
		Bottom	3	Johnson and Brun (1966)
Grape	All	Entire	0	Grncarevic and Radler (1967)
Orange	—	Equator	74	Rokach (1953)
		Ends	67	Rokach (1953)

stomata are much less common on fruits than on leaves of most plants. Stomate density also varies among different positions on the fruit. Haas (1936) observed that stomata were distributed evenly over small avocado fruits, but more fruit enlargement occurred near the stem end than at the stylar end, resulting in a lower density of stomata near the stem end. Stomata also became less functional as the fruit aged.

It is not likely that stomata exert as much control over gas exchange and water relations in fruits as they do in leaves. Haas and Klotz (1935) found that water loss of the stem half of ripened Valencia oranges exceeded that of the stylar half even though stomatal frequency was greater on the stylar half. Perhaps cuticular transpiration is always significant in oranges, even if the stomate density is relatively high compared to that in other fruits. It is not known whether stomata of all fruits function normally although Johnson and Brun (1966) found that stomata of banana fruits opened when light was available, even in fruits shipped from the tropics. More research is required to determine if fruit stomata have any real significance in fruit physiology and growth.

C. ACCUMULATION OF OSMOTICALLY ACTIVE SUBSTANCES IN FRUITS

Fruit growth, like growth of other tissues, requires turgor pressure, and many environmental and physiological conditions that limit growth may be related to effects on turgor pressure. Two factors, tissue water potential and osmotic potential, are primarily responsible for the regulation of turgor pressure. Short-term fluctuations in water potential cause immediate changes in cell turgor and are largely responsible for diurnal changes in fruit volume. Osmotic potentials of fruit tissues also change, but the rate of change is much less pronounced. Fluctuations in the concentration of osmotically active substances can occur in several ways. For example, changes in tissue water content alter the cell sap concentration. Neither the rate of supply of soluble carbohydrates to the tissue nor the rate of conversion of soluble carbohydrates into structural substances and other insoluble materials is constant.

Many osmotically active chemicals derived from the products of photosynthesis occur in fruit tissue. Kliewer (1966) examined the composition of grapes and identified 9 sugars and 23 organic acids. R. L. Clements (1964a,b) examined varietal and seasonal changes in organic acids of citrus fruits, and R. L. Clements and Leland (1962) studied seasonal changes in free amino acids in Valencia oranges. While no attempt will be made to summarize these reports, they are mentioned here as examples of research showing that variations exist in cell sap composition, and that these differences can have effects on osmotic relationships of the tissue. Inorganic compounds can also influence osmotic potential, but their importance in fruits does not seem to be appreciable.

Most of the osmotically active substances which accumulate in fruits are derived from carbohydrates entering through the vascular system. While translocation of carbohydrates to fruits has relatively little direct relationship with water deficits and reproductive growth, translocation is important in establishing the osmotic potential of cells and in providing structural material required when turgor pressure increases cell size. Sucrose is the major carbohydrate translocated to orange fruits (Kriedemann, 1969a) and probably in most other fruits, but sorbitol is translocated into apples and plums (Bieleski, 1969; Hansen, 1967; Williams *et al.,* 1967). Rates of dry matter increase in growing fruits can be phenomenally high. H. F. Clements (1940), in his classic experiments on fruit growth in sausage trees (*Kigelia*), found that the dry weight of clusters of four fruits increased as much as 32.6 gm per day. He also calculated that from 500 to 5700 gm of water were transported to the fruit each day. Although more recent research suggests that the phloem sap was more concentrated than Clements' calculations showed, his inability to account for all the water entering the fruit still has not been fully explained. The strength of the carbohydrate sink (i.e., fruit size and growth rate) also can affect the rate of carbohydrate translocation. Wardlaw (1965) observed that reductions in the number of grains in wheat ears decreased the import of ^{14}C compounds from the second leaf to the ear but had no effect on movement from the flag leaf to the ear. Lenz (1967) observed that navel orange fruits on rooted cuttings competed more successfully than the vegetative structures for photosynthate. Vegetative growth was less when cuttings bore several fruits compared with nonbearing cuttings. Partial defoliation reduced fruit growth less than it decreased vegetative growth. Swarbrick and Luckwill

(1953), citing data of A. H. Lees, suggested that high rainfall during the summer months may cause vegetative growth at the expense of carbohydrate availability for fruit bud formation for the following year. Position of fruits on trees may affect the concentration of osmotically active substances. Sites and Reitz (1949) found that the percent soluble solids in mature Valencia oranges was highest for more exposed fruits and for fruits located near the top of the tree. Soluble solids were lowest in fruits from the northeast sector of the tree.

Changes in carbohydrate translocation during ripening may be relatively small, particularly during the early stages of ripening. Kriedemann (1969b) found that lemons continued to import ^{14}C-labeled photosynthate until harvest. During an 8-wk period after labeling, most ^{14}C moved from the pericarp region to the vesicles. ^{32}P-labeled phosphate continued to move into tomato fruits at all stages of maturity, but ^{14}C-glucose and other ^{14}C-labeled photosynthates were translocated into the fruit only until about the tenth day after color change began (McCollum and Skok, 1960). It was concluded that harvesting just before the full-ripe stage would not impair fruit quality. While the abscission zone may affect translocation in tomatoes Stosser et al. (1969) observed that the abscission zone formed between the fruit and pedicel of sour cherries did not form through the vascular bundles, which apparently were responsible for continued fruit attachment.

Osmotically active substances in fruits may result directly from fruit photosynthesis. Lovell and Lovell (1970) found that 60% of the CO_2 fixed by the carpel of dwarf pea pods was transported to the enclosed seeds. They concluded that carpel photosynthesis may be a major source of carbohydrate for the seeds. Pea pods contain two photosynthetically active regions, one exposed primarily to the outside environment, and the other lining the internal seed cavity and utilizing CO_2 from respiring seeds (Flinn and Pate, 1970). The calculated contributions of pods, stipules, and leaflets to the total carbon in seeds is shown in Fig. 6. Pod photosynthesis became significant about 12 days after anthesis. Wardlaw (1967) found that photosynthesis of the wheat ear was reduced by water stress, but not as rapidly as leaf photosynthesis. Ear respiration was not affected by water stress. Evans and Rawson (1970) studied photosynthesis and respiration of components of the ear of wheat and concluded that grain photosynthesis accounted for 33 to 42% of the gross ear photosynthesis. As much as 76% of the total grain requirements for assimilates during early growth was provided by ear photosynthesis, and varieties that have awns produced more photosynthate than varieties without awns. Photosynthetic fixation of CO_2 and dark fixation are known to occur in other fruits of angiosperms (Bean and Todd, 1960; Hansen, 1970a,b; Kriedemann, 1968) and in

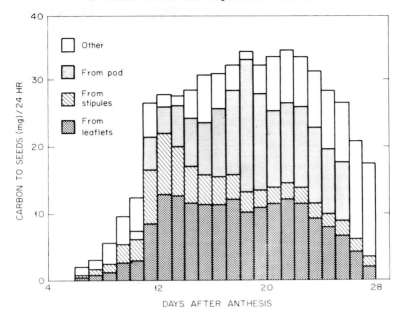

Fig. 6. Contributions of carbon by pods, stipules, and leaflets to growing seeds at the first fruiting node of pea plants. The estimated supply from each source includes photosynthetically fixed carbon, carbon available through mobilization of dry matter, and in the case of pods, carbon reassimilated from respiring leaves. The unhatched areas represent the amount of carbon known to be needed for seed growth but not supplied by organs at the fruiting node. From Flinn and Pate (1970).

strobili of conifers (Dickmann and Kozlowski, 1970). While photosynthesis in small fruits may be important, it is apparent that the photosynthetic capacity of large fruits is not great in comparison to the total amount of assimilate required.

A number of researchers have reported osmotic potential (or osmotic pressure) measurements in fruits and factors related to changes in osmotic potential. Smock (1941) found that partial girdling of stems of young apple fruits increased the osmotic potential (solution became more dilute) measured later in the season. Other experiments have indicated that girdling the trunk of fruit trees may increase the carbohydrate supply to fruits and reduce their osmotic potential because no carbohydrates are translocated to the roots. Rokach (1953) and Kaufmann (1970b) observed that loss of water from orange fruits reduced the osmotic potential in vesicle and pericarp tissues. Kerr and Anderson (1944) examined the osmotic potential of cottonseeds in growing bolls. Osmotic potential increased sharply about 16 days after flowering when fiber elongation stopped and secondary thickening began.

Differences in osmotic potential exist among fruit tissues and at different stages of development. In growing navel orange fruits, osmotic potential is more negative at the stem end than at the stylar end (Fig. 7). Ting (1969) reported on the distribution and local differences in concentrations of soluble solids, total acids, and sugars of mature Valencia oranges. Because the vesicles are isolated from the vascular network in oranges, osmotic potential remained stable diurnally and after girdling of the stem near the fruit, but osmotic potential in the pericarp tissues changed more readily (Kaufmann, 1970b). Osmotic potential of grape berries decreased from —8.7 bars to —26.4 bars as the fruits matured while osmotic potential of the cane sap remained constant. Mature fruits had a more negative osmotic potential at the apex end than at the stem end (Meynhardt, 1964).

IV. FRUIT DISORDERS RELATED TO WATER

Rupture of the outer portions of fruits (termed splitting, cracking, or checking) is common in a number of fruits. In nearly every case, fruit cracking is related to high availability of soil water and high relative

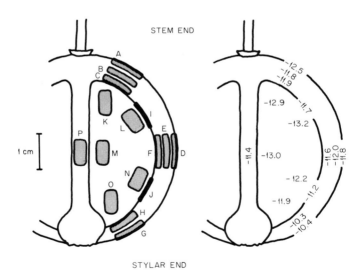

Fig. 7. Longitudinal sections of naval oranges 5.4–5.7 cm in diameter showing tissue sampling positions (left) and osmotic potentials (right). Sites A–H are in the pericarp, I and J are in the endocarp, K–O are vesicles, and P is in the central axis. Osmotic potentials were measured on samples collected at 8:00 A.M. when the exocarp water potential (near site D) was —5.5 bars. From Kaufmann (1970b).

humidities, and frequently to low temperatures. For example, avocado fruits checked after rain (Haas, 1936), and end-cracking of prunes occurred when irrigation followed a period of water stress (Uriu *et al.,* 1962). Randhawa *et al.* (1958) found that splitting of lemon fruits was most frequent when rains, accompanied with low temperatures and high relative humidities, followed drought. Observations by the writer indicate that navel oranges also split when the night temperature is cool and relative humidity is high. Rain and high relative humidity are associated with cherry fruit cracking (Gerhardt *et al.,* 1945; Levin *et al.,* 1959), but Bullock (1952) observed that more cherries immersed in water cracked at high temperatures than at low temperatures. Ripe cherries are particularly susceptible to cracking. High humidities are correlated with splitting of grape berries (Meynhardt, 1964) and cracking of apples (Verner, 1935).

Fruit splitting or cracking appears to be related to high internal pressures which cause tensile forces in the outer tissues to exceed the breaking point. In some fruits, cracking is related to lower than normal osmotic potentials which would cause high turgor pressures. For example, grape berries that split had a lower osmotic potential than those remaining intact (Meynhardt, 1964). In apples, osmotic potential was more negative in regions of the fruit where the skin was modified than in regions having normal skin (Verner, 1935). Osmotic potential of tissue beneath sound skin having low color was −17.2 bars; beneath moderately colored, sound skin the osmotic potential was −18.8 bars; and beneath highly colored skin −19.6 to −20.3 bars. Tissue beneath russeted skin with high color, where cracking was common, had osmotic potentials of −22.1 bars. Cracking may be more localized in apples than in other fruits. Verner observed cracking in an uninfected portion of a fruit having soft rot and concluded that cracking depended on rapid enlargement of restricted portions of tissue rather than of the fruit as a whole.

While osmotic and turgor effects are important in fruit cracking, other factors can also be significant. Kertesz and Nebel (1935) found that the water-holding capacity of macerated, sweet cherry tissue was greater than that of sour cherries, which are less susceptible to cracking. Treatment of the macerated pulp with pectic, amylolytic, and proteolytic enzymes sharply reduced water retention by sweet cherry pulp and had no effect on sour cherry pulp, indicating that colloidal effects contributed to water retention in the sweet cherries. Therefore swelling by colloidal imbibition may complement the turgor pressure associated with low osmotic potentials, resulting in more frequent splitting of sweet cherries. An abscission zone forms between the pedicel and fruit in sour cherries, but not in sweet cherries (Stosser *et al.,* 1969). It is not clear whether the lack of an abscission zone in sweet cherries affects fruit cracking under natural conditions. It has also been observed (Kaufmann, 1969, 1970a) that the pericarp tissue of oranges is under the greatest mechanical tension when the temperature is low and the water potential is high (see Fig. 5). Since fruit splitting occurs when environmental conditions are likely to result in high tension in the pericarp, splitting in oranges and other citrus fruits probably occurs because the tensile strength of the pericarp tissue is exceeded at some weak point. In navel oranges, splitting in almost all cases begins at an irregularity of the pericarp periphery at the navel on the stylar end. Fruits having very small navels are less susceptible to splitting than those with large navels. Perhaps the effects of water potential and temperature on tissue strength observed in citrus also occur in other fruits which split or crack. Resistance to cracking may not be a fruit characteristic in some cases, however. Reynard (1951) observed that fruits on tomato plants infected with *Fusarium* wilt or weakened by excessive soil moisture showed less than normal amounts of cracking. Perhaps turgor pressure is reduced in fruits under these conditions.

Certain other fruit disorders may be related to water relations. Cahoon *et al.* (1964) found that oleocellosis, a spotting of citrus pericarp with oil from ruptured oil glands, was higher during harvest when soil moisture and relative humidity were high and temperature was low. Handling of fruit caused more damage because the fruit water content and turgor were highest under these conditions. Tukey (1959) found that the high relative humidity caused by enclosing apple fruits in bags resulted in russeting. He suggested that cuticle development was altered by the high humidity. Corking disorder in apples (water-soaking followed by brown coloration of tissue near the core) was associated with water deficiency (Burrell, 1934).

REFERENCES

Abrol, Y. P., and McIlrath, W. J. (1966). Studies on the dehydration of seeds during fruit development in *Lycopersicon esculentum* Mill. var. Marglobe. *Indian J. Plant. Physiol.* 9, 66.

Aljibury, F. K., and May, D. (1970). Irrigation schedules and production of processed tomatoes on the San Joaquin Westside. *Calif. Agr.* 24(8), 10.

Alvim, P. de T. (1960). Moisture stress as a requirement for flowering in coffee. *Science* 132, 354.

Alvim, P. de T. (1964). Tree growth periodicity in tropical climates. *In* "The Formation of Wood in Forest Trees" (M. H. Zimmermann, ed.), pp. 479–496. Academic Press, New York.

Anderson, D. B., and Kerr, T. (1943). A note on the growth behavior of cotton bolls. Plant Physiol. 18, 261.

Bain, J. M. (1958). Morphological, anatomical, and physiological changes in the developing fruit of the Valencia orange, *Citrus sinensis* (L.) Osbeck. *Aust. J. Bot.* 6, 1.

Bartholomew, E. T. (1926). Internal decline of lemons. III. Water deficit in lemon fruits caused by excessive leaf evaporation. *Amer. J. Bot.* **13**, 102.

Bean, R. C., and Todd, G. W. (1960). Photosynthesis and respiration in developing fruits. I. $C^{14}O_2$ uptake by young oranges in light and in dark. *Plant Physiol.* **35**, 425.

Ben-Yehoshua, S. (1969). Gas exchange, transpiration, and the commercial deterioration in storage of orange fruit. *J. Amer. Soc. Hort. Sci.* **94**, 524.

Bieleski, R. L. (1969). Accumulation and translocation of sorbitol in apple phloem. *Aust. J. Biol. Sci.* **22**, 611.

Brown, D. S. (1953). The effects of irrigation on flower bud development and fruiting in the apricot. *Proc. Amer. Soc. Hort. Sci.* **61**, 119.

Bullock, R. M. (1952). A study of some inorganic compounds and growth promoting chemicals in relation to fruit cracking of Bing cherries at maturity. *Proc. Amer. Soc. Hort. Sci.* **59**, 243.

Burrell, A. B. (1934). The effect of irrigation on the occurrence of a form of the cork disease and on the size of apple fruits. *Proc. Amer. Soc. Hort. Sci.* **30**, 415.

Cahoon, G. A., Grover, B. L., and Eaks, I. L. (1964). Cause and control of oleocellosis on lemons. *Proc. Amer. Soc. Hort. Sci.* **84**, 188.

Campbell, C. A., McBean, D. S., and Green, D. G. (1969a). Influence of moisture stress, relative humidity and oxygen diffusion rate on seed set and yield of wheat. *Can. J. Plant Sci.* **49**, 29.

Campbell, C. A., Pelton, W. L., and Nielson, K. F. (1969b). Influence of solar radiation and soil moisture on growth and yield of Chinook wheat. *Can. J. Plant. Sci.* **49**, 685.

Chaney, W. R., and Kozlowski, T. T. (1969a). Seasonal and diurnal changes in water balance of fruits, cones, and leaves of forest trees. *Can. J. Bot.* **47**, 1407.

Chaney, W. R., and Kozlowski, T. T. (1969b). Seasonal and diurnal expansion and contraction of fruits of forest trees. *Can. J. Bot.* **47**, 1033.

Chaney, W. R., and Kozlowski, T. T. (1969c). Diurnal expansion and contraction of leaves and fruits of English Morello cherry. *Ann. Bot. (London)* [N. S.] **33**, 991.

Clements, H. F. (1940). Movement of organic solutes in the sausage tree, *Kigelia africana*. *Plant Physiol.* **15**, 689.

Clements, R. L. (1964a). Organic acids in citrus fruits. I. Varietal differences. *J. Food Sci.* **29**, 276.

Clements, R. L. (1964b). Organic acids in citrus fruits. II. Seasonal changes in the orange. *J. Food Sci.* **29**, 281.

Clements, R. L., and Leland, H. V. (1962). Seasonal changes in the free amino acids in Valencia orange juice. *Proc. Amer. Soc. Hort. Sci.* **80**, 300.

Coit, J. E., and Hodgson, R. W. (1918). The June drop of Washington navel oranges, a progress report. *Calif., Agr. Exp. Sta., Bull.* **290**, 203.

Cooper, W. C., Peynado, A., Furr, J. R., Hilgeman, R. H., Cahoon, G. A., and Boswell, S. B. (1963). Tree growth and fruit quality of Valencia oranges in relation to climate. *Proc. Amer. Soc. Hort. Sci.* **82**, 180.

Coyne, D. P. (1968). Differential response of stylar elongation in tomatoes to soil moisture levels. *HortScience* **3**(1), 39.

Davenport, D. C., Uriu, K., and Hagan, R. M. (1969). Unpublished data.

Derera, N. F., Marshall, D. R., and Balaam, L. N. (1969). Genetic variability in root development in relation to drought tolerance in spring wheats. *Exp. Agr.* **5**, 327.

de Vries, H. A. M. A. (1968). Development of the structure of the normal smooth cuticle of the apple "Golden Delicious." *Acta Bot. Neer.* **17**, 229.

Dickmann, D. I., and Kozlowski, T. T. (1970). Photosynthesis by rapidly expanding green strobili of *Pinus resinosa*. *Life Sci.* **9**, Part II, 549.

Dubetz, S., and Mahalle, P. S. (1969). Effect of soil water stress on bush beans *Phaseolus vulgaris* L. at three stages of growth. *J. Amer. Soc. Hort. Sci.* **94**, 479.

Elfving, D. C., and Kaufmann, M. R. (1971). Unpublished data.

Erickson, L. C., and Brannaman, B. L. (1960). Abscission of reproductive structures and leaves of orange trees. *Proc. Amer. Soc. Hort. Sci.* **75**, 222.

Erickson, L. C., and Richards, S. J. (1955). Influence of 2,4-D and soil moisture on size and quality of Valencia oranges. *Proc. Amer. Soc. Hort. Sci.* **65**, 109.

Evans, L. T., and Rawson, H. M. (1970). Photosynthesis and respiration by the flag leaf and components of the ear during grain development in wheat. *Aust. J. Biol. Sci.* **23**, 245.

Feldstein, J., and Childers, N. F. (1965). Effects of irrigation on peaches in Pennsylvania. *Proc. Amer. Soc. Hort. Sci.* **87**, 145.

Flinn, A. M., and Pate, J. S. (1970). A quantitative study of carbon transfer from pod and subtending leaf to the ripening seeds of the field pea (*Pisum arvense* L.). *J. Exp. Bot.* **21**, 71.

Flocker, W. J., and Lingle, J. C. (1961). Field applications of tensiometers and soil moisture blocks as criteria for irrigation of canning tomatoes. *Proc. Amer. Soc. Hort. Sci.* **78**, 450.

Fulton, J. M. (1970). Relationships among soil moisture stress, plant populations, row spacing and yield of corn. *Can. J. Plant Sci.* **50**, 31.

Gamble, S. J. (1962). Ray Klackle irrigates his apples every year for annual bearing. *Mich. State Hort. Soc. Annu. Rep.* **92**, 69.

Gerhardt, F., English, H., and Smith, E. (1945). Cracking and decay of Bing cherries as related to the presence of moisture on the surface of the fruit. *Proc. Amer. Soc. Hort. Sci.* **46**, 191.

Grimes, D. W., Dickens, L., Anderson, W., and Yamada, H. (1967). Irrigation and nitrogen for cotton. *Calif. Agr.* **21**(11), 12.

Grimes, D. W., Miller, R. J., and Dickens, L. (1970). Water stress during flowering of cotton. *Calif. Agr.* **24**(3), 4.

Grncarevic, M., and Radler, F. (1967). The effect of wax components on cuticular transpiration—model experiments. *Planta* **75**, 23.

Haas, A. R. C. (1936). Growth and water relations of the avocado fruit. *Plant Physiol.* **11**, 383.

Haas, A. R. C., and Klotz, L. J. (1935). Physiological gradients in citrus fruits. *Hilgardia* **9**, 179.

Hales, T. A., Mobayen, R. G., and Rodney, D. R. (1968). Effects of climatic factors on daily 'Valencia' fruit volume increases. *Proc. Amer. Soc. Hort. Sci.* **92**, 185.

Hall, D. M. (1966). A study of the surface wax deposits on apple fruit. *Aust. J. Biol. Sci.* **19**, 1017.

Hansen, P. (1967). ^{14}C-studies on apple trees. I. The effect of the fruit on the translocation and distribution of photosynthates. *Physiol. Plant.* **20**, 382.

Hansen, P. (1970a). ^{14}C-studies on apple trees. V. Translocation of labelled compounds from leaves to fruit and their conversion within the fruit. *Physiol. Plant.* **23**, 564.

Hansen, P. (1970b). ^{14}C-studies on apple trees. VI. The influence of the fruit on the photosynthesis of the leaves, and the relative photosynthetic yields of fruits and leaves. *Physiol. Plant.* **23**, 805.

Hartmann, H. T., and Panetsos, C. (1961). Effect of soil moisture deficiency during floral development on fruitfulness in the olive. *Proc. Amer. Soc. Hort. Sci.* **78**, 209.

Hendrickson, A. H., and Veihmeyer, F. J. (1951). Irrigation experiments with grapes. *Calif., Agr. Exp. Sta., Bull.* **728**, 1–31.

Hilgeman, R. H. (1966). Effect of climate of Florida and Arizona on grapefruit fruit enlargement and quality; apparent transpiration and internal water stress. *Proc. Fla. State Hort. Soc.* **79**, 99.

Hodgson, R. W. (1917). Some abnormal water relations in citrus trees of the arid Southwest and their possible significance. *Univ. Calif. Publ. Agric. Sci.* 3, 37.

Hoff, J. E., and Dostal, H. C. (1968). A method for determining gas flow characteristics in apple fruits. *Proc. Amer. Soc. Hort. Sci.* **92**, 763.

Hoffman, J. C. (1967). Morphological variations of snap bean pods associated with weight loss and wilting. *Proc. Amer. Soc. Hort. Sci.* **91**, 294.

Jackson, D. I. (1969). Effects of water, light, and nutrition on flower-bud initiation in apricots. *Aust. J. Biol. Sci.* **22**, 69.

Johnson, B. E., and Brun, W. A. (1966). Stomatal density and responsiveness of banana fruit stomates. *Plant Physiol.* **41**, 99.

Jones, W. W., and Cree, C. B. (1965). Environmental factors related to fruiting of Washington navel oranges over a 38-year period. *Proc. Amer. Soc. Hort. Sci.* **86**, 267.

Kaufmann, M. R. (1968). Unpublished data.

Kaufmann, M. R. (1969). Relation of slice gap width in oranges and plant water stress. *J. Amer. Soc. Hort. Sci.* **94**, 161.

Kaufmann, M. R. (1970a). Extensibility of pericarp tissue in growing citrus fruits. *Plant Physiol.* **46**, 778.

Kaufmann, M. R. (1970b). Water potential components in growing citrus fruits. *Plant Physiol.* **46**, 145.

Kaufmann, M. R. (1970c). Unpublished data.

Kerr, T., and Anderson, D. B. (1944). Osmotic quantities in growing cotton bolls. *Plant Physiol.* **19**, 338.

Kertesz, Z. I., and Nebel, B. R. (1935). Observations on the cracking of cherries. *Plant Physiol.* **10**, 763.

Klepper, B. (1968). Diurnal pattern of water potential in woody plants. *Plant Physiol.* **43**, 1931.

Kliewer, W. M. (1966). Sugars and organic acids of *Vitis vinifera*. *Plant Physiol.* **41**, 923.

Kozlowski, T. T. (1968). Diurnal changes in diameters of fruits and tree stems of Montmorency cherry. *J. Hort. Sci.* **43**, 1.

Kriedemann, P. E. (1968). Observations on gas exchange in the developing sultana berry. *Aust. J. Biol. Sci.* **21**, 907.

Kriedemann, P. E. (1969a). ^{14}C translocation in orange plants. *Aust. J. Agr. Res.* **20**, 291.

Kriedemann, P. E. (1969b). ^{14}C distribution in lemon plants. *J. Hort. Sci.* **44**, 273.

Lenz, F. (1967). Relationships between the vegetative and reproductive growth of Washington navel orange cuttings (*Citrus sinensis* L. Osbeck). *J. Hort. Sci.* **42**, 31.

Levin, J. H., Hall, C. W., and Deshmukh, A. P. (1959). Physical treatment and cracking of sweet cherries. *Mich., Agr. Exp. Sta. Quart. Bull.* **42**, 133.

Lombard, P. B., Stolzy, L. H., Garber, M. J., and Szuszkiewicz, T. E. (1965). Effects of climatic factors on fruit volume increase and leaf water deficit of citrus in relation to soil suction. *Soil Sci. Soc. Amer., Proc.* **29**, 205.

Lovell, P. H., and Lovell, P. J. (1970). Fixation of CO_2 and export of photosynthate by the carpel in *Pisum sativum. Physiol. Plant.* **23**, 316.

McCollum, J. P., and Skok, J. (1960). Radiocarbon studies on the translocation of organic constituents into ripening tomato fruits. *Proc. Amer. Soc. Hort. Sci.* **75**, 611.

McIlrath, W. J., Abrol, Y. P., and Heiligman, F. (1963). Dehydration of seeds in intact tomato fruits. *Science* **142**, 1681.

Massey, P. H., Jr., Eheart, J. F., Young, R. W., and Mattus, G. E. (1962). The effect of soil moisture, plant spacing, and leaf pruning on the yield and quality of broccoli. *Proc. Amer. Soc. Hort. Sci.* **81**, 316.

Maurer, A. R., Ormrod, D. P., Scott, N. J. (1969). Effect of five soil water regimes on growth and composition of snap beans. *Can. J. Plant Sci.* **49**, 271.

Meynhardt, J. T. (1964). Some studies on berry-splitting of Queen of the Vineyard grapes. *S. Afr. J. Agr. Sci.* **7**, 179.

Modlibowska, I. (1961). Effect of soil moisture on frost resistance of apple blossom, including some observations on "ghost" and "parachute" blossoms. *J. Hort. Sci.* **36**, 186.

Nakata, S., and Suehisa, R. (1969). Growth and development of *Litchi chinensis* as affected by soil-moisture stress. *Amer. J. Bot.* **56**, 1121.

Possingham, J. V., Chambers, T. C., Radler, F., and Grncarevic, M. (1967). Cuticular transpiration and wax structure and composition of leaves and fruit of *Vitis vinifera. Aust. J. Biol. Sci.* **20**, 1149.

Prokofév, A. A., and Kholodova, V. P. (1968). Features of the changes in water content in ripening seeds. *Fiziol. Rast.* **15**, 1022.

Protsenko, D. F., Shmatkó, I. G., and Rubanyuk, E. A. (1968). Drought resistance of winter wheats in relation to their amino acid content. *Fiziol. Rast.* **15**, 680.

Randhawa, G. S., Singh, J. P., and Malik, R. S. (1958). Fruit cracking in some tree fruits with special reference to lemon (*Citrus limon*). *Indian J. Hort.* **15**, 6.

Reuther, W., and Crawford, C. L. (1945). Irrigation experiments with Deglet Noor dates. *Date Growers Inst.* **22**, 11.

Reynard, G. B. (1951). Inherited resistance to radial cracks in tomato fruits. *Proc. Amer. Soc. Hort. Sci.* **58**, 231.

Richards, S. J., and Warneke, J. E. (1969). Sprinkler and furrow irrigation management for lemon trees under coastal climatic conditions. *Proc. Int. Citrus Symp., 1st, 1968* Vol. 3, p. 1749.

Richards, S. J., Warneke, J. E., and Weeks, L. V. (1960). Irrigation of avocados. *Yearb. Calif. Avocado Soc.* **44**, 73.

Robins, J. S., and Domingo, C. E. (1953). Some effects of severe soil moisture deficits at specific growth stages in corn. *Agron. J.* **45**, 618.

Rokach, A. (1953). Water transfer from fruits to leaves in the Shamouti orange tree and related topics. *Palestine J. Bot. (Rehovot)* **8**, 146.

Salter, P. J., and Drew, D. H. (1965). Root growth as a factor in the response of *Pisum sativum* L. to irrigation. *Nature (London)* **206**, 1063.

Salter, P. J., and Goode, J. E. (1967). "Crop Response to Water at Different Stages of Growth," Res. Rev. No. 2. Commonw. Agr. Bur., Farnham Royal, Bucks, England.

Schreiber, H. A., Stanberry, C. O., and Tucker, H. (1962). Irrigation and nitrogen effects on sweet corn row numbers at various growth stages. *Science* **135**, 1135.

Simons, R. K. (1963). Anatomical studies of apple fruit abscission in relation to irrigation. *Proc. Amer. Soc. Hort. Sci.* **83**, 77.

Sites, J. W., and Reitz, H. J. (1949). The variation in individual Valencia oranges from different locations of the tree as a guide to sampling methods and spot-picking for quality. I. Soluble solids in the juice. *Proc. Amer. Soc. Hort. Sci.* **54**, 1.

Slatyer, R. O. (1957). The influence of progressive increase in total soil moisture stress on transpiration, growth, and internal water relationships of plants. *Aust. J. Biol. Sci.* **10**, 320.

Smock, R. M. (1941). Studies on bitter pit of the apple. *Cornell Univ., Agr. Exp. Sta., Mem.* **234**, 1.

Stockton, J. R., Carreker, J. R., and Hoover, M. (1967). Irrigation of cotton and other fiber crops. *In* "Irrigation of Agricultural Lands" (R. M. Hagan, H. R. Haise, and T. W. Edminster, eds.), Monogr. No. 11, pp. 661–673. Amer. Soc. Agron., Madison, Wisconsin.

Stosser, R., Rasmussen, H. P., and Bukovac, M. J. (1969). A histological study of abscission layer formation in cherry fruits during maturation. *J. Amer. Soc. Hort. Sci.* **94**, 239.

Sturkie, D. G. (1934). A study of lint and seed development in cotton as influenced by environmental factors. *J. Amer. Soc. Agron.* **26**, 1.

Swarbrick, T., and Luckwill, L. C. (1953). Fruit culture. The factors governing fruit-bud formation. *In* "Science and Fruit" (T. Wallace and R. W. Marsh, eds.), pp. 99–109. Univ. of Bristol, Bristol, England.

Taylor, S. A., Haddock, J. L., and Pedersen, M. W. (1959). Alfalfa irrigation for maximum seed production. *Agron. J.* **51**, 357.

Ting, S. V. (1969). Distribution of soluble components and quality factors in the edible portion of citrus fruits. *J. Amer. Soc. Hort. Sci.* **94**, 515.

Tukey, L. D. (1959). Observations on the russeting of apples growing in plastic bags. *Proc. Amer. Soc. Hort. Sci.* **74**, 30.

Tukey, L. D. (1964). A linear electronic device for continuous measurement and recording of fruit enlargement and contraction. *Proc. Amer. Soc. Hort. Sci.* **84**, 653.

Turrell, F. M., Monselise, S. P., and Austin, S. W. (1964). Effect of climatic district and of location in tree on tenderness and other physical characteristics of citrus fruit. *Bot. Gaz.* **125**, 158.

Uriu, K., Hansen, C. J., and Smith, J. J. (1962). The cracking of prunes in relation to irrigation. *Proc. Amer. Soc. Hort. Sci.* **80**, 211.

Uriu, K., Werenfels, L., Post, G., Retan, A., and Fox, D. (1964). Cling peach irrigation. *Calif. Agr.* **18**(7), 10.

Verner, L. (1935). A physiological study of cracking in Stayman Winesap apples. *J. Agr. Res.* **51**, 191.

Vittum, M. T., Alderfer, R. B., Janes, B. E., Reynolds, C. W., and Struchtmeyer, R. A. (1963). Crop response to irrigation in the Northeast. *N. Y., Agr. Exp. Sta., Bull.* **800**, 1–66.

Wardlaw, I. F. (1965). The velocity and pattern of assimilate translocation in wheat plants during grain development. *Aust. J. Biol. Sci.* **18**, 269.

Wardlaw, I. F. (1967). The effect of water stress on translocation in relation to photosynthesis and growth. I. Effect during grain development in wheat. *Aust. J. Biol. Sci.* **20**, 25.

Williams, M. W., Martin, G. C., and Stahly, E. A. (1967). The movement and fate of sorbitol-C^{14} in the apple tree and fruit. *Proc. Amer. Soc. Hort. Sci.* **90**, 20.

Zimmerman, L. H. (1958). Castorbeans: A new oil crop for mechanized production. *Advan. Agron.* **10**, 257.

CHAPTER 4

PROTOPLASMIC RESISTANCE TO WATER DEFICITS

Johnson Parker

NORTHEASTERN FOREST EXPERIMENT STATION, FOREST SERVICE
U. S. DEPARTMENT OF AGRICULTURE, HAMDEN, CONNECTICUT

I. INTRODUCTION

"Protoplasmic resistance" might be considered the ability of the protoplasm to survive a serious reduction in water content. Of course, plant dehydration ordinarily entails much more than protoplasmic desiccation. A great variety of defenses against soil and atmospheric drought occur in the plant kingdom. There are many mechanisms for water retention in vascular plants subjected to drought that delay water loss and, in fact, allow water absorption and metabolism to continue as the plant desiccates.

Yet, at least in severe cases of drought, it is the protoplasm itself that is placed in danger of injury. Here in the protoplasm one finds great variability of resistance, both in different plant taxa and in different seasons. It may be an oversimplification to equate resistance with the ability of protoplasm to withstand dehydration. In any case, protoplasmic resistance has been the focal point for much of the research in drought resistance in the past forty years. It is the purpose of this review to try to clarify, or at least to summarize, some of the current concepts on this subject.

II. DESICCATION EFFECTS ON PROTOPLASM

In determining the effects of desiccation on the plant cell, two things should be measured: the amount of desiccation and the effect on the plant. The amount of desiccation is usually measured in terms of either plant water content or the vapor pressure of the ambient air; the effect is determined with some criterion of survival. From these data one may derive a "lethal level." This is usually considered to be a level at which approximately half of the cells of test plants are dead and half are alive. We shall consider the two subjects—measurement of stress and measurement of the effects—in Section II,A. In Sections II,B and C, two other stress-producing agents are discussed, heat and cold. All three factors—desiccation, heat, and cold—seem to have something in common, although it seems that they cannot be precisely equated.

A. LETHAL LEVEL DETERMINATIONS

This subject has been taken up previously (Parker, 1968). Nevertheless, it is of value to mention here some of the practical difficulties encountered in quantifying desiccation resistance. Perhaps the most serious problem is determination of water status. This has also been discussed by Cowan and Milthorpe (1968) and Barrs (1968).

1. Quantitative Water Deficit Assessment

In the study of desiccation resistance (definitions by Parker, 1969), it has been difficult to quantify results to permit comparison of the resistance of different species. Determination of the lethal level is often made in terms of relative water content (RWC). RWC is considered to be the amount of water contained relative to what the plant organ or tissue can hold when fully turgid (Cowan and Milthorpe, 1968).

Originally, Stocker (1929) expressed water content as the "saturation deficit," namely, the difference between saturation and fresh weight divided by the difference between saturation and dry weight. Various revisions have been made in this formula, and it has gone by different names. But little new has really been added. Recently, the term relative turgidity has been used by a number of authors. This expression, proposed by Weatherly (1950), implies the measurement of turgidity, but there is none. It was intended only as a relative value, and is really another way of expressing water content in mathematical terms. Walter (1963) criticized relative turgidity as "incomprehensible," and Weatherly (1965) agreed that the term might best be dropped.

Most of these formulas, including those used to calculate "critical saturation deficit" (e.g., Parker, 1968, 1969), make use of the saturation weight. This term was introduced to escape the difficulty of basing total water content on the uncertain value of fresh weight. Unfortunately, saturation weight too has its inaccuracies. As Barrs (1968) has pointed out, it does not represent a simple hydration end point; also, dry weight, on which it is based, can change with time. These difficulties and others led Walter (1963, 1965) to use "hydrature" to express the moisture "status" of the cell. Hydrature is supposed to represent the water vapor pressure of the cytoplasm or vacuole as compared to that of free water at the same temperature and pressure. It is usually measured by the cryoscopic method in terms of osmotic pressure (ψ_s) of the expressed sap.

The assumption has to be made that ψ_s of the expressed sap is a measure of ψ_s of the living protoplasm. This seems to be basically correct. However, the tonoplast (the membrane separating the large central vacuole from the cytoplasm) is complex and may not be a simple semipermeable membrane. Active loading of ion-water molecules into the vacuome is known to take place and may be one of the specific functions of the tono-plast (Criddle, 1969). Also, the amount of relatively free water in the protoplasm of the amoeba, for example, may vary from one location to another (Allen and Francis, 1965).

Other indices of water deficit have been suggested. Among these is water potential (ψ_{cell}). Some writers give the impression that ψ_{cell} quantitatively assesses the water deficit, although Barrs (1968) adds that both water content and ψ_{cell} are needed to express the water deficit "completely." Since ψ_{cell} changes much more than osmotic pressure (ψ_s) during the day in a typical crop plant, ψ_{cell} seems to be a more sensitive measure of the water deficit than ψ_s. There is no question, for example, that a slight drop (increasing negativity) in ψ_{cell} has an effect on growth (Boyer, 1968) as well as on photosynthesis, even though the stomata remain open (Boyer, 1965). But the question here is whether ψ_{cell} values can be related to survival itself, assuming no change in ψ_s.

Suppose the plant has a turgor pressure (ψ_p) such that it counter-balanced an unusually low ψ_s. Then ψ_{cell} would be 0 in the formula $\psi_{cell} = \psi_s + \psi_p$; for example, $0 = -40 + (+40)$ in which units are atmospheres. The effect on the protoplasm could conceivably be injurious even though ψ_{cell} was 0.

Another way of illustrating this point is that if there were a "low" ψ_{cell}, such as probably occurs in trees on hot sunny days with low humidity, there could be a turgor pressure that counterbalanced the ψ_{cell}. Assuming that ψ_s was 5 atm, then we might have $-20 = -5 + (-15)$ and there would probably be no injurious effect on the cell. Yet the theorists would expect injury on the basis of the low ψ_{cell} value.

Turgor pressures of an opposite sign (as in the above example) from that usually encountered are theoretically possible, although considered rare (Kramer *et al.*, 1966). Such turgor pressures more likely occur in semidesert plants that dry to low levels than in most mesophytic plants. The main difficulty here is that the turgor values mentioned above cannot be measured directly, except in certain specialized cases described by Burstrom *et al.* (1967). Ordinarily, ψ_p is derived from the above formula knowing ψ_{cell} and ψ_s.

There is evidence to support the notion of lack of effect of a very low ψ_{cell} on viability. Respiration was found to be unaffected, in situations where ψ_{cell} either was very low or was counterbalanced by applied pressure (Flowers and Hanson, 1969). In other words, respiratory inhibition did not result from lowered ψ_{cell} itself, but from some other effect of water on solute concentration. Under such circumstances, at least, ψ_s is probably a better measure than ψ_{cell} of the water deficit.

Besides the theoretical difficulties of using ψ_{cell} for estimating water deficit, there have long been practical difficulties in measuring it. Barrs (1968) described the various ways of measuring ψ_{cell} and many of the attendant problems. The thermocouple psychrometer method of determining ψ_{cell} has, at least previous to about 1968, required elaborate instrumentation and strict temperature control. The procedure was thus difficult to use, especially in the field (Knipling and Kramer, 1967). Recently, however, a type of miniature thermocouple psychrometer has been shown applicable to field use and gives good agreement with other methods (Wiebe *et al.*, 1970). The so-called dye or Chardakov method evidently is relatively inaccurate for determining ψ_{cell}, but requires no elaborate equipment and can be used for rough work in both laboratory and field (Knipling and Kramer, 1967).

The invention of the pressure chamber (Scholander *et al.*, 1965) made available a simple field means of measuring ψ_{cell} (presumably) and is in wide use today (e.g., Turner and Waggoner, 1968; DeRoo, 1969). However, in woody plants with large, gas-filled vessels, gross errors can be involved (Kaufmann, 1968). Also, in measuring the ψ_{cell} of leaves, the pressure chamber may really determine the pressure potential and not water potential, because of the resistance to water flow between leaf cells and xylem (Boyer, 1967). Boyer concluded that for accurate determinations of leaf ψ_{cell}, the pressure chamber measurements must be calibrated from measurements made with a thermocouple psychrometer.

In spite of difficulties of measurement, it is apparent that in the field plants can attain very low ψ_{cell} values without injury. For example, Jarvis and Jarvis (1963) found that at the "critical tolerance" or "desiccation tolerance" level, spruce leaves attained a ψ_{cell} of -950 J/hg* at a relative turgidity (RT) of 40, pine -700 at RT of 38, birch -500 at RT of 40, and aspen, the lowest in both drought tolerance and avoidance, -400 at RT of 54. Such low ψ_{cell} values (requiring over 200 lb/in.^{2} to obtain stem exudation) occur even when the soil moisture is not far below field capacity (ca -6 atm) (Pierpoint, 1967). All of this suggests that at least moderately low ψ_{cell} values in themselves are not deleterious to protoplasm.

2. Viability Determinations

In measuring desiccation resistance, determination of tissue survival is probably the most important problem. It is usually treated as if it were the least important. Either growth or protoplasmic streaming might seem satisfactory criteria of life. When growth or streaming occurs, the cells certainly are alive, but lack of growth or streaming is no assurance that cells are dead. Growth can, of course, resume after a dormant period, dormancy persisting in spite of optimum growth conditions. Protoplasmic streaming is open to the same criticism.

In a summary of various life criteria, Alexandrov (1964) pointed out that each viability indicator, be it vital staining, protoplasmic streaming, photosynthesis, or one of several others, may indicate a different lethal level in terms of final tissue water content (Fig. 1). In our work with heat

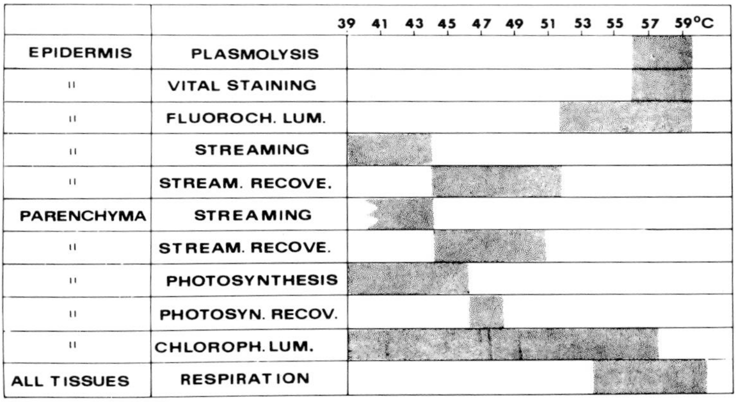

		39 41 43 45 47 49 51 53 55 57 59°C
EPIDERMIS	PLASMOLYSIS	
"	VITAL STAINING	
"	FLUOROCH. LUM.	
"	STREAMING	
"	STREAM. RECOVE.	
PARENCHYMA	STREAMING	
"	STREAM. RECOVE.	
"	PHOTOSYNTHESIS	
"	PHOTOSYN. RECOV.	
"	CHLOROPH. LUM.	
ALL TISSUES	RESPIRATION	

Fig. 1. Appearance of signs of injury in cells of rosette leaves of *Campanula persicifolia* after heating 5 min at temperatures shown. Fluoroch. lum., fluorochrome luminescence; stream. recov., recovery from cessation of protoplasmic streaming; photosyn. recov., recovery from depression of photosynthesis; chloroph. lum., decrease of chlorophyll luminescence. Original from various authors; summarized by Alexandrov (1964), by permission of the *Quarterly Review of Biology*.

* Joules per hectogram: 1.0 J/hg = 0.1 bar or about 0.1 atm.

resistance, photosynthesis was more readily stopped by heat than was any other process measured, including respiration, tetrazolium chloride reduction, or vacuolar retention of neutral red (Parker, 1972).

So far as I know, the tetrazoliums when applied to living cells are only reduced through the mediation of respiratory enzymes. At least three different enzyme systems, succinic, isocitric, and malic dehydrogenases, can influence tetrazolium reduction (Novikoff, 1955). However, the high correlation between the amount of freezing injury and the ability to reduce tetrazolium chloride is related to disruption of oxidative phosphorylation and intracellular localization of substrates and cofactors, rather than denaturation of specific dehydrogenases (Steponkus and Gregg, 1968).

Since methylene blue is reduced to leucomethylene blue (colorless) it is not as valuable an indicator of the cells' living condition as the tetrazoliums, which become colored when reduced. Although a number of strong reducing agents are known that will reduce tetrazolium chloride, it seems that tetrazolium reduction by cells into which it penetrates is a good indication that such cells are alive (Parker, 1953a,b; Larcher and Eggarter, 1960; Monk and Wiebe, 1961).

Other viability tests have been used in stress research. Natural or autofluorescence was stated by Larcher (1963) to be a good indication of viability. According to him, vacuolar fluorescence disappears when the cell dies. In our experience the test seemed unreliable in comparison to tetrazolium reduction. Of course, microorganisms can reduce the tetrazoliums, but the effect is usually recognizable, appearing in the solution as a reddish color or flocculent surrounding the tissue. As for contamination of the tissue itself, Steponkus and Lanphear (1967a) found that although bacteria were present, contamination accounted for less than 2% of the total reduction. In using respiration rate as a life criterion, invasion by microorganisms can be a more serious problem; also, respiration may increase following injury (see Sections II,B,4 and C,4).

Electrical resistivity methods have been used by some investigators in judging injury and viability of tissues and whole plants. It was known from Osterhout's work in 1922 (de Plater and Greenham, 1959) that in water of high electrolyte content, the low frequency resistance of a tissue to an alternating current ultimately decreases with injury and falls to a low value on death. De Plater and Greenham (1959) used a portable AC bridge for determining the impedance of the electrical system at 1 kc, 10 kc, and 1 Mc/sec frequencies. In one series of studies it was found that virus-infected tubers had a lower capacitance than healthy tubers (Greenham et al., 1952). Results were estimated in terms of an equivalent resistance in parallel with a capacitance.

Subsequently, it was found that the injury in a freshly bruised region of an apple can be quantitatively measured by means of the resistance ratio (low frequency resistance/high frequency resistance), since this ratio decreased progressively and significantly as severity of bruising increased (Greenham, 1966a). Mechanical injury of cell membranes whose resistance can be measured (Greenham, 1966b) was the probable cause of the decrease.

In trees, electrical resistance techniques using AC currents for determining both frost hardiness and viability have shown promise (Wilner, 1967). Electrical resistance in tissues injured by desiccation during winter greatly increased in the spring whereas that of normal shoots decreased. Desiccation alone may cause an increase in resistance (de Plater and Greenham, 1959).

A telemetric system, capable of monitoring large numbers of electric measurements in a digital form suitable for computer analysis, can also be used to test winter hardiness of woody plants (Wilner and Brach, 1970).

B. Similarities of Desiccation and Freezing Injury

More research has probably been done on freezing resistance, injury, and related problems than on desiccation resistance and its related problems. The apparent connection between the mechanism of cold resistance and that of desiccation resistance prompted the following summary:

1. Mechanical Considerations

Previous to about the mid-nineteenth century, similarities between effects of freezing and drying were evidently not suspected. Perhaps this was because they seemed to be totally unrelated processes. Freezing was thought to cause expansion and compression of tissues, whereas drying was believed to cause shrinkage and separation of certain tissues (reviewed by Parker, 1963). However, as von Sachs (1874, p. 703) and Maximov (1914) pointed out, in certain cases at least, freezing could have a dehydrative effect on cells. Since then, good correlations have been found between the amount of drying and the extent to which temperature can be lowered before cells are killed as well as between the amount of plasmolysis that cells can withstand in solution and the amount of drying they can withstand in dry air (Siminovitch and Briggs, 1953). Plasmolysis occurs as the result of osmotic loss of water, while freezing (at least extracellular freezing) also occurs at the expense of cell water.

Freezing may take place outside the cells (extracellularly) or within the cells (intracellularly). Extracellular freezing is sometimes referred to as intercellular. It is most likely that extracellular freezing causes protoplasmic dehydration. This may result in removal or disturbance of the bound or icelike water, normally associated with protoplasmic proteins or lipoproteins. This, in turn, may result in denaturation and perhaps aggregation (Section III,B,3,b).

Intracellular freezing, once initiated, usually proceeds in a fraction of a second (flash freezing) and is almost invariably destructive (reviewed by Levitt, 1956; Parker, 1963; Asahina, 1966; Mazur, 1969). Probably the relatively free water of the cell is reoriented to form ordinary ice (Section III,B,2). Freezing does not remove all of the bound water of proteins; the amount removed depends on the temperature at which freezing occurs. The lower the freezing point, the less unfrozen water is left in proteins (Bull and Breese, 1968).

Both intra- and extracellular freezing appear to be dehydrative processes, although extracellular freezing is more gradual and results in a dehydration similar to that in dry air. Intracellular freezing is usually destructive, although cases of recovery have been observed. Freezing of isolated mitochondria can cause fragmentation into spherical, membrane-bound vesicles (Cunningham, 1964). Some cases of recovery may have occurred because ice did not form in the protoplasm but rather between the cell wall and shrunken protoplast (Asahina, 1956).

According to Mazur (1969), freezing may be harmful since one or several of the following may occur: (1) attainment of a critical electrolyte concentration, (2) removal of stabilizing, bound, or icelike water, (3) alteration in spatial relations of macromolecules resulting in formation of bonds such as S—S, or (4) cell shrinkage. Levitt (1967) argued against (1) and favored (3). He (1969) distinguished between denaturation, which can be reversible, and aggregation (S—S bonds form), which is not reversible. Some investigators like Meryman (1967), Heber (1968), and Mazur (1969) seem to favor (1), although Mazur expressed dissatisfaction with all these theories. Mazur (1970) concluded that in some types of cells injury is caused by increases in solute concentration and in others by intracellular freezing. Mazur criticized (3) on several grounds. Little information is available for (4) although Luyet (1967) indicated the importance of protoplasmic shrinkage and resulting distortion during freezing. Meryman (cited by Mazur, 1970) suggested that damage to frozen erythrocytes results not from concentrated electrolytes per se but from the inability of the red cells to shrink below 55% of their normal volume without injury.

Comparisons of freezing and dehydration injury are made difficult because of different degrees of injury and various effects produced by different speeds of cooling or drying, or of subsequent warming or rehydration. Furthermore, there may be more than one mechanism in plants to combat these stresses. Some workers use experimental plants which harden only to about $-12°C$ while others use those which harden to at least $-180°C$. There must be differences between these kinds of plants.

Results from studies on insect physiology have sometimes paralleled those in plants. For example, there is a relation between desiccation and cold resistance; very drought-resistant insects can be cooled to the temperature of liquid helium and still survive (Asahina, 1966). Seasonal increases and decreases in certain sugars and polyhydric alcohols in insects (Tanno, 1967) closely resemble seasonal fluctuations in plants. Yet at the same time there is evidently a strictly protoplasmic factor in some animal cells that changes with hardening (Asahina, 1966).

2. Seasonal Similarities

In overwintering plants of cold climates there is normally an increase in cold resistance in autumn. This seems to render tissues progressively more resistant to other kinds of stress as well as to cold. I. W. Bailey (1954) pointed out that besides being more cold resistant, cambial initials of certain trees in the winter condition are more resistant to intense illumination or application of hypertonic solutions than they are in the growing season. It appeared in Parker and Philpott's (1961) work that tissues collected in winter and prepared for electron microscopy gave better results than those obtained in summer because of greater resistance of membranes of the winter specimens to the dehydration process. This was before the use of prefixation in glutaraldehyde was known. When glutaraldehyde was used, such seasonal differences were not so apparent, apparently because it protects the membranes (Parker, 1972).

In fern sporophytes that remain green throughout the winter, Kappen (1964) found a marked increase in desiccation resistance in early winter when there was a rise in cold resistance (Fig. 2). However, in some species an increase in desiccation resistance occurred in the spring before the main increase in cold and desiccation took place in autumn (Fig. 2). This may be because of the decrease in relatively free water in the cell in early summer as the new growth matures.

If freezing resistance in cold-climate trees is high in winter and low in summer, then desiccation resistance should follow similar trends. Results of Pisek and Larcher (1954) with *Pinus cembra, Picea excelsa, Rhododendron ferrugineum,* and certain other evergreens showed this to be the case. However, in a species that develops no true winter cold resistance, e.g., *Olea europea,* even though its cold resistance increases slightly in winter, desiccation resistance may not change at all with the season (Larcher, 1963). The increased winter resistance was believed to be caused by a lowered cell freezing point, not by true protoplasmic resistance.

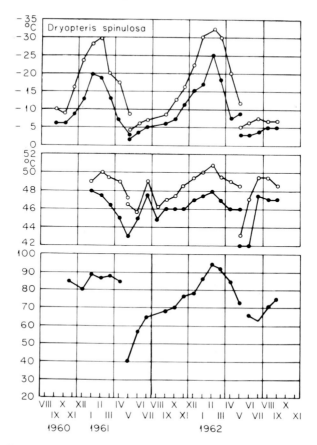

Fig. 2. Frost resistance (top), heat resistance (middle), and desiccation re-sistance (bottom) in the fern *Dryopteris spinulosa* over a 3-yr period. Months given as Roman numerals. Open circles: vital resistance level (10% injured); darkened circles: lethal resistance level (90% injured). From Kappen (1964) by permission of the author and Flora (Jena).

Evidently this freezing point change, which might indicate increased osmotic pressure, has no influence on desiccation resistance.

3. Species Similarities

Species that are cold hardy should also be drought hardy if the theory outlined above applies. There is, however, at least one problem here. Plants tested for drought hardiness in the growing season should not be compar-able to those tested in the winter season. As Levitt (1956) pointed out, hardiness to cold is not comparable to hardiness to drought at a time when little or no cold hardiness is attained. It appears that woody evergreens such as *Camellia,* which are not very cold resistant in winter, also possess rather low desiccation resistance (Parker, 1966).

Species differences in cold resistance are most readily demonstrated in winter. In summer, when hardiness is not developed, differences in cold resistance of woody species have been difficult to demonstrate (Parker, 1963). However, species differences in summer frost resistance of common forest tree species evidently do occur (Pomerleau and Ray, 1957) and perhaps play an important role in species composition in areas subject to summer frosts (R. Bailey, 1960).

Some very winter-hardy species, such as certain of the Betulaceae, can withstand unusually low temperatures as well as drying in the frozen state (Krasavtsev, 1967). Freeze-drying appears to be more severe in its effect on the cell than freezing alone (Hanafusa, 1969). When twigs of certain birches were freeze-dried in a specially designed apparatus, then thawed and remoistened, they survived as indicated by growth (Krasavtsev, 1967). Krasavtsev stated that twigs were "almost" dried. This is important since when the remaining water was removed, the twigs did not recover. Twigs were still viable when only 8.5% moisture remained in the tissues. Other species appeared to require somewhat more water to survive the treatment. Oppenheimer (reviewed by Parker, 1968) observed that *Ceterach officinarum* required a certain small amount of water to survive drying treatment in spite of its great desiccation resistance. Presumably, this plant should also be very cold resistant. It is known that desiccation-resistant mosses and liverworts of western Greenland can resist temperatures of $-26°C$ (Biebl, 1967–1968).

The effect of dehydration of leaf cells on the Hill-reaction in the photosynthetic process, as well as on photophosphorylation, was found to be identical to that caused by freezing (Santarius and Heber, 1967). The critical limit for dehydration of protoplasmic structures seemed to be nearly 10 to 15% of total water content.

4. Respiratory Phenomena

Respiration rate, either as O_2 absorbed or CO_2 released, has been extensively studied in connection with drought resistance, although not primarily as it relates to cold resistance. Some 40 years ago it was found that twigs of the more frost-resistant apple varieties exhibited a relatively small respiratory rise upon warming from a frozen condition to somewhat above 0°C (DeLong *et al.,* 1930). Less hardy varieties exhibited a larger rise. Although these results have been questioned, they seem to support the findings of Montfort and Hahn (1950), who showed that respiration of desiccation-resistant plants rose less than those of less resistant plants. In such experiments so many environmental and internal factors affect respiration that results are difficult to duplicate unless stringently controlled environmental conditions are maintained and the previous treatment of the plant is uniform for separate experiments.

Studies on plant respiration during dehydration indicate that a respiratory rise occurs with some species but not with others (Brix, 1962). Also it appears that the time course of imposition of drought is important to the type of response (Slatyer, 1967). On the other hand, Mooney (1969) concluded that the rate of desiccation did not necessarily alter the basic rate response of respiration of certain species of *Pistacia* and *Quercus* to drought. Mooney found an initial rise in respiration in these Mediterranean woody plants, then a decline as the water content continued to decrease. He thought this initial rise could be the result of lowered resistance to gaseous diffusion as intercellular spaces dried out. Increases in respiratory rates during drying have also been attributed to (1) uncoupling in the respiratory system (Zholkevich, 1961), (2) an increase in respiratory substrate as starch is hydrolyzed to sugars (Brix, 1962), and (3) protein reconstitution during the restitution phase (Stocker, 1960; Henkel, 1967).

C. SIMILARITIES OF DESICCATION AND HEAT INJURY

Heat and desiccation resistance have long been suspected of having mechanisms in common (e.g., Levitt, 1956), but research has not always revealed such a connection [e.g., Schölm (1968) with freshwater algae]. Some studies with tracheophytous plants (Kappen, 1964; Schwarz, 1969) indicate certain relationships do exist here.

1. Mechanical Considerations

As early as 1877 it was known that seeds resistant to prolonged exposure to dry air also were capable of withstanding high temperatures. Dry seeds, capable of germination when moistened, survived boiling water and even 120°C (Just, 1877). But if seeds were moistened and allowed to swell, they could not withstand such temperatures.

These findings suggest that when the living protoplasm of seeds is hydrated it becomes heat-sensitive. One might then ask whether certain heat and drying treatments are similar in their effects on cells. There is some evidence that they are. Kaloyereas (1958) concluded that heat stability of pine leaf chlorophyll could be used as a measure of drought resistance in different varieties of *Pinus taeda*. According to Majumber (1969), the chlorophyll index (a measure of chlorophyll heat stability) can be used as a measure of adaptation to overwintering in various woody plants. Results are sometimes difficult to duplicate because speeds and duration of heating or drying influence the resulting injury.

The most obvious cause of heat injury is protoplasmic denaturation. This occurs at least within certain ranges, speeds, and durations of heating. It should be understood that there is at least one time factor here. For example, a temperature of 57°C for 1 min usually has about the same effect as 54°C for about 5 min. These relationships were expressed by Lepeshkin (1935):

$$T = a - b \log Z$$

where T is heat-killing temperature, Z time of heating, and a and b are constants establishing position and slope of the relationship. Interestingly, the killing time for protoplasm and the coagulation time for protein sols have a similar logarithmic relation (Lepeshkin, 1935).

One peculiarity of protoplasm is that its Q_{10} (the number of times the rate of a process increases with a 10°C rise in temperature) is very high.

Respiratory Q_{10} values in the biokinetic (not killing) zone are commonly about 2 or 3 for most plants. But when a tissue is heated enough to injure it seriously, Q_{10} values may be much higher (Lorenz, 1939). Q_{10} values based on speed of killing for heat-treated *Catalpa* twigs were about 360 in the 54°–65°C range, although values for twigs of other tree species tested were lower. Lorenz mentioned that the Q_{10} of the marine colonial animal *Tubularia crocea* may be over 3000. On the other hand, the Q_{10} based on CO_2 released from the tissue were never more than about 10 (Parker, 1970).

Electron microscope studies (Parker, 1970) suggest that chloroplast lamellae are broken down by just enough heat to stop photosynthesis in *Picea pungens* leaves in summer (57°C for 15–30 sec). *Juniperus virginiana* was somewhat more heat resistant. A time-dependent breakdown was evident in *Juniperus* (Figs. 3 and 4). The more severe the damage, the more rapid was the subsequent breakdown of all cell membranes. In severe cases of heat injury only the starch grains were distinguishable within the cell walls. In this connection it is known that chlorophyll–protein complexes in maize chloroplasts are readily denatured by heat. This process is pH-sensitive. Most denaturation occurs at pH 5 (Nagy and Faludi-Daniel, 1967).

It might be expected that pH would be important for chloroplast membranes since interactions between carboxyl and amino groups help to establish and maintain the characteristic conformation of the protein. Both these interactions and the shape of the molecule vary with pH (Bartley *et al.,* 1968). Extremes of pH, like heating, can result in denaturation.

Molecular-level studies suggest that during heat denaturation water of hydration (bound or icelike water) is probably disrupted. One bit of evidence for this is the positive correlation that exists between resistance to

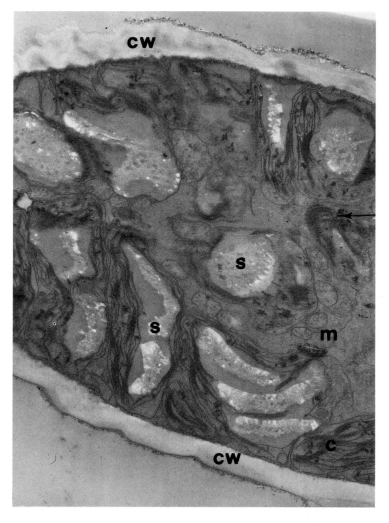

Fig. 3. Portion of a chlorenchyma cell from *Juniperus virginiana* leaf 10 min after heating 2 min at 57°C. Chloroplast lamellae beginning to disintegrate (arrow). S, starch at chloroplast center; cw, cell wall; m, mitochondria; c, chloroplast (intact). Epoxy resin embedded, April, 1969; prefixed in glutaraldehyde, KMnO₄ stain. Magnification: × 9760.

heat and resistance to ethyl alcohol in *Amoeba* (Sopina, 1968). It is generally accepted that ethyl alcohol in sufficient concentration removes the water of hydration from proteins.

Furthermore, adsorbed ions, partly intercalated in the hydration water

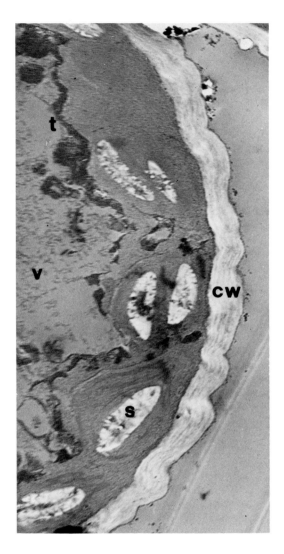

Fig. 4. Portion of a chlorenchyma cell from *Juniperus virginiana* leaf 2 hr after heating 2 min at 57°C. S, starch at chloroplast center; cw, cell wall; v, vacuole; t, tonoplast (here disrupted; strictly, this is a $KMnO_4$-accumulating layer about 0.5 μ thick commonly occurring against the vacuolar membrane in many conifer species). Note that chloroplast lamellae are largely disintegrated. Plasmalemma just to left of cw is a row of dark-staining particles. Epoxy resin embedded, April, 1969; prefixed in glutaraldehyde, $KMnO_4$ stain. Magnification: $\times$ 7000.

(Ling, 1969), would probably be affected by heat treatment. When transfer RNA was denatured by heat, it took up manganese ions more readily than before. This suggests that sites for this ion were not previously available (Molin and Bekker, 1967).

Heilbrunn (1958) found evidence that lipid droplets appeared in the cytoplasm after heating. As a result, he favored the idea that membrane lipids melted at high temperatures. Levitt (1956), however, suggested that intermolecular forces should be strong enough to maintain orientation of the membrane lipids, even above the liquefaction temperature for such lipids in the free state. Although this may be correct, results of Nir and Klein (1970) seem to support Heilbrunn's conclusions.

The importance of membranes to the theory of heat injury is suggested by a number of recent investigations. For example, activity of the enzyme galactosidase, in salmon liver tissue, declined when temperatures were raised from 15 to 30°C (Gatt, 1969). The enzyme was suspected of being lysosomal in origin. Lysosomes (Gahan, 1968), similar to mitochondria, may rupture as a result of this warming treatment, resulting in leakage of their enzymes and subjecting of the protoplasm to foreign proteins. If these foreign substances are not inactivated, the cell is destroyed (Gatt, 1969).

Obviously, high temperatures can injure enzymes. Less severe heat can dislodge certain enzymes like ATPase from the mitochondrial wall (Green and MacLennan, 1969). Some enzymes can be heat hardened. Activity of acid phosphatase extracted from cucumber seedling leaves was decreased as tissues were heat hardened (Feldman *et al.,* 1966). At the same time, the thermostability of the phosphatase was increased and persisted after dialysis of the phosphatase. This indicates that the stability of the enzyme is not dependent on the presence of a protective dialyzable substance. Protective agents are known that can stabilize certain proteins against heat denaturation. The addition of the sodium salt of pyridinedicarboxylic acid (PDA) to human serum albumin prevents the solution from becoming turbid when heated to between 60° and 95°C (Mishiro and Ochi, 1966). These investigators reported that PDA occurs naturally in germinating bacterial spores. In other research heat stability of certain dehydrogenases was enhanced by an interaction between succinic acid and the enzyme on which the flavine–quinone is formed (Giovenco *et al.,* 1967). An enzyme–substrate interaction was found to increase heat hardiness of the enzymes as well as decrease the tendency of the enzymes to aggregate (Tomita and Kim, 1966). Further results concerning resistance of enzymes to stress are discussed by Todd in Chapter 5 of this volume.

Addition of auxin or 2,4-D to plant cells can decrease the effects of heat in bringing about coagulation of cytoplasmic proteins. This increased protein stability may be the result of conversion of —SH groups to intramolecular S—S bonds (Morré, 1970). Note that these are intramolecular, not intermolecular; the latter is involved in aggregation, according to Levitt (1969). These changes, in turn, may be related in some way, still obscure, to the redox balance between ascorbic acid and glutathione, a balance that may be important to growth.

2. Seasonal Similarities

Seasonal similarities between effects of heat and desiccation on protoplasm have not been very clear in the past. Some reports indicate certain parallels (Lange, 1955, 1959, 1961). For example, in the moss *Barbula gracilis* a peak in heat resistance occurred in early summer (75°–93°C for 0.5 hr). This was apparently the result of dry conditions at that time when desiccation resistance was presumably high, because a decline in heat resistance followed rains in July and August (Lange, 1955).

In certain evergreens (e.g., top graph, Fig. 5), heat resistance increased in spring, peaked in early summer, and then declined in late summer or autumn with a subsequent increase to a winter maximum (Lange, 1961). Lange concluded that there was no exact parallel between heat and cold resistance in the woody plants he studied, yet heat resistance reached a peak in winter in species of *Erica* (Fig. 5) and *Asarum* when one would expect peaks in cold resistance at the same time. In other plants, including mosses and ferns, Lange (1967) sometimes found a relation between heat and cold resistance and sometimes none (Fig. 5).

A study by Parker (1970) showed that heat resistance in various species of evergreen and deciduous forest trees declined in spring, remained low during the summer, increased in September, and reached a winter peak at least by December. Shortening day length seemed associated with the early September increase, since warm weather continued well into September. These trends resemble those for cold resistance (e.g., Parker, 1959, 1962).

Kappen (1964) showed certain similarities between heat resistance and desiccation resistance in fern sporophytes (see Fig. 2). Desiccation and cold resistance were also related, but not as strongly as desiccation resistance and heat resistance. Heat resistance of pines and rhododendron may be controlled by the plant's inner rhythm and influenced, in its onset at least, by day length (Schwarz, 1969). Schwarz found that heat resistance increased in autumn with cold resistance, but plants became fully cold-hardy only when subjected to low temperature.

3. Species Similarities

Plant species possessing both great desiccation- and heat resistance are usually very primitive (Parker, 1968). Some investigators seem to have assumed that heat- and desiccation resistances in higher plants can be

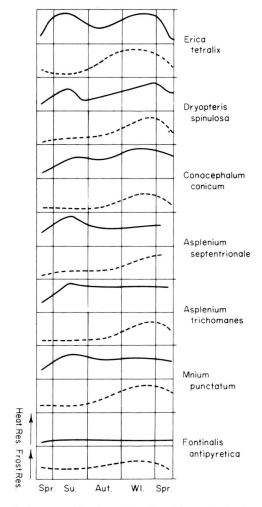

Fig. 5. Seasonal changes of heat- and frost resistance in the leaves of hibernating plants near Göttingen, Germany; *Erica tetralix* (phanerogam), *Dryopteris spinulosa, Conocephalum conicum, Asplenium septentrionale, A. trichomanes* (ferns), *Mnium punctatum, Fontinalis antipyretica* (mosses). Abscissa: seasons (time). Ordinates: heat resistance (—), and frost resistance (---). From Lange (1967) by permission of the author and Pergamon Press Ltd.

equated (e.g., Jameson, 1961), evidently based on Levitt's (1956) theory. As an approximate rule this seems true for many species, provided, of course, we are dealing with desiccation resistance, not overall drought resistance. It has been shown that *Juniperus virginiana* and related *Juniperus* species are unusually desiccation resistant (Parker, 1966). *Juniperus virginiana* is also more heat resistant than other gymnosperms tested (Parker, 1970).

4. Respiratory Phenomena

Heating would be expected to cause a respiratory rise, both in O_2 consumption and CO_2 release. But it has been found that this rise following warming is greater in some species of trees than in others and varies widely with season (Parker, 1970). Respiration of *Picea pungens* leaves is much higher following heating than are those of *Juniperus virginiana*. In fact the respiratory rise in *Picea* in autumn can be 10 times the normal versus only 3 times in *Juniperus,* measured at 24°C after heating to 57°C for 2 min. Since the extent of this rise varies widely with season, relationships to stored foods and/or hardiness are suggested.

Heat-sensitive species often have a higher respiratory rate after heating than do more resistant species (Parker, 1970). This corresponds to findings of Montfort and Hahn (1950), mentioned above (Section II,B,4). Since increased heat normally results in a respiratory rise anyway, whether injury occurs or not, it is difficult to separate the effect from one of injury.

III. NATURAL MECHANISMS OF PROTOPLASMIC RESISTANCE

A basic understanding of the causes of protoplasmic resistance of plants to desiccation continues to elude us even today. Relatively little research effort has been devoted to investigations of stress resistance as compared to studies of growth hormones, solute translocation, and photosynthesis in green plants. This is perhaps because investigators have been more interested in knowing what makes the plant grow than in understanding factors preventing this growth. In addition, many investigations of protoplasm had to await the introduction of methods that only in the past fifteen years or so have solved long-standing problems.

A. PROBLEMS AT THE CELLULAR LEVEL

The problems of desiccation-, heat-, and cold resistance were at first attacked with the means at hand—mainly the light microscope and some crude analytical procedures. Work with the microscope was limited, partly because of an inadequate understanding of the meaning of staining reactions. In spite of such difficulties, the reviews of Küster (1935), Guilliermond (1941), and others attest to the enormous amount of cytological information accumulated by that time. At the same time, analytical methods were often hampered by impurities in plant extracts that interfered with results, or by a lack of understanding of what was being measured. The analysis of sugars serves as a good example of this. Only when paper chromatography became widely used was it realized that most plants contained sugars other than glucose, fructose, and sucrose.

This brief review does not attempt to cover thoroughly the subjects considered below. Rather, it is intended to give the reader some insight into mechanisms of protoplasmic resistance as they relate to problems of heat-, desiccation-, and cold-resistance in plants and certain other organisms.

1. Protoplasm

Protoplasm, as a subject for study, has received intensive investigation in the past few years. This is partly because of development of new investigative and analytical methods. But there have been difficulties in the new methods. For example, the transition from light to electron microscopy has left us with a legacy of terms difficult to apply to the things we are seeing with the electron microscope.

The terminology of studies on protoplasm has always been somewhat vague. There has even been disagreement on a basic definition for protoplasm. Strasburger in 1882, in disagreement with von Mohl, included the nucleus as well as the cytoplasm in protoplasm (Anonymous, 1961). Most English-speaking physiologists think of the cell contents as consisting of (1) cytoplasm, (2) nucleus or nuclei, and (3) vacuome (one or more vacuoles); the protoplasm is thought to consist of (1) and (2) together. This terminology is used here. Plant physiologists usually exclude the large central vacuole from this definition of protoplasm, but it becomes a problem of how big the vacuole has to be before it is excluded. Those writing in German usually exclude the nucleus from protoplasm or *"Protoplasma"* (Höfler, 1960).

One of the most interesting characteristics of cytoplasm is its ability to stream. Only in recent years have we begun to understand how this occurs (e.g., Kamiya, 1960; Kamiya and Kuroda, 1966; Wohlfarth-Bottermann, 1968; Stockem *et al.,* 1969). In amoebae and slime molds, the so-called pressure flow theory seems to fit the facts at least as well as any (Wohlfarth-Bottermann, 1968). The motive force in streaming appears to originate from contractile elements, such as myosin, which obtain their energy from adenosine triphosphate (ATP) as in muscle (Kamiya, 1960). In the cyclosis of plant cells, on the other hand, a different mechanism seems to exist. Sol–gel interfaces are essential for generation of the streaming force, possibly because of the waving action of flagellalike threads protruding from the gel into the sol. In *Acetabularia* (the "mermaid's wineglass," Chlorophyta) protoplasmic streaming takes place in the form of many slender channels along a fibrous structure inside a thin ectoplasmic layer (Kamiya and Kuroda, 1966). This is reminiscent of the fibrillar appearance of slime in sieve tubes of vascular plants as seen with the light microscope (Parker, 1965) or the electron microscope (Evert *et al.*, 1966). However, this comparison may be totally misleading, since with slime or P-protein we are dealing with components requiring magnifications of about 80,000 to reveal hollow tubules (Evert and Deshpande, 1969).

I mention these apparently extraneous observations to emphasize the complexity and delicacy of protoplasm in the living form. It would be expected that stresses of various kinds would affect such mechanisms. Studies on cytoplasmic streaming in overwintering trees should help clarify the various states of protoplasm in the resting and active state. But only a few such studies have been made.

Cytoplasmic streaming in the cambium of trees at a standard test temperature is generally more rapid in the spring than at other times of the year (Thimann and Kaufman, 1958). This may be related to increased auxin levels in spring over other seasons resulting in reduced cytoplasmic viscosity. Streaming ceases at about $-1°C$, but can be slowly reinitiated in winter by warming. Streaming is observed in summer in *Robinia* bark cortical cells as a series of mitochondria and other organelles moving in a thin peripheral layer along the cell wall (Siminovitch *et al.*, 1967). This, in general, is similar to protoplasmic streaming in cambial initials of trees (I. W. Bailey, 1954).

In summer the nucleus is nearly transparent and thus difficult to see in *Robinia* bark (Siminovitch *et al.*, 1967). In winter, streaming usually ceases; the nucleus can be readily seen at that time, is usually centrally located, and possesses radiating cytoplasmic strands (Fig. 6). Other changes visible in the light microscope occur. Chloroplasts migrate to the center of the cell in autumn in leaves of species of *Picea,* whereas those of cold-adapted Pinaceae leaves migrate to the cell periphery (Parker, 1963).

Cytoplasm of these Pinaceae in the winter condition stains a darker blue with the protein stain, bromphenol blue–$HgCl_2$, than it does in the summer condition, indicating more sites for dye attachment in winter (Parker, 1960) (Fig. 7). This could result from protein augmentation in the cytoplasm (Siminovitch *et al.*, 1967) or from increased numbers of —SH radicals (Levitt *et al.*, 1962). When the protoplasm of sea urchin eggs becomes cold-hardy following fertilization (Asahina and Tanno, 1963), there may be an increase in —SH radicals (Kawamura and Dan, cited by Asahina and Tanno, 1963).

It was an old theory that cambial cells turned from a gel to a sol state in spring (reviewed by Parker, 1963). But it seemed that the main gel–sol change was taking place in the vacuome, not in the cytoplasm as many

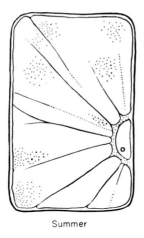
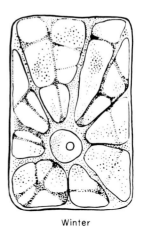

Summer   Winter

Fig. 6. Diagrammatic representation of summer and winter aspects of cells of living bark (fixation with alcohol-acetic acid). Magnification: × 1800. From Siminovitch *et al.* (1967), with permission of the author and the Institute of Low Temperature Science, Hokkaido University.

have supposed (Parker, 1960) (Fig. 7). This brings up the question of changes in the cytoplasmic viscosity.

Whether cytoplasm becomes more viscous or less viscous following cold hardening still remains in doubt. There is evidence that the endoplasm, the often-streaming inner part of cytoplasm, becomes more viscous in winter, while the ectoplasm, or outer part of the cytoplasm in which the chloroplasts are usually anchored, becomes less viscous (Levitt and Siminovitch, 1940, 1941). This appears to be in conflict with the conclusions of Kessler and Ruhland (1938), who observed, by centrifuging chloroplasts *in situ,* that the "protoplasm" was more viscous in winter than in summer. The speed with which the chloroplasts sank in the cytoplasm under the force of the centrifuge was taken as a measure of viscosity. Results of Franke (1962) with ferns have supported those of Kessler and Ruhland, although they are open to the same criticism: the density of chloroplasts may be altered with change of season. Certainly, the large starch grains present in summer within the chloroplasts either disappear or are much reduced in size by winter. Also there occur layers of endoplasmic reticulum through which such bodies must pass during centrifugation.

During drought hardening, perhaps similar to the cold hardening mentioned above, Stocker and Ross (cited by Stocker, 1960) found that protoplasmic viscosity slowly increased at first during dehydration, then on

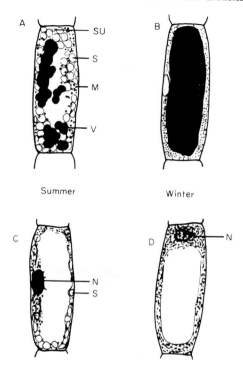

Fig. 7. A and B, axial phloem parenchyma cells of *Pinus strobus* bark fixed in Fleming's fluid, dehydrated in alcohol, and hematoxylin-stained. A and C, summer condition; B and D, winter condition. C and D, cells similar to A and B, but stained with bromphenol blue–HgCl₂. v, vacuolar material (black in A and B); m, mitochondrion; s, starch; su, sudan-staining particle (here unstained); n, nucleus. From Parker (1960), by permission of Protoplasma, Vienna.

irrigation of the plant it declined relatively rapidly. These results were based on the plasmolysis time method. It may be questioned whether the speed of plasmolysis is a true function of viscosity alone. Henkel (cited by Stocker, 1960) found that viscosity, measured by time plasmolysis and centrifugation, was increased as in Stocker's experiments. Elasticity, measured by centrifugation, was also increased on hardening. The effect of protoplasmic adhesion to the cell wall was supposedly avoided by treatment with weak hypotonic sucrose solutions. A third characteristic of hardening plants, according to Henkel, is that the imbibitional water of the protoplasm increases. As this occurs, protoplasmic structure is theoretically strengthened.

Another way of studying changes in cytoplasmic viscosity is by observing the morphology of plasmolyzed cytoplasm during hardening. The shape of the plasmolyzed protoplast is taken as a measure of viscosity changes. In winter the protoplast often rounds up evenly, whereas in summer it forms concavities between the plasmalemma and the cell wall. The latter form was taken by Levitt and Siminovitch (1940, 1941) as evidence of a greater ectoplasmic viscosity in summer. Also, as the cytoplasm pulled away from the cell wall, the Hechtian strands were seen to break more readily in summer. This suggested lesser ductility, if not viscosity, at that time (Scarth, 1944). The difficulty of the term "viscosity" is that cytoplasm is a non-Newtonian fluid and does not obey the usual laws of solutions. Ductility implies elasticity or stretchability.

In the final analysis we find ourselves involved in both interpretation and semantics. Cytoplasmic adhesion to the cell wall may be involved in "concave" and "convex" plasmolysis. Many of the terms used are hard to define. These problems may take on new significance when more information accumulates on how growth and hardiness are interrelated.

Another peculiarity of cold-hardened plant cells is their increased water permeability (Scarth and Levitt, 1937). Similarly, sea urchin eggs increase in water permeability as they become frost-hardy (Asahina and Tanno, 1963). In plants it appears that when cells are rendered increasingly water-permeable, ice is more likely to form outside them and continue to form there as cooling proceeds. This results in a lowered chance of ice forming intracellularly and thus destructively. The advantage of such a permeability change to the droughted cell is, however, not clear.

There is new evidence that the permeability of protoplasm of various tree species is no different in dormant than in nondormant buds (Badanova and Vartapetyan, 1967). If correct, permeability may not be a factor involved in hardiness changes. Stadelmann (1969) discussed water permeability at length and concluded that little is known about the site of action of any factor which might cause a change in permeability; for example, whether it influences the sieve effect of cell membranes or alters the capacity of membranes to act as a lipid solvent. He concluded that the sieve mechanism seems more sensitive to damage, whereas the lipid arrangement (in membranes) is less sensitive and also more easily repaired following injury.

2. Vacuome

The vacuome in most plant cells frequently occurs as a single large entity. Since it usually constitutes a large proportion of the cell volume (e.g., Fig. 7) and has a high water content, it is of the greatest importance to desiccation resistance. It can be both an asset and a liability. As an asset it acts as a reservoir for the cytoplasm during diurnal and seasonal fluctuations in water content; as a liability it shrinks under severe dehydration and may place the cytoplasm under great mechanical stress.

Rouschal (1938) discovered that certain species of ferns possessed a vacuome that entirely solidified during dehydration. This apparently prevented cytoplasmic distortion during cell dehydration and resulted in greatly improved desiccation resistance. Iljin (Parker, 1968) favored the idea that damage to the cell by dehydration is produced by mechanical distortion. Samygin and Mateeva (1967a) concluded that both mechanical rupture of cell membranes and dehydrative denaturation were important to injury, but that the latter was probably the more important of the two.

In forest trees, especially in the Pinaceae of northern climates, the vacuome of cortical cells of twigs can solidify completely to a true gel in winter (Parker, 1960). The solid matter of the vacuome can then be picked out with a micromanipulator needle. In summer, on dehydration, there is also a tendency for the vacuome to solidify, but there appears to be less solute material available, and droplets of solidified matter form in the vacuome instead of a single homogeneous mass (Fig. 7,A). Changes in vacuolar composition are indicated by the reaction of the vacuome to neutral red, since droplets form in summer (staining red) whereas the vacuome stains homogeneously red in winter.

I. W. Bailey (1954) made a thorough study of the vacuome in cambial initials of a number of tree species. The February condition in *Robinia pseudoacacia* commonly includes numerous spherical or oval vacuoles varying in number from about three or four large ones to hundreds of small ones. In an active cambium, on the other hand, protoplasmic strands extend mainly in the long axis of the cell, sometimes moving across a single large vacuole or along the cell periphery without crossing the vacuole. Presumably, the fractionated state of the vacuome in winter is related to the cell's greater resistance. However, this seems very different from our observations in bark cortical cells mentioned above.

B. Problems at the Molecular Level

The tremendous proliferation of molecular biology in recent years has done much to forward an understanding of stress injury to the living cell. Possibly too much emphasis has been placed on structural effects and not enough on the attractive and electrical properties of protoplasm related to charge transfer, energy bands, and biopotentials (Szent-Györgyi, 1968). Denaturation of cell proteins, for example, is probably more complex than has been supposed. In general, it appears to be at least a two-stage process and may give rise in some proteins to an unfolded conformation which then may undergo further change. The result may be formation of an aggregate, precipitate, gel, or refolded, nonnative conformation (Bradbury and King, 1969). In any case, it is this phenomenon of protein breakdown or change that seems to be the crux of the problem of cell stress injury.

1. Membrane Systems

Central to the problem of protoplasmic resistance to water stress is the membrane, a feature reviewed previously in Volume I of this series (Crafts, 1968). A number of additional points could be mentioned.

An early concept held that protoplasm was a single unit consisting of some kind of sol–gel system bounded by inner and outer membranes. While this may be true for certain portions of the cytoplasm, the cell membrane including the endoplasmic reticulum (ER) and membranes of various organelles is now believed to form the main structural portion of the living cell. At least it is this structure that appears to undergo disruption under stress. For example, Schnepf's (1961) studies in pea roots showed that Golgi bodies often broke down during dehydration. Likewise, the protoplasmic membranes of yeast cells were injured by freeze-thawing, resulting in leakage of certain components of the enzyme mechanism (Souzu, 1967). Nir and Klein (1970) concluded that dehydration of root cells of Zea mays caused certain changes in the structure of mitochondrial cristae which prevented their staining with osmium. Since lipid droplets were found in the cytoplasm of dehydrated cells, it seemed possible that lipid components were displaced from the membranes by drying.

Heating germinating sporangiospores and conidiospores of certain fungi at 52°C for 2.5 min resulted not only in disrupted mitochondria but also in disorganization of nuclei and aggregation of ribosomes (Baker and Smith, 1970). They found withdrawal of the plasmalemma from the spore wall, but this may be the result of dehydration before embedding. In our work on evergreen leaves (Parker, 1970) chloroplast lamellae appeared particularly sensitive to heat (57°C for about 1 min), and mitochondrial cristae also readily broke down. As mentioned before (Section II,C,1) there is a time factor involved in breakdown following heat treatment (Figs. 3 and 4).

It is now more certain that a class of structural proteins exists in plant cytoplasm. Their molecular weights vary from 20,000 to as much as 1,000,000 (Criddle, 1969), although agreement is lacking on the definition of a structural protein (Criddle and Willemot, 1969). Most estimations of molecular weight of chloroplastic and mitochondrial structural proteins range from 22,000 to 25,000 (Criddle, 1969). There is now increasing evidence that certain enzymes are united with these membranes. For example, structural protein has been observed to combine stoichiometrically with cytochromes a, b, and c, and with cytochrome oxidase (Criddle, 1969). Goldberger *et al.* (1962) found that a complex formation between isolated cytochrome b and structural protein resulted in a change of the oxidation-reduction potential of the enzyme so that the complex behaved similarly to that of the intact mitochondrion. Other evidence shows that certain of the dehydrogenases are closely associated with structural proteins, since mutations affecting structural proteins also affect the dehydrogenases (Criddle, 1969). A number of dehydrogenases and cytochromes are listed by Sjöstrand (1969) as occurring in mitochondrial membranes. There is now also evidence that the succinoxidase enzyme complex occurs on the mitochondrial cristae (Kalina *et al.*, 1969).

The exact structural composition of the typical cell membrane, whether of endoplasmic reticulum, nucleus, chloroplast, and so on, is still much in doubt and may vary widely in these different entities (e.g., Weiss, 1969; Sjöstrand, 1969). The "unit membrane" consisting of protein and lipid layers is accepted by some, but many doubts have been expressed as to its architecture. In fact, Branton (1969) concluded that, in a sense, our knowledge of membranes stands where our knowledge of DNA stood in 1945.

At least it seems clear that the cell membrane surfaces are hydrophilic and the interiors are hydrophobic (Criddle, 1969). Both proteins and lipids contain hydrophobic groups that appear to be buried in the membrane matrix. The protein is supposed to be held to the lipid by a number of different kinds of forces. These include covalent binding, electrostatic binding, polarization interaction, dispersion interaction, hydrophobic binding, and still others (Chapman, 1969). It has been suggested that the native structures of membranes, as in globular proteins, are primarily determined by hydrophobic rather than electrostatic interactions (Tria and Barnabei, 1969). The hydrophobic "interaction" (not to say "bond") refers to the tendency of nonpolar groups (as in adjacent peptide chains) to adhere to one another in an aqueous environment by van der Waals' forces and solvent effects.

With osmium stain–fixative, a typical mitochondrial envelope appears under the electron microscope as two lines. Each of these two lines or membranes in the mitochondrial envelope is considered a unit membrane. Each unit membrane appears with high resolution electron microscopy as one line with osmium stain but as two lines with $KMnO_4$. These two parts of the unit membrane may represent the two lipid layers (Finean, 1961, p. 86). Yet when the lipid is removed before fixation, the two layers still appear. This suggests that the lipid and protein exist in the same location. Possibly the lipid is entangled with the irregular bends of the polypeptide, although the lipid ends are still oriented as in the two-layered lipid micelle (Benson, cited by Tria and Barnabei, 1969).

Similar problems exist in interpretation of the chloroplastic thylakoid membrane. Some suggest that it consists of a globular protein framework surrounded by lipid, whereas others consider it to be made up of a central layer of lipid covered on both sides by globular subunits (Boardman, 1968).

Researchers sometimes do not realize the very disturbing effects of fixation, dehydration, and embedding on electron microscopy. A number of effects, including especially oxidation, polymerization, dehydration, and denaturation, can disrupt structures in various ways during preparation or even before preparation of specimens. This has probably accounted for different interpretations of membrane structures. Early electron microscopy of sieve tubes suggested that mature cells were largely devoid of structure, but this was probably because $KMnO_4$ is not a good stain–fixative for sieve tube slime and destruction of proteins during polymerization of the embedding medium may have occurred. Prefixation in glutaraldehyde has been a marked improvement and Evert and Deshpande (1969), for example, have revealed the great complexity of sieve tube contents.

Some years ago Frey-Wyssling pointed out the presence of mosaic structures or arrays of micellar or globular subunits (reviewed by Branton, 1969). Negative staining of mitochondria suggested that 90 Å-wide subunits occur, each composed of a head piece, stalk, and base piece (Fernández-Morán, cited by Folkers, 1967). Head pieces appeared to contain membrane ATPases. The base pieces may make up the continuum of the membrane, but there is much disagreement here (Branton, 1969). Some membranes may, in fact, be one particle thick, made of fused or nesting subunits. Particles seen may be lipoprotein macromolecules, all of the same size (Green and Tzagaloff, 1966, quoted by Branton, 1969).

Sjöstrand (1969) considered the basic membrane to be a two-dimensional array of globular protein molecules and lipoprotein complexes with the lipid molecules partially embedded in the protein molecules and partially distributed among the globular lipoprotein complexes. He thought that there were fairly large areas consisting almost entirely of lipid molecules in a bilayer extending between areas where globular proteins were accumulated. This situation may exist, for example, in the plasma membrane. He offered some serious criticisms of the widely accepted unit membrane idea and pointed not only to the frequent danger of artifacts in electron microscopy of membranes, but to the dynamic quality of these membranes. He held that some denaturation occurs during dehydration or fixation of material so that certain membrane features seen by some workers must be artifacts. Other artifacts are produced by pretreatment of material.

Further light has been shed on membrane structure by freeze-etching procedures for electron microscopy. Such techniques suggest that enzymatically active proteins occur here and there, inserted in the membrane. The freeze-etching process itself seems to break the hydrophobic bonds down the middle of the membrane, thus splitting it into two sheets (Fig. 8), while the hydrophilic portions are not much affected by freezing because of dipole interactions between hydrophilic portions. Branton (1969) concluded that the bimolecular leaflet best explains the unspecialized properties of such membranes. Unfortunately, freeze-etching, too, can produce artifacts. Nečas et al. (1969) found that a nonetchable layer in frozen-etched protoplasts was artificial. This appears as a thin layer on the surface of the yeast protoplast.

The character of membranes can also be inferred through studies on permeability. Some solutes appear to penetrate these membranes according to the laws of substances passing through lipids. Many workers support the two-pathway theory, in which the pores of a membrane, such as the plasmalemma, control passage of that portion of the total amount of a penetrating, small-molecular-weight substance that does not enter by lipid solution (Stadelmann, 1969). The permeation of a nonelectrolyte such as urea seems explained by the sieve theory, whereas certain lipid-soluble substances appear to pass through the lipid fraction.

It should be mentioned in this review that many kinds of lipids occur in membranes. Phospholipids are important in stabilizing and organizing subunits; they may also function in electron transport (Criddle and Willemot, 1969).

Membrane proteins appear to have certain properties which resemble those of the α helical muscle proteins like paramyosin (Neville, cited by Anonymous, 1969). It is probable that the entire unit membrane hy-

a   b

Fig. 8. Molecular configurations that could account for the freeze-etch results of membranes. Lipids shown as circles with two tails; proteins as irregular coils. Particles seen in the fracture faces are interpreted as lipid micelles (cluster of 6 lipid molecules in a) in equilibrium with a lipid bilayer (a), or (b) protein elements (heavy curled lines) in the hydrophobic membrane interior. In each case (a or b), the fractured membrane is on the left. From Branton (1969), by permission of the author and the *Annu. Rev. of Plant Physiology*, Palo Alto, Calif.

pothesis is in need of major revision, as Stadelmann (1969) stated, since there are so many kinds of membranes, thicknesses, and so on, which cannot be brought into agreement with the original concept.

2. Water in Protoplasm

To write of water in protoplasm implies that water is simply water and protoplasm is a sort of homogeneous mass. This is, of course, not true. Water can exist in many different bond arrangements and protoplasm is anything but a homogeneous mass. Even the idea of protoplasm having a fine architecture is doubtful. The structure that seems so positively set under the electron microscope is, in life, constantly changing. Many of the organelles are not only vibrating under Brownian movement, but electrostatic charges are changing, ions are exchanging positions on the polypeptides, vacuoles are enlarging or shrinking, materials are being excreted, protoplasm is streaming, and so on.

A number of peculiarities of water relate to the problem of protoplasmic resistance to water stress. For example, the cohesiveness of water under tension permits (along with other forces) the rise of water to the tops of tall trees (e.g., Kozlowski, 1961). In fact, water has many peculiarities which are puzzling to the physical chemist as well as to the physiologist.

The unusual colligative properties of water are mainly the result of strong attraction between its molecules. This attraction is associated with a structural feature called the hydrogen bond (H—) (e.g., Pauling, 1952; Frank, 1970). We might expect water to exist commonly as a gas at standard temperature and pressure (STP) since H_2S, CH_4, NH_3, and HF tend to do so. But it does not; it ordinarily remains liquid under these circumstances, usually with some vaporization. This might be expected on the basis of the boiling points of these compounds (Table I). The boiling point of HF is higher than that of H_2S because of H—F————H—F

TABLE I

COMPARISON OF MELTING AND BOILING POINTS OF WATER
AND ITS CLOSE RELATIVES

Compound	Melting point (°C)	Boiling point (°C)
H_2O	0.000	100.000
HF (or HxFx)	−92.3	19.4
H_2S	−82.9	−61.8
NH_3	−77.7	−33.35
CH_4	−184.0	−161.5

hydrogen bonding. Likewise the boiling point of NH_3 is higher than that of CH_3 because of H_3N————HNH_2 hydrogen bonding.

Water not only binds to itself (Crafts, 1968) but also forms hydrates with many other compounds and elements. In an aqueous medium the proton (H^+) is known to be combined with one or more molecules of water and is often represented as the hydronium ion (H_3O^+). However, it can be further hydrated to $H_9O_4^+$ (Lehninger, 1970). Also, ions such as Mg^{2+} and Al^{3+} polarize water in their immediate neighborhood (e.g., Bernal, 1965; Ling, 1969). In the case of ions associated with proteins one has to consider both the water of hydration of these ions and the fact that ions have different properties within the layers of icelike water adjacent to the peptide chains than without (Ling, 1969).

From the energy point of view water is considered at the bottom of the scale (Szent-Györgyi, 1968). This is because electrons reach the minimum of their energy by being oxidized; in this case, coupled to oxygen to form water. In spite of this, the water molecule remains somewhat polarized. This polarization allows it to make subsidiary bonds to adjacent oxygens in other water molecules in addition to the regular covalent H—O bonds. These subsidiary bonds are weaker than the regular covalent bonds, yet stronger than the normal van der Waals' forces (Pauling, 1952). Water binds to hydrophilic groups of the peptide chain as amino, imino, carboxyl, and SH groups. Adsorptive as well as H-bonding forces are involved here (Bernal, 1965). There may well be other forces (Derjaguin, 1965).

The electrostatic properties of water as well as of its ion solutes should not be underestimated. One unusual feature of water containing ions in solution is illustrated as follows: When a glass plate is submerged in water to which certain salts are added, a layer of ions tends to form along this glass surface in an immovable layer of adsorbed water. Another ion layer with an opposite charge forms over the first layer, thus forming a system similar to a parallel plate capacitor (condenser). This constitutes the Helmholtz double layer across which there is a potential called the zeta (ζ) potential (Gucker and Meldrum, 1944). Now if a capillary tube is filled with water containing the same ions, one might obtain such a layer on the inside surfaces. The charged ions in the immovable layer should not move, but the charged ions in the movable layer may migrate toward one or another electrode, assuming electrodes are present. Such a flow may occur through pores or capillaries in membranes. The phenomenon has been termed electroosmosis.

Lipid molecules forming a membrane orient themselves so that their ionic or head groups (circles in Fig. 6) are directed toward the bulk aqueous phase. The electrostatic charge on the head groups attracts counter ions (of opposite charge) in the adjacent icelike or aqueous phase.

The total potential difference here is considered the ψ-potential (nothing to do with the water potential discussed in Section II,A,1), whereas the potential between the counter ions and the charge in the bulk aqueous phase constitutes the ζ-potential (Chapman, 1969). Neutralization of the ζ-potential on lipoproteins by certain salts, for example, may allow the membrane to aggregate with other such molecules, normally repelled by similar electrostatic charges.

In the case of a simple hydrophobic sol (containing no hydration water), one can discharge the micelles and obtain coagulation. With a hydrophilic sol, containing both hydration water and charged ions, the situation is more complex. Certain electrolytes such as $(NH_4)_2SO_4$, besides removing ions from the sol can also acquire large quantities of hydration water and thus dehydrate the sol as well, resulting in coagulation. This has been termed "salting out" (Meyer and Anderson, 1952). It may be one of the causes of protoplasmic coagulation resulting from dehydration and the attendant increase in concentration of certain native ions initiated by drought.

The electrostatic properties of peptide chains are also complex. The charge on one reactive group influences other parts of the peptide chain; the chain as a whole can transmit these effects along its axis (Ling, 1969). A consequence of the stable water lattice surrounding the peptides is that it accelerates the diffusion-controlled transfer of protons along the macromolecule (Berendsen and McCulloch, 1960). The means by which ions are adsorbed on proteins must be related to the icelike arrangement of water as well as to the peculiarities of adsorbing points on the protein (Ling, 1969).

Usually water is considered to exist in cells either bound or free. However, as brought out by Crafts (1968), the existence of truly free water in the living cell is unlikely if not impossible. It now appears that there must be several layers of icelike water around proteins and that these layers are more strongly bound than was at first suspected. In a colloidal clay particle, addition of a monolayer of water increases its thickness about 4 Å. As many as 10 monolayers can be added as the clay particle expands to 40 Å (Bernal, 1965). This is, of course, water in an oriented form, but as mentioned above, not the same as the crystal arrangement of pure frozen water.

It may be of interest to mention here a new form of water discovered in small capillaries, called ordered, anomalous, or polywater (Lippincott et al., 1969; Castellion et al., 1970). As the distance from the capillary wall increases, both the amount of polywater and the extent of ordering is supposed to decrease. This water is considered by these authors to be a true polymer, and is bound by O—H—O—H linkages, in contrast to the various forms of ice in which the linkage may be shown as $H—O\cdots H$. On the other hand, there is the possibility that polywater is not a true polymer and that its peculiar properties are produced by the presence of particulate matter in the water (Kurtin et al., 1970). These investigators point out that even traces of sol-formers will entirely alter the properties of water.

Two sets of ice appear to be associated with protein: internal and external, neither identical with ice of free water (Bernal, 1965). In skeletal muscle there is a minimum of two phases of ordered or icelike water (Hazelwood et al., 1969). About 30% of the protein weight consists of icelike water arranged in certain places.

As mentioned before, icelike water is believed to exist in layers on the protein surface. Layers may be 10–20 Å thick (Bernal, 1965). In water layers about 100 Å thick, molecules are not so firmly bound, but are influenced by their proximity to the protein. Even at distances up to 4000 Å the protein exerts some influence on the water shell (Bernal, 1965).

In living cells, bound water may be arranged on peptides in rows to accord with the repeated occurrence of amino acid residues protruding laterally from the peptide chain (Berendsen and McCulloch, 1960). In this way, one hexagon of ice corresponds to each amino acid residue to form a chain paralleling the peptide and perhaps spiraling with it.

In killed plant tissue, such as cabbage leaves, there is more relatively free water than in living cells of the same material. Furthermore, different amounts of water are released from plant cells killed under different conditions (Macovscki and Marineanu, 1967).

Protein-bound water does not, however, appear to be like ordinary ice, but has been called cubic, cuboidal, or of the iceberg form (e.g., Hanafusa, 1967). It has also been termed Ice Ic (Eisenberg and Kauzmann, 1969). It differs in structural arrangement from ordinary Ice I or from one of the pressure forms of ice (Ice II to Ice VIII). Ice bound to proteins may be in one of the pressure forms, perhaps up to Ice VI (Berendsen and McCulloch, 1960).

Ice Ic differs mainly from Ice I in arrangement of the hexagonal rings. In both Ice I and Ic the O atoms are arranged in puckered layers containing hexagonal rings with the "chair" conformation. However, in Ic, unlike I, the stacking of the rings is such that the rings formed by six O's in one layer and six in the next layer also have the chair conformation (Eisenberg and Kauzmann, 1969). In the meantime, the H atoms are disordered in the same way in both types of ice. Also, the O—H distance is about 0.97 Å in Ic and 1.01 Å in I; these variations are probably not significantly different (Honjo and Shimaoka, cited by Eisenberg and Kauzmann, 1969). This O–H distance should not be taken as a measure of the spacing between adjacent water molecules. Instead, there is a spacing in Ice I between adjacent oxygens of 2.76 Å (Eisenberg and Kauzmann, 1969). This is important in estimating the number of ice layers in small cellular spaces, assuming Ice Ic has a similar spacing as I.

Water possesses a great propensity to form strongly induced dipoles (Ling, 1969). The restriction of water within protein chains can be likened to that of iron filings in a magnetic field (Fig. 9). The space between protein chains (at least in muscle cells) averages 16.9 Å. This allows room enough for six layers of water molecules (Fig. 9). When ions are introduced into this system, a situation occurs analogous to the addition of the nails in the diagram. The ions are severely restricted in rotational tendency by the surrounding bound water.

The idea of bound water is at least sixty years old but its existence has been frequently ignored by physiologists. It may even be implicated in the action of auxin on growth (Van Overbeek, cited by Vaadia *et al.*,

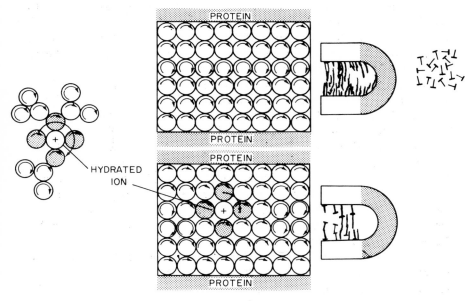

Fig. 9. Diagrammatic representation of the adsorption of water molecules as polarized multilayers on proteins. On entering such a system, the hydrated ion shown to the left suffers severe rotational restriction. A simple model of this effect is shown on the right where the restricted orientation of small nails in the field of the horseshoe magnet filled with iron filings is shown. On the upper right, the nails (representing ions) are not included with the filings (representing H_2O), while on the lower right they are included. From Ling (1969), by permission of the author.

1961) and in the action of certain hormones on animal eggs (Hays and Leaf, 1962). Some of the curious permeability properties of trout eggs are better explained if one assumes the presence of structural water. Hays and Leaf suggest that the limiting permeability, even to water itself, is water in an icelike state in or on the membrane. Alteration of the membrane pores by hormones may result from a change in bonding of the water molecules.

It was estimated that after freeze-drying egg albumin, about 100 molecules of water were left per protein molecule (Hanafusa, 1969). Freeze-drying seemed to result in a "phase change" of solvent water in certain globular proteins like catalase, whereas unfolding occurred in fibrous proteins (Hanafusa, 1967).

Although such water as that adsorbed to proteins is in an icelike state, it is not as firmly set as ordinary Ice I, if Berendsen and McCulloch (1960) are correct. Also, this icelike water seems to be readily exchangeable with water in the surrounding medium. This was proved by using $H_2^{18}O$. Furthermore, partial removal of the water of hydration did not greatly reduce this mobility (Vartapetyan, 1965).

Considerable information has now been obtained on the nature of the linkages tending to hold protein chains in certain configurations. These linkages include (1) electrostatic interactions (—COO^- and —NH_3^+ groups), (2) hydrogen bonding, (3) interaction between nonpolar side chains caused by mutual repulsion of the solvent, (4) Van der Waals' interactions, and (5) disulfide linkages (Bartley et al., 1968). Yet even these may not be enough to maintain the configuration without the help of bound water in proteins as well as lipids.

One would thus expect that stabilization of bound water by such substances as sugars, polyols, or other compounds of water-binding character might be of benefit to the protein in preventing denaturation during dehydration. Obviously, the problem of organic binding to proteins is complex. For a review on binding of neutral molecules, metals, organic ions, and other substances to proteins the reader is referred to Steinhardt and Reynolds (1970).

3. Theories of Resistance

As we have seen (Sections II,B and C), hardiness of certain species of plants to desiccation can sometimes be related to hardiness to either cold or heat. The relationships among these various stress factors allow us to borrow results of research from widely separated fields. In fact, much of the theory in this field of resistance comes to us from studies outside botany.

a. Carbohydrates. It has been known for over fifty years that sugars increase in woody plants capable of hardening. The antiquity of this knowledge has possibly tended to degrade it in the eyes of modern investigators. Sugars have, nevertheless, been shown to be cryoprotective agents. For example, they can protect cold-sensitive enzymes from injury due to freezing (Ullrich and Heber, 1957; Heber, 1958, 1968).

Other close relatives of sugars such as the polyhydric alcohols are known to be protective. Bovine and human spermatozoa can be protected from freezing with glycerol (Sherman, 1962). Glycerol can protect overwintering forms of insects from cold (Salt, 1961). Other polyhydric alcohols, occurring naturally in some plants, are known to increase during hardening (Sakai, 1960). Various organic acids, however, do not seem to change much with the season (Parker, 1963).

The rise of the trisaccharide raffinose from negligible quantities to levels similar to those of sucrose during hardening suggests a relationship between sugars and hardening (Parker, 1958). Certainly there is nothing magical about this, but it suggests that galactose is being made available to sucrose to form raffinose. If so, one might speculate that protein–galactose compounds may also be forming. It has been shown, for example, that a glycoprotein is formed in the gel of a slime mold (*Physarum polycephalum*) and is absent in the sol (Sheen *et al.,* 1969).

Henze's data (1959) pointed to autumn increases in a glycoprotein fraction in fruit tree bark. This fraction was mainly an arabinose water-soluble protein combination, although other sugars including glucose, galactose, ribose, and rhamnose appeared in protein hydrolyzates. We found that water soluble proteins in *Hedera* contained arabinose, galactose, and ribose (Parker, 1963), but we did not find a good relation between changes in water-soluble proteins and hardiness. Two types of sugar–protein complexes have been found in hardened *Hedera* leaves: one covalently linked as a glycoprotein and the other more loosely linked by H-bonding. Furthermore, protein from cold-acclimated tissue exhibited a higher sugar-binding capacity than protein from nonacclimated tissue (Steponkus, 1969).

Whatever the sugar relationships with proteins may be, there is now enough evidence to justify concluding that sugars and their close relatives, such as polyols, may serve as protective agents. This idea is based on the following: (1) Sugars always increase on hardening in all overwintering woody plants of cold climates (Parker, 1963), and even in certain insects or insect instars (Tanno, 1967); certain specific sugars, like raffinose and sucrose, are especially closely related to cold-hardiness changes (Parker, 1958, 1959, 1962). (2) Sugars like glucose can protect various kinds of living cells (Webb, 1965). (3) Sugars can, in theory at least, replace bound water or some phase of water adjacent to sensitive proteins [however, Hanafusa (1967) did not find evidence of this in enzymes]. (4) Sugars can theoretically influence water binding because of the affinity of their —OH radicals to water and perhaps because of their structural similarity to that of water (discussed above). (5) Monosaccharides have been shown to bind directly to certain proteins (Giles and McKay, 1962) and protein–sugar complexes have been found in hardened plants (Steponkus, 1969). (6) The production of sucrose is important to cold hardening (Steponkus, 1968; Heber, 1968). (7) Amylases usually increase in droughted plants with a resultant decline in starch and a rise in sugars (Vaadia *et al.,* 1961), although this seems refuted by Takaoki (1968), who found that α- and β-amylase decreased in cowpeas (*Vigna sinensis*) as soil moisture declined.

This protection phenomenon may also apply to desiccation resistance; Webb's work (1965) with viruses, bacteria, and mammalian tumor ascites strongly suggests that it does. The protective effects of inositol and glucose have been demonstrated in these organisms and tissues. When glucose was methylated, its protective effect was lost (Webb, 1965). One is, therefore, tempted to think as Webb did, that these substances replace water somewhere in the protein–water complex of the protoplasm.

As described in the previous section, water as ordinary ice (Ice I) is pictured as a series of linked hexagons, the joints of the hexagons consisting of O atoms. In Ice Ic, the hexagon is pictured as having the chair shape (Fig. 10). Glucose can—and most commonly does—also occur in the cyclic form. It is shown in Fig. 10 as β-D-($+$)-glucose with the chair shape. Since both ice and glucose form chair-shaped hexagons we might suppose that the sugar hexagon could fit into the space left by the loss of the ice hexagon. However, examination of Fig. 10 shows that the ice hexagon is considerably larger than that of glucose. This would seem to be an impasse. Yet, if we make models of ice and glucose, as in Fig. 10, it can be shown that we can superimpose the glucose on the ice in such a way that four of the O's of the glucose OH's, as well as the carbon of the glucose side chain, fit very nearly in the corners of the ice hexagon. In this way a glucose could occupy the space left by the loss of 6 HOH's in a hexagon.

The fact that methylation of glucose blocks its protective action (Webb, 1965) suggests that this either prevents fit or that the methyl group repels water rather than absorbing it. One could readily picture that two glycerol molecules arranged in two V's ($<>$) would have a similar spatial arrangement as one glucose. Actually, the V's fit best parallel to each other.

The protective effect of such compounds as glycerol and ethylene glycol against freezing has now been demonstrated in plant cells. In some cases they were more effective as protective agents than sucrose, perhaps because of better cell penetration (Samygin and Mateeva, 1967b). Experiments

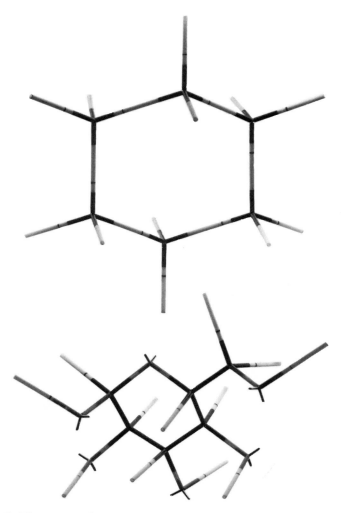

Fig. 10. The comparative structure of Ice Ic (top) and β-D-(+)-glucose (below) is shown with bonds between atoms as bars. In ice, O's occur at the joints and have a covalent radius equal to the length of the black portion. The atomic center of the H's is shown by a line on some of the bars. H's are arranged as in Ice II to illustrate how H's may be associated with O's, although H's in Ice Ic are disordered. Nearest neighbor distance in Ice Ic is 2.75 Å.

The glucose model (below) is in the same scale as ice. C's or O's occur at the joints of the model. H's and covalent radii are indicated as above. Each O has two short prongs, representing two pairs of unshared electrons. The white part of the bond extending out from the H's represents the extent of the radius of van der Waals' forces. Photographed by John Howard, U. S. Forest Service, from models by the author. Ice model based on drawings from Eisenberg and Kauzmann (1969), by permission of Oxford University Press.

with mixed solutions of sucrose and glycerol or with sucrose and ethylene glycol showed that penetrating substances protected the cells not only by raising the concentration of the intercellular solution, but also by affecting the stability of the protoplasm as it relates to water removal to extracellularly formed ice (Samygin and Mateeva, 1967b).

Common to many of the protective compounds like glycerol, inositol, and various sugars, besides their similar structural arrangements, is the presence of numerous —OH radicals. These radicals bind the surrounding water. For this reason, and because the organic molecule is more stable than water alone, these compounds may have a stabilizing influence. Ling (1969) indicates that the rotational tendency of water molecules as well as the forces between water molecules are changed by the introduction of ions. While sugars and polyols are nonpolar, they are known to have unusual water-binding effects.

Mazur (1970) considers that most additives like glycerol and dimethyl sulfoxide protect proteins because of solution effects. Although the effects of such additives are explicable on the basis of molarity and solubility (according to Mazur), protection by macromolecules such as polyvinylpyrrolidone (mol. wt. 40,000) is not.

A number of other organic, nonproteinaceous compounds have protective action *in vivo,* including dimethylformamide, dimethylacetamide, and N-methylpyrrolidinone (Roberts, 1969). Such compounds are characterized by their high affinity for water and form strong H bonds with both themselves and water.

b. Proteins. Since about 1950 there has been great interest in proteins as this relates to cytoplasmic protection to cold stress (Siminovitch and Briggs, 1949; Siminovitch *et al.,* 1967; Jung *et al.,* 1967; Levitt, 1967; Gaff, 1966; Roberts, 1969). In studies of drought hardening, Henkel (1967, 1970) concluded that plant tolerance to desiccation primarily depends on a high rate of protein synthesis during dehydration.

Since changes in sugars do not always exactly correspond to cold hardiness changes, a new factor seemed needed. The discovery that water-soluble proteins in *Robinia pseudoacacia* increase in autumn and decline in spring (Siminovitch and Briggs, 1949) did much to establish the importance of proteins to cold hardening. However, much of the theory was based on this one species, largely because other overwintering woody plants did not yield sufficient water-soluble protein for analysis. This was because water-soluble proteins were readily precipitated by tanninlike compounds during homogenization of tissues.

Total proteins can be shown to increase in the bark of a number of woody plant species in autumn, although much less so than in that of *Robinia* (Parker, 1958). Protein gains, in turn, appear to be preceded by a pronounced rise in RNA (Siminovitch *et al.,* 1967). However, in the less-hardy, woody evergreen vine *Hedera helix,* gains in sucrose, raffinose, and anthocyanins seemed better related to cold hardiness than did an increase in water soluble protein (Parker, 1962). It is often true that these sugar-hardiness relationships are not clear-cut, and this has led many workers to doubt the validity of a hardiness association. But it should be kept in mind that whole tissues are being extracted and that it is frequently uncertain whether the sugar increase is vacuolar or cytoplasmic, or is in fact occurring in some tissue other than that tested for hardiness.

Proteins, too, are not always well related quantitatively to cold-hardiness changes. Also, as Levitt (1969) said, both the older and newer electrophoresis methods have consistently failed to discover the appearance of new proteins or the loss of old ones during hardiness changes. Most research indicates that all the separable bands in water-soluble protein increase on cold hardening. On the other hand, Craker *et al.* (1969) claim to have obtained evidence of the appearance and disappearance of specific protein bands separated by electrophoresis following hardiness changes. But their data suggest that most bands simply change somewhat in density with the season. During drought hardening of wheat leaves, certain peroxidase bands disappeared and new ones appeared; at the same time, total peroxidase increased (Stutte and Todd, 1969). Meanwhile, water-soluble proteins drastically declined under water stress in these experiments, the opposite of what might have been expected from cold-hardiness studies.

In the final analysis, both proteins and sugars seem involved in hardiness changes. Steponkus and Lanphear (1967b) concluded that in *Hedera helix* there might be two factors in cold hardening: one, a predisposing condition, and the other a light-generated promoter. Studies with ^{14}C-sugars suggested that the promoter could be translocated from illuminated leaves to darkened leaves and that the promoter could be sucrose. Steponkus (1969) concluded that protein–sugar combinations are produced during hardening.

Although sugars are effective as protective agents in both cold- and drought hardening, two proteinlike compounds were found to be much more effective than sugars in protecting chloroplasts from either drying or freezing (Heber, 1968). Tumanov (1969) concluded that the stable state in which the organism could withstand freezing was obtained by cessation of growth, accumulation of protective substances, increase in cell membrane permeability to water, and optimal gelatinization of the protoplast. However, growth cessation is not necessarily related to hardening. Cox and Levitt (1969) found that the ability of cabbage leaves to harden increased as growth rate increased.

If new proteins are the main cause of hardiness, then one would expect that membranes or other protein-containing structures, as "seen" with the electron microscope, might change following hardening. Most studies have not, however, revealed any striking changes. In *Pinus strobus* leaves, chloroplast lamellae appeared to be the same in winter as in summer (Parker and Philpott, 1961). In *Fraxinus americana* bark, Srivastava (1966) reported similar findings. But in the work of Parker and Philpott there was some evidence of a reticulum occurring in winter in the cytoplasm that was less, or absent, in summer. In fact, light microscope studies suggest the presence of a reticulumlike structure in pine bark cells in winter (Fig. 7D). Mouravieff (1969) found that after drought hardening the endoplasmic reticulum was easier to see than before hardening. More recent work by the author (unpublished) does nothing to substantiate the idea of an increase in reticulum but rather in the total mass of ground protoplasm, also suggested by results of Srivastava (1966).

One could suppose that the separation of S—S bonds during hardening resulted in an increase in —SH proteins and this might be the source of increased water soluble protein (Levitt, 1967). However, the —SH levels decreased after an initial increase and continued to decrease as hardening proceeded (Levitt, 1967).

There are other difficulties with the —SH theory of hardening. If small amounts of sugar are present, no change in —SH content of chloroplasts occurred during dehydration; yet in the absence of sugar, —SH increased, apparently because of denaturation (Santarius and Heber, 1967). In no case could oxidation to S—S be observed following drying or freezing as Levitt's theory predicted. SH $\leftrightharpoons$ S—S changes are certainly complex: they may be involved in auxin-growth effects as well as in protein stability under heat stress (Morré, 1970). It would seem that a simple solution to the hardiness phenomenon based on such changes is not at present forthcoming.

Levitt (1967) found mercaptoethanol capable of binding to —SH sites and of increasing frost-hardening. According to his theory, —SH radicals in adjacent proteins are prevented by this binding to form S—S bridges. Mazur (1969), doubting the validity of this theory, pointed out that S—S bond formation is not ordinarily considered the cause of denaturation (Levitt's "aggregation" when S—S bonds form), but it is only one of the steps resulting in irreversible aggregation of a previously denatured protein.

Another theory of hardening is that new proteins substitute for already existing ones (Roberts, 1969). Replacement of proteins in such complex structures as are found in chloroplast membranes seems unlikely, although Roberts points out that isozymic forms of certain enzymes can be added or deleted in the cell under certain circumstances.

During fertilization of certain egg cells (Asahina and Tanno, 1963), the hardiness change is so rapid that some sort of bond or structural change is indicated. When heat-hardening occurs (Section II,C) sensitive proteins of the cell are stabilized so that they become less affected by injurious agents such as $CdCl_2$, ethyl ether, sodium azide, among others (Lomagin et al., 1963). This perhaps supports Heber's theory (1968) that protection by sugars or proteins of sensitive membrane parts represents protection against increasing ion concentrations.

The appearance of higher levels of amino acids and amides in cold- or drought-injured cells than in normal ones may not be related to a protective mechanism, but may simply be the result of some protein breakdown. This might account for results of Saunier et al. (1968) in droughted seedlings of Larrea divaricata. There may also be an increase in NH_4^+ during drought, as a result of protein hydrolysis. This can be toxic, perhaps producing salting out of proteins. Henkel (1964) suggested that plants which retain their capacity to control the NH_4^+ level by incorporation of it into amino acids appear to be drought adapted. Less-tolerant plants show enhanced formation of ammonium salts (Henkel, 1967, 1970).

In a study of Carex pachystylis, Hubac et al. (1969) found that plants grown under dry conditions contained large amounts of proline whereas watered plants contained hardly any. No other amino acid showed this remarkable difference, although arginine and γ-aminobutyric acid were also higher in droughted plants. They cited other workers who found somewhat similar results in other plants. Parker (1972) found more proline in roots of Quercus and Acer seedlings that were not watered from mid-June until late July than in roots of thoroughly watered seedlings. The reason for this increase is uncertain. Proline may originate from glutamic acid but may also serve as a storage compound for nitrogen (Hubac et al., 1969).

IV. CONCLUSIONS

In conclusion, it is apparent that there are many problems to be solved in explaining the protoplasmic resistance mechanisms of plants. Probably a number of different things can fail in the life mechanism of cells under drought, cold, or heat stress, Much probably depends on species as well as on speed and extent of this stress. Injuries may be associated with loss of icelike water in or adjacent to lipoproteins, changes in ion interactions, distortion and even separation of certain membrane components, and irreversible alteration in lipoprotein membranes. Abnormal leakage through these membranes may result in general cell autolysis.

Mechanisms for prevention of injury may include changes in the physical nature of the vacuome, increased elasticity, and perhaps gelation of the ground cytoplasm, and production of membrane-protective agents such as sugars, polyols, peptides, proteins, and still other substances. Control of toxic compounds formed during drying may also be important.

REFERENCES

Alexandrov, V. Ya. (1964). Cytophysiological and cyto-ecological investigations of resistance of plant cells toward the action of high and low temperature. *Quart. Rev. Biol.* **39,** 35.

Allen, R. D., and Francis, D. W. (1965). Cytoplasmic contraction and the distribution of water in the amoeba. *Symp. Soc. Exp. Biol.* **19,** 259–271.

Anonymous. (1961). "Webster's New International Dictionary." G. & C. Merriam, Springfield, Massachusetts.

Anonymous. (1969). Membrane proteins. *Nature (London)* **221,** 614.

Asahina, E. (1956). The freezing process of plant cell. *Contrib. Inst. Low Temp. Sci., Hokkaido Univ.,* **10,** 83.

Asahina, E. (1966). Freezing and frost resistance in insects. In "Cryobiology" (H. T. Meryman, ed.), pp. 451–484. Academic Press, New York.

Asahina, E., and Tanno, K. (1963). A remarkably rapid increase of frost resistance in fertilized egg cells of the sea urchin. *Exp. Cell Res.* **31,** 223.

Badanova, K. A., and Vartapetyan, B. B. (1967). The permeability to water of dormant bud protoplasm of woody plants. *Dokl. Akad. Nauk SSSR* **176,** 476.

Bailey, I. W. (1954). "Contributions to Plant Anatomy." Chronica Botanica, Waltham, Massachusetts.

Bailey, R. (1960). Summer frosts: A factor in plant range and timber succession. *Wis. Acad. Rev.* **7,** 153.

Baker, J. E., and Smith, W. L., Jr. (1970). Heat-induced ultrastructural changes in germinating spores of *Rhizopus stolonifer* and *Monilinia fructicola. Phytopathology* **60,** 869.

Barrs, H. D. (1968). Determination of water deficits. In "Water Deficits and Plant Growth" (T. T. Kozlowski, ed.), Vol. 1, pp. 235–368. Academic Press, New York.

Bartley, W., Birt, L. M., and Banks, P. (1968). "The Biochemistry of the Tissues." Wiley, New York.

Berendsen, H. J. C., and McCulloch, W. S. (1960). Structure of water in biological tissue. "Neurophysiology," Quart. Prog. Rep. Res. Lab. Elec., Massachusetts Institute of Technology, Cambridge, Massachusetts.

Bernal, J. D. (1965). The structure of water and its biological implications. *Symp. Soc. Exp. Biol.* **19,** 17–32.

Biebl, R. (1967–1968). Über Wärmehaushalt und Temperaturresistenz arktischer Pflanzen in Westgrönland. *Flora (Jena)* **157,** 327.

Boardman, N. K. (1968). The photochemical systems of photosynthesis. *Advan. Enzymol.* **30,** 1.

Boyer, J. S. (1965). Effects of osmotic water stress on metabolic rates of cotton plants with open stomata. *Plant Physiol.* **40,** 229.

Boyer, J. S. (1967). Leaf water potentials measured with a pressure chamber. *Plant Physiol.* **42,** 133.

Boyer, J. S. (1968). Relationship of water potential to growth of leaves. *Plant Physiol.* **43,** 1056.

Bradbury, J. H., and King, N. D. R. (1969). Denaturation of proteins: Single or multiple step process? *Nature* (*London*) 223, 1154.

Branton, D. (1969). Membrane structure. *Annu. Rev. Plant Physiol.* 20, 209.

Brix, H. (1962). The effect of water stress on the rates of photosynthesis and respiration in tomato plants and loblolly pine seedlings. *Physiol. Plant.* 15, 10.

Bull, H. B., and Breese, K. (1968). Protein hydration. II. Specific heat of egg albumin. *Arch. Biochem. Biophys.* 128, 497.

Burstrom, H., Uhrström, I., and Wurscher, R. (1967). Growth, turgor, water potential and Young's modulus in pea internodes. *Physiol. Plant.* 20, 213.

Castellion, G. A., Grabar, D. G., Hession, J., and Burkhard, H. (1970). Polywater: Methods for identifying polywater columns and evidence for ordered growth. *Science* 167, 865.

Chapman, D. (1969). The physico-chemical approach to the study of lipoprotein interactions. *In* "Structural and Functional Aspects of Lipoproteins in Living Systems" (E. Tria and A. M. Scanu, eds.), pp. 3–36. Academic Press, New York.

Cowan, I. R., and Milthorpe, F. L. (1968). Plant factors influencing the water status of plant tissues. *In* "Water Deficits and Plant Growth" (T. T. Kozlowski, ed.), Vol. 1, pp. 137–193. Academic Press, New York.

Cox, W., and Levitt, J. (1969). Direct relation between growth and frost hardening in cabbage leaves. *Plant Physiol.* 44, 923.

Crafts, A. S. (1968). Water structure and water in the plant body. *In* "Water Deficits and Plant Growth" (T. T. Kozlowski, ed.), Vol. 1, pp. 23–47. Academic Press, New York.

Craker, L. E., Gusta, L. V., and Weiser, C. J. (1969). Soluble proteins and cold hardiness of two woody species. *Can. J. Plant Sci.* 49, 279.

Criddle, R. S. (1969). Structural proteins of chloroplasts and mitochondria. *Annu. Rev. Plant Physiol.* 20, 239.

Criddle, R. S., and Willemot, J. (1969). Mitochondrial structural lipoprotein. *In* "Structural and Functional Aspects of Lipoproteins in Living Systems" (E. Tria and A. M. Scanu, eds.), pp. 173–199. Academic Press, New York.

Cunningham, W. P. (1964). Oxidation of externally added NADH by isolated corn root mitochondria. *Plant Physiol.* 39, 699.

DeLong, W. A., Beaumont, J. H., and Willaman, J. (1930). Respiration of apple twigs in relation to winter hardiness. *Plant Physiol.* 5, 509.

de Plater, C. V., and Greenham, C. G. (1959). A wide-range AC bridge for determining injury and death. *Plant Physiol.* 34, 661.

Derjaguin, B. V. (1965). Recent research into the properties of water in thin films and in microcapillaries. *Symp. Soc. Exp. Biol.* 19, 55–60.

DeRoo, H. C. (1969). Water stress gradients in plants and soil-root systems. *Agron. J.* 61, 511.

Eisenberg, D., and Kauzmann, W. (1969). "The Structure and Properties of Water." Oxford Univ. Press, London and New York.

Evert, R. F., and Deshpande, B. P. (1969). Electron microscope investigation of sieve-element ontogeny and structure in *Ulmus americana. Protoplasma* 68, 403.

Evert, R. F., Murmanis, L., and Sachs, I. B. (1966). Another view of the ultrastructure of *Cucurbita* phloem. *Ann. Bot.* (*London*) [N. S.] 30, 563.

Feldman, N. L., Kamentseva, I. E., and Yurashevskaya, K. N. (1966). Acid phosphatase thermostability in the extracts of cucumber and wheat seedling leaves after heat hardening. *Tsitologiya* 8, 755.

Finean, J. B. (1961). "Chemical Ultrastructure in Living Tissues." Thomas, Springfield, Illinois.

Flowers, T. J., and Hanson, J. B. (1969). The effect of reduced water potential on soybean mitochondria. *Plant Physiol.* **44,** 939.

Folkers, K. (1967). Research on coenzyme Q. *In* "Phenolic Compounds and Metabolic Regulation" (B. J. Finkle and V. C. Runeckles, eds.), pp. 95–119. Appleton, New York.

Frank, H. S. (1970). The structure of ordinary water. *Science* **169,** 635.

Franke, I. (1962). Untersuchungen über den Einfluss des Frostes auf Blattparenchymzellen von *Polypodium vulgare. Protoplasma* **55,** 63.

Gaff, D. F. (1966). The sulfhydryl-disulphide hypothesis in relation to desiccation injury of cabbage leaves. *Aust. J. Biol. Sci.* **19,** 291.

Gahan, P. B. (1968). Lysosomes. *In* "Plant Cell Organelles" (J. B. Pridham, ed.), pp. 228–238. Academic Press, New York.

Gatt, S. (1969). Thermal lability of beta galactosidase from pink salmon liver. *Science* **164,** 1422.

Giles, C. H., and McKay, R. B. (1962). Studies in hydrogen bond formation. XI. Reactions between a variety of carbohydrates and proteins in aqueous solutions. *J. Biol. Chem.* **237,** 3388.

Giovenco, M. A., Giordano, M. G., Caiafa, P., Giovenco, S., and Magni, G. (1967). Activation of succinate dehydrogenase by succinate. *Ital. J. Biochem.* **16,** 43.

Goldberger, R., Pumphrey, A., and Smith, A. (1962). Studies on the electron transport system. XLVI. On the modification of the properties of cytochrome B. *Biochim. Biophys. Acta* **58,** 307.

Green, D. E., and MacLennan, D. H. (1969). Structure and function of the mitochondrial cristael membranes. *BioScience* **19,** 213.

Greenham, C. G. (1966a). Bruise and pressure injury in apple fruits. *J. Exp. Bot.* **17,** 404.

Greenham, C. G. (1966b). The relative electrical resistances of the plasmalemma and tonoplast in higher plants. *Planta* **69,** 150.

Greenham, C. G., Norris, D. O., and Brock, R. D. (1952). Some electrical differences between healthy and virus-infected potato tubers. *Nature (London)* **169,** 973.

Gucker, F. T., and Meldrum, W. B. (1944). "Physical Chemistry." American Book Co., New York.

Guilliermond, A. (1941). "The Cytoplasm of the Plant Cell" (Transl. by L. R. Atkinson). Chronica Botanica, Waltham, Massachusetts.

Hanafusa, N. (1967). Denaturation of enzyme protein by freeze-thawing. *In* "Cellular Injury and Resistance in Freezing Organisms" (E. Asahina, ed.), Vol. II, pp. 35–50. Inst. Low Temp. Sci., Hokkaido.

Hanafusa, N. (1969). Denaturation of enzyme protein by freeze-thawing and freeze-drying. *In* "Freezing and Drying of Microorganisms," pp. 117–129. Univ. of Tokyo Press, Tokyo.

Hays, R. M., and Leaf, A. (1962). Permeability of the isolated toad bladder to solutes and its modification by vasopressin. *J. Gen. Physiol.* **45,** 933.

Hazlewood, C. F., Nichols, B. L., and Chamberlain, N. F. (1969). Evidence for the existence of a minimum of two phases of ordered water in skeletal muscle. *Nature (London)* **222,** 747.

Heber, U. (1958). Ursachen der Frostresistenz bei Winterweizen. I. Die Bedeutung der Zucker für die Frostresistenz. *Planta* **52,** 144.

Heber, U. (1968). Freezing injury in relation to loss of enzyme activities and protection against freezing. *Cryobiology* **5**, 188.

Heilbrunn, L. B. (1958). The viscosity of protoplasm. *Protoplasmatologia* **2**, Sect. C, Part 1, 1–109.

Henckel, P. A. (1964). Physiology of plants under drought. *Annu. Rev. Plant Physiol.* **15**, 363.

Henckel, P. A. (1967). Trends in the development of physiology of drought- and salt-resistant plants. *Izv. Akad. Nauk SSSR, Ser. Biol.* **1**, 46.

Henckel, P. A. (1970). Role of protein synthesis in drought resistance. *Can. J. Bot.* **48**, 1235.

Henze, J. (1959). Untersuchungen über die Kohlenhydrat-Eiweiss-Haushalt der Rinde in seiner Beziehung zur Frostresistenz von Obstgehölzen. *Z. Bot.* **47**, 42.

Höfler, K. (1960). Meiosomes and groundplasm. *Protoplasma* **52**, 295.

Hubac, C., Guerrier, D., and Ferran, J. (1969). Résistance à la sécherese du *Carex pachystylis* (J. Gay) plante du désert du Neger. *Oecol. Plant.* **4**, 325.

Jameson, D. A. (1961). Heat and desiccation resistance of tissue of important trees and grasses of the pinyon-juniper type. *Bot. Gaz.* **122**, 174.

Jarvis, P. G., and Jarvis, M. S. (1963). The water relations of tree seedings. IV. Some aspects of the tissue water relations and drought resistance. *Physiol. Plant.* **16**, 501.

Jung, G. A., Shih, S. C., and Shelton, D. C. (1967). Influence of purines and pyrimidines on cold hardiness of plants. III. Associated changes in soluble proteins and nuclei acid content and tissue pH. *Plant Physiol.* **42**, 1653.

Just, L. (1877). Ueber die Einwirkung hoher Temperaturen auf die Erhaltung der Keimfähigkeit der Samen. *Beitr. Biol. Pflanz.* **2**, 311.

Kalina, M., Weaver, B., and Pearse, A. G. E. (1969). Fine structural localization of succinoxidase complex on the mitochondrial cristae. *Nature (London)* **221**, 479.

Kaloyereas, S. A. (1958). A new method of determining drought resistance. *Plant Physiol.* **33**, 232.

Kamiya, N. (1960). Physics and chemistry of protoplasmic streaming. *Annu. Rev. Plant Physiol.* **11**, 323.

Kamiya, N., and Kuroda, K. (1966). Some observations of protoplasmic streaming in *Acetabularia*. *Bot. Mag.* **79**, 706.

Kappen, L. (1964). Untersuchungen über den Jahresverlauf der Frost-, Hitze- und Austrocknungsresistenz von Sporophyten einheimischer Polypodiaceen (*Filicinae*). *Flora (Jena)* **155**, 123.

Kaufmann, M. R. (1968). Evaluation of the pressure chamber technique for estimating plant water potential of forest tree species. *Forest Sci.* **14**, 369.

Kessler, W., and Ruhland, W. (1938). Weitere Untersuchungen über die inneren Ursachen der Kälteresistenz. *Planta* **28**, 159.

Knipling, E. B., and Kramer, P. J. (1967). Comparison of the dye method with the thermocouple psychrometer for measuring leaf water potentials. *Plant Physiol.* **42**, 1315.

Kozlowski, T. T. (1961). The movement of water in trees. *Forest Sci.* **7**, 177.

Kramer, P. J., Knipling, E. B., and Miller, L. N. (1966). Terminology of cell-water relations. *Science* **153**, 889.

Krasavtsev, O. A. (1967). Frost hardening of woody plants at temperatures below zero. *In* "Cellular Injury and Resistance in Freezing Organisms" (E. Asahima, ed.), Vol. II, pp. 131–138. Inst. Low Temp. Sci., Hokkaido.

Kurtin, S. L., Mead, C. A., Mueller, W. A., Kurtin, B. C., and Wolf, E. D. (1970). "Polywater": A hydrosol? *Science* 167, 1720.

Küster, E. (1935). "Die Pflanzenzelle." Parey, Berlin.

Lange, O. L. (1955). Untersuchungen über die Hitzeresistenz der Moose in Beziehung zu ihrer Verbreitung. I. Die Resistenz starck ausgetrockneter Moose. *Flora (Jena)* 142, 381.

Lange, O. L. (1959). Untersuchungen über Wärmehaushalt und Hitzeresistenz Mauretanischer Wüsten- und Savannenpflanzen. *Flora (Jena)* 147, 595.

Lange, O. L. (1961). Die Hitzeresistenz einheimischer immer- und wintergrüner Pflanzen im Jahresablauf. *Planta* 56, 666.

Lange, O. L. (1967). Investigations on the variability of heat-resistance in plants. *In* "The Cell and Environmental Temperature" (A. S. Troshin, ed.), pp. 131–141. Pergamon, Oxford.

Larcher, W. (1953). Schnellmethode zur Unterscheidung lebender von toten Zellen mit Hilfe der Eigenfluoreszenz pflanzlicher Zellsäfte. *Mikroskopie* 8, 299.

Larcher, W. (1963). Zur Frage des Zusammenhanges zwischen Austrocknungsresistenz und Frosthärte bei Immergrünen. *Protoplasma* 57, 569.

Larcher, W., and Eggarter, H. (1960). Anwendung des Triphenyltetrazoliumchlorids zur Beurteilung von Frostschäden in verschiedenen Achsengeweben bei *Pirus*-Arten, und Jahresgang der Resistenz. *Protoplasma* 51, 595.

Lehninger, A. L. (1970). "Biochemistry." Worth Publ., New York.

Lepeshkin, W. W. (1935). Zur Kenntnis des Hitzetodes des Protoplasmas. *Protoplasma* 23, 349.

Levitt, J. (1956). "The Hardiness of Plants." Academic Press, New York.

Levitt, J. (1962). A sulfhydryl-disulfide hypothesis of frost injury and resistance in plants. *J. Theor. Biol.* 3, 355.

Levitt, J. (1967). The mechanism of hardening on the basis of the SH $\rightleftarrows$ SS hypothesis of freezing injury. *In* "Cellular Injury and Resistance in Freezing Organisms" (E. Asahina, ed.), Vol. II, pp. 51–61. Inst. Low Temp. Sci., Hokkaido.

Levitt, J. (1969). Growth and survival of plants at extremes of temperature—a unified concept. *Symp. Soc. Exp. Biol.* 23, 385–448.

Levitt, J., and Siminovitch, D. (1940). The relation between frost resistance and the physical state of protoplasm. I. The protoplasm as a whole. *Can. J. Res., Sect. C* 18, 550.

Levitt, J., and Siminovitch, D. (1941). The relation between frost resistance and the physical state of protoplasm. II. The protoplasmic surface. *Can. J. Res., Sect. C* 19, 9.

Levitt, J., Sullivan, C. Y., and Johansson, N. O. (1962). A new factor in frost resistance. III. Relation of SH increase during hardening to protein, glutathione, and glutathione oxidizing activity. *Plant Physiol.* 37, 266.

Ling, G. N. (1969). A new model for the living cell: A summary of the theory and recent experimental evidence in its support. *Int. Rev. Cytol.* 26, 1–64.

Lippincott, E. R., Stromberg, R. R., Grant, W. H., and Cessac, G. L. (1969). Polywater. Vibrational spectra indicate unique stable polymeric structure. *Science* 164, 1482.

Lomagin, A. G., Antropova, R. A., and Ilmete, A. (1963). The influence of heat-hardening on the resistance of plant cells to different injurious agents. *Inst. Ser. Monogr. Pure Appl. Biol., Div. Zool.* 34, 180.

Lorenz, R. W. (1939). High temperature tolerance of forest trees. *Minn., Agr. Exp. Sta., Tech. Bull.* 141, 1–25.

Luyet, B. (1967). On the possible biological significance of some physical changes encountered in the cooling and the rewarming of aqueous solutions. *In* "Cellular Injury and Resistance in Freezing Organisms" (E. Asahina, ed.), Vol. II, pp. 1–20. Inst. Low Temp. Sci., Hokkaido.

Macovschi, E., and Marineanu, I. (1967). The structure theory, the state of protoplasm water and the problem of plant sap origin. *Rev. Roum. Biochim.* 4, 235.

Majumber, S. K. (1969). Stability of "chlorophyll index" as a measure of adaptation for over-wintering. *Plant Physiol., Suppl.* 44, 17.

Maximov, N. A. (1914). Experimentelle und kritische Untersuchungen über das Gefrieren und Erfrieren der Pflanzen. *Jahrb. Wiss. Bot.* 53, 327.

Mazur, P. (1969). Freezing injury in plants. *Annu. Rev. Plant Physiol.* 20, 419.

Mazur, P. (1970). Cryobiology: The freezing of biological systems. *Science* 168, 939.

Meryman, H. T. (1967). The relationship between dehydration and freezing injury in the human erythrocyte. *In* "Cellular Injury and Resistance in Freezing Organisms" (E. Asahina, ed.), Vol. II, pp. 231–244. Inst. Low Temp. Sci., Hokkaido.

Meyer, B. S., and Anderson, D. B. (1952). "Plant Physiology." Van Nostrand-Reinhold, Princeton, New Jersey.

Mishiro, Y., and Ochi, M. (1966). Effect of dipicolinate synthesis during sporogenesis and germination of bacteria on the heat denaturation of proteins in human and bovine sera. *Nature (London)* 211, 1190.

Molin, Yu. N., and Bekker, Zh. M. (1967). Proton relaxation studies of the t-RNA Mn^{2+} interaction during heat denaturation. *Biofizika* 12, 337.

Monk, R. W., and Wiebe, H. H. (1961). Salt tolerance and protoplasmic salt hardiness of various woody and herbaceous ornamental plants. *Plant Physiol.* 36, 478.

Montfort, C., and Hahn, H. (1950). Atmung and Assimilation als dynamische Kennzeichen abgestufter Trockenresistenz bei Farnen und höheren Pflanzen. *Planta* 38, 503.

Mooney, H. A. (1969). Dark respiration of related evergreen and deciduous Mediterranean plants during induced drought. *Bull. Torrey Bot. Club* 96, 550.

Morré, D. J. (1970). Auxin effects on the aggregation and heat coagulability of cytoplasmic proteins and lipoproteins. *Physiol. Plant.* 23, 38.

Mouravieff, I. (1969). Sur les caractères protoplasmiques de cellules épidermiques foliaires soumises à la influence d'une déshydration progressive: Expériences avec la tétracycline comme fluorochrome vitale. *Physiol. Veg.* 7, 191.

Nagy, A. H., and Faludi-Daniel, A. (1967). On the nature of the binding forces stabilizing carotenoid-protein and chlorophyll-protein complexes *in vivo. Photosynthetica* 1, 69.

Nečas, O., Kopecká, M., and Brichta, J. (1969). Interpretation of surface structures in frozen-etched protoplasts of yeasts. *Exp. Cell Res.* 58, 411.

Nir, I., and Klein, S. (1970). The effect of water stress on mitochondria of root cells. *Plant Physiol.* 45, 173.

Novikoff, A. B. (1955). Histochemical and cytochemical staining methods. *In* "Analytical Cytology" (R. C. Mellors, ed.), Chapter 2, pp. 1–63. McGraw-Hill, New York.

Parker, J. (1953a). Some applications and limitations of tetrozolium chloride. *Science* 118, 770.

Parker, J. (1953b). Criteria of life: Some methods of measuring viability. *Amer. Sci.* 41, 614.

Parker, J. (1958). Changes in sugars and nitrogenous compounds of tree barks from summer to winter. *Naturwissenschaften* **45**, 139.

Parker, J. (1959). Seasonal variations in sugars of conifers with some observations on cold resistance. *Forest Sci.* **5**, 56.

Parker, J. (1960). Seasonal changes in the physical nature of the bark phloem parenchyma cells of *Pinus strobus*. *Protoplasma* **52**, 223.

Parker, J. (1962). Relationships among cold hardiness, water-soluble protein, anthocyanins, and free sugars in *Hedera helix*. *Plant Physiol.* **37**, 809.

Parker, J. (1963). Cold resistance in woody plants. *Bot. Rev.* **29**, 123.

Parker, J. (1965). Strand characteristics in sieve tubes of some common tree species. *Protoplasma* **60**, 86.

Parker, J. (1966). Leaf water retention versus desiccation resistance as the cause of drought resistance in leaves of woody evergreens. *Advan. Front. Plant Sci.* **15**, 157.

Parker, J. (1968). Drought-resistance mechanisms. *In* "Water Deficits and Plant Growth" (T. T. Kozlowski, ed.), Vol. 1, pp. 195–231. Academic Press, New York.

Parker, J. (1969). Further studies on drought resistance in woody plants. *Bot. Rev.* **35**, 317.

Parker, J. (1970). The effect of heat on survival, respiration, and ultrastructure in tree twigs. *Plant. Physiol., Suppl.* **46**, 22.

Parker, J. (1971). Effects of defoliation and drought on root food reserves in sugar maple seedlings. *U. S. Dep. Agr., Forest Ser., Res. Pap. NE* **169**, 1–8.

Parker, J. (1972). Unpublished data.

Parker, J., and Philpott, D. E. (1961). An electron microscopic study of chloroplast condition in summer and winter in *Pinus strobus*. *Protoplasma* **53**, 575.

Pauling, L. (1952). "College Chemistry." Freeman, San Francisco, California.

Pierpoint, G. (1967). Direct measurement of internal moisture deficits in trees. *Forest Chron.* **43**, 145.

Pisek, A., and Larcher, W. (1954). Zusammenhang zwischen Austrocknungsresistenz und Frosthärte bei Immergrünen. *Protoplasma* **44**, 30.

Pomerleau, R., and Ray, R. G. (1957). Occurrence and effects of summer frost in a conifer plantation. *Forest Res. Div. (Can.). Tech. Note* **51**, 1–15.

Roberts, D. W. A. (1969). Some possible roles for isozymic substitutions during cold hardening in plants. *Int. Rev. Cytol.* **26**, 303–328.

Rouschal, E. (1938). Eine physiologische Studie an *Ceterach offiicianarum* Wild. *Flora (Jena)* **132**, 305.

Sakai, A. (1960). The frost-hardening process of woody plant. VIII. Relation of polyhydric alcohols to frost hardiness. *Contrib. Inst. Low Temp. Sci., Hokkaido Univ., Ser. B* **18**, 15.

Salt, R. W. (1961). Principles of insect cold-hardiness. *Annu. Rev. Entomol.* **6**, 55.

Samygin, G. A., and Mateeva, N. M. (1967a). Dehydration of protoplasts as a cause of cell death in freezing. *Dokl. Akad. Nauk SSSR.* **174**, 1219.

Samygin, G. A., and Mateeva, N. M. (1967b). Protective effect of glycerine and other substances which easily penetrate protoplasts during freezing of plant cells. *Fiziol. Rast.* **14**, 1048.

Santarius, K. A., and Heber, U. (1967). Das Verhalten von Hill-Reaktion und Photophosphorylierung isolierter Chloroplasten im Abhängigkeit vom Wassergehalt. II. Wasserentzug über $CaCl_2$. *Planta* **73**, 109.

Saunier, R. E., Hull, H. M., and Ehrenreich, J. H. (1968). Aspects of the drought tolerance in creosote bush (*Larrea divaricata*). *Plant Physiol.* 43, 401.

Scarth, G. W. (1944). Cell physiological studies of frost resistance. *New Phytol.* 43, 1.

Scarth, G. W., and Levitt, J. (1937). The frost-hardening mechanism of plant cells. *Plant Physiol.* 12, 510.

Schnepf, E. (1961). Über Veränderungen der plasmatischen Feinstruktur während des Welkens. *Planta* 57, 156.

Scholander, P. F., Hammel, H. T., Bradstreet, E. D., and Hemmingsen, E. A. (1965). Sap pressure in vascular plants. *Science* 148, 339.

Schölm, H. E. (1968). Untersuchungen zur Hitze und Frostresistenz einheimischer Süsswasseralgen. *Protoplasma* 65, 97.

Schwarz, W. (1969). Der Einfluss der Photoperiode auf die Frosthärte und Hitzeresistenz von Zirben und Alpenrosen. *Ber. Deut. Bot. Ges.* 82, 109.

Sheen, S. J., Gailey, F. B., Miller, D. M., Anderson, J. D., Bargmann, T. J., and Carter, D. A. (1969). Sol-gel differences in plasmodia of the acellular slime mold, *Physarum polycephalum*. Research Reports. *BioScience* 19, 1003.

Sherman, J. K. (1962). Preservation of bull and human spermatozoa by freezing in liquid nitrogen vapour. *Nature (London)* 194, 1291.

Siminovitch, D., and Briggs, D. R. (1949). The chemistry of the living bark of the black locust tree in relation to frost hardiness. I. Seasonal variations in protein content. *Arch. Biochem.* 23, 8.

Siminovitch, D., and Briggs, D. R. (1953). The validity of plasmolysis and desiccation tests for determining the frost hardiness of bark tissue. *Plant Physiol.* 28, 15.

Siminovitch, D., Gfeller, F., and Rheaume, B. (1967). The multiple character of the biochemical mechanism of freezing resistance of plant cells. *In* "Cellular Injury and Resistance in Freezing Organisms" (E. Asahina, ed.), Vol. II, pp. 93–117. Inst. Low Temp. Sci., Hokkaido.

Sjöstrand, F. S. (1969). Morphological aspects of lipoprotein structures. *In* "Structural and Functional Aspects of Lipoproteins in Living Systems" (E. Tria and A. M. Scanu, eds.), pp. 73–128. Academic Press, New York.

Slatyer, R. O. (1967). "Plant-water Relationships." Academic Press, New York.

Sopina, V. A. (1968). The role of the nucleus and cytoplasm in the heredity of heat-resistance in amoeba. *Genetika* 4, 82.

Souzu, H. (1967). Location of polyphosphate and polyphosphatase in yeast cells and damage to the protoplasmic membrane of the cell by freeze-thawing. *Arch. Biochem. Biophys.* 120, 344.

Srivastava, L. M. (1966). On the fine structure of the cambium of *Fraxinus americana* L. *J. Cell Biol.* 31, 79.

Stadelmann, E. J. (1969). Permeability of the plant cell. *Annu. Rev. Plant Physiol.* 20, 585.

Steinhardt, J., and Reynolds, J. A. (1970). "Multiple Equilibria in Proteins." Academic Press, New York.

Steponkus, P. L. (1968). Cold acclimatation of *Hedera helix*—a two-step process. *Cryobiology* 4, 276.

Steponkus, P. L. (1969). Protein-sugar interactions during cold acclimatation. *Proc. Int. Bot. Congr., 11th, 1969* p. 209.

Steponkus, P. L., and Gregg, J. M. (1968). Effect of freezing on dehydrogenase activity and reduction of triphenyl tetrazolium chloride. *Plant Physiol.* 43, S-5.

Steponkus, P. L., and Lanphear, F. O. (1967a). Refinement of the triphenyl tetrazolium chloride method of determining cold injury. *Plant Physiol.* 42, 1423.

Steponkus, P. L., and Lanphear, F. O. (1967b). Light stimulation of cold acclimatation: Production of a translocatable promoter. *Plant Physiol.* 42, 1673.

Stockem, W., Wohlfarth-Bottermann, K. E., and Haberey, M. (1969). Pinocytose und Bewegung von Amöben. V. Mitteilung. Konturveränderung und Faltungsgrad der Zelloberfläche von *Amoeba proteus. Cytobiologie* 1, 37.

Stocker, O. (1929). Das Wasserdefizit von Gefässpflanzen in verschiedenen Klimazonen. *Planta* 7, 382.

Stocker, O. (1960). Physiological and morphological changes in plants due to water deficiency. *In* "Plant-water Relationships in Arid and Semi-arid Conditions," Rev. Res., pp. 63–104. UNESCO, Paris.

Stutte, C. A., and Todd, G. W. (1969). Some enzyme and protein changes associated with water stress in wheat leaves. *Crop Sci.* 9, 510.

Szent-Györgyi, A. (1968). "Bioelectronics." Academic Press, New York.

Takaoki, T. (1968). Relationship between drought tolerance and aging in higher plants. II. Some enzyme activities. *Bot. Mag.* 81, 297.

Tanno, K. (1967). Freezing injury in fat-body cells of the poplar sawfly. *In* "Cellular Injury and Resistance in Freezing Organisms" (E. Asahina, ed.), Vol. II, pp. 245–257. Inst. Low Temp. Sci., Hokkaido.

Thimann, K. V., and Kaufman, D. (1958). Cytoplasmic streaming in the cambium of white pine. *In* "Physiology of Forest Trees" (K. V. Thimann, ed.), pp. 479–492. Ronald Press, New York.

Tomita, G., and Kim, S. S. (1966). Inhibition of heat-denatured taka-amylase A by substrate. *Experientia* 22, 392.

Tria, E., and Barnabei, O. (1969). Lipoproteins and lipopeptides of cell membranes. *In* "Structural and Functional Aspects of Lipoproteins in Living Systems" (E. Tria and A. M. Scanu, eds.), pp. 143–171. Academic Press, New York.

Tumanov, I. I. (1969). Physiology of frost-resistant plants. *Izv. Akad. Nauk SSSR, Ser. Biol.* 4, 469; *Biol Abstr.* p. 16385 (1970).

Turner, N. C., and Waggoner, P. E. (1968). Effects of changing stomatal width in a red pine forest on soil water content, leaf water potential, bole diameter, and growth. *Plant Physiol.* 43, 973.

Ullrich, H., and Heber, U. (1957). Über die Schutzwirkung der Zucker bei der Frostresistenz von Winterweizen. *Planta* 48, 724.

Vaadia, Y., Raney, F. C., and Hagan, R. M. (1961). Plant water deficits and physiological processes. *Annu. Rev. Plant Physiol.* 12, 265.

Vartapetyan, B. B. (1965). Water relations of plants in experiments with heavy isotope O^{18}. *In* "Water Stress in Plants" (B. Slavik, ed.), pp. 72–79. Junk, Publ., The Hague.

von Sachs, J. (1874). "Lehrbuch der Botanik," 4th ed., Vol. III. Physiologie. Engelmann, Leipzig.

Walter, H. (1963). Zur Klärung des spezifischen Wasserzustandes im Plasma und in der Zellwand. *Ber. Deut. Bot. Ges.* 76, 40.

Walter, H. (1965). Zur Klärung des spezifischen Wasserzustandes im Plasma. III. *Ber. Deut. Bot. Ges.* 78, 104.

Weatherly, P. E. (1950). Studies in the water relations of the cotton plant. I. The field measurement of water deficits in leaves. *New Phytol.* 49, 81.

Weatherley, P. E. (1965). The state and movement of water in the leaf. *Symp. Soc. Exp. Biol.* 19, 157–184.

Webb, S. J. (1965). "Bound Water in Biological Integrity." Thomas, Springfield, Illinois.

Weiss, L. (1969). The cell periphery. *Int. Rev. Cytol.* **26**, 63–105.

Wiebe, H. H., Brown, R. W., Daniel, T. W., and Campbell, E. (1970). Water potential measurements in trees. *BioScience* **20**, 225.

Wilner, J. (1967). Changes in electric resistance of living and injured tissues of apple shoots during winter and spring. *Can. J. Plant Sci.* **47**, 469.

Wilner, J., and Brach, E. J. (1970). Comparison of radio telemetry with another electric method for testing winter injury of outdoor plants. *Can. J. Plant Sci.* **50**, 1.

Wohlfarth-Bottermann, K. E. (1968). Dynamik der Zelle. *Mikroskopie* **23**, 71.

Zholkevich, V. N. (1961). Energy balance of respiring plant tissues under various conditions of water supply. *Fisiol. Rast.* **8**, 407.

CHAPTER 5

WATER DEFICITS AND ENZYMATIC ACTIVITY

Glenn W. Todd

DEPARTMENT OF BOTANY AND PLANT PATHOLOGY, OKLAHOMA STATE UNIVERSITY,
STILLWATER, OKLAHOMA

I. INTRODUCTION

Many metabolic reactions occurring in cells take place in a water milieu. Even those reactions which take place at surfaces are dependent upon a liquid phase so that molecules can move from one place to another.

This dependency on a liquid medium is obvious, and as water is the liquid used by cells, most metabolic reactions cease in the absence of water. As discussed elsewhere in these volumes it is apparent that during drying important changes in metabolic reactions and overall physiological functions occur in most cells and tissues long before an air-dry state is reached (Crafts, 1968). The rate attained by any particular metabolic reaction in a cell depends on the presence and amount of reactants and enzymes. The capability of the enzyme will depend upon its local environment of hydration, ions, temperature, cofactors, inhibitors, etc. In addition many enzymes function as part of a complex structure in association with other enzymes, proteins, lipids, or other macromolecules.

Because of the great variations in susceptibility to drying of different plants or organs (Parker, 1968) there have been many attempts to associate these differences with particular subcellular features including the enzymes themselves. Many of these studies concern the presence or absence of certain enzymes although, less frequently, studies of enzyme function *in situ* have been made. Information concerning enzymes in the extracted state may be of less direct help in interpretation because of our ignorance of how the enzyme was situated in the intact cell. Dehydration of tissue which causes such dramatic shifts in overall metabolism may lead to changes in enzymatic capabilities which are not a measure of the effect of dehydration on the enzyme, but rather on the degradation or inactivation of the enzyme by other enzymes or chemical substances. Thus, while all these phases are interrelated, for purposes of discussion we shall examine separately the effects of dehydration on levels of tissue enzymes and on functioning of subcellular organelles carrying on multiple reactions. In addition, attention will be given to effects of dehydration *per se* on the enzymes themselves and on their functions.

II. EFFECT OF WATER DEFICITS ON ENZYME ACTIVITY FOUND IN THE "SOLUBLE" FRACTION

Most studies have utilized plant extracts for the enzyme sources following a desiccation treatment. A relatively small number of studies have been made and many of those were conducted before we were aware of the problems of extracting proteins from different types of plant material. For example, extraction of leaves from cotton (*Gossypium hirsutum*) plants yields virtually no protein after the usual procedure of homogenization in buffered solutions followed by centrifugation unless some agent such as polyvinylpyrrolidone or Carbowax is added during extraction, whereas such additions appear to be unnecessary with wheat leaves (Todd, 1969). If tissues develop more phenolic- or tannin-type compounds as a result of water deficit then an apparent decrease in activity of a certain enzyme may simply be the result of poor extraction or of protein coagulation during extraction of the desiccated tissue. Interpretation of the significance of lowered amounts of enzyme may also be somewhat difficult. A certain enzyme may be present in quantities greater than actually required for plant survival so that a reduction in amount of enzyme may not be of any consequence, whereas loss of enzyme below a certain critical level may be extremely important.

Loss of an enzyme may not be critical to survival of the tissue if new enzyme can be synthesized as soon as the tissue is rehydrated. Unfortunately, little work has been done on enzymes required for *de novo* synthesis of RNA or protein. There are indications that moisture-stressed tomato (*Lycopersicon esculentum*) leaves retain the ability to incorporate ^{32}P into RNA but the rate of destruction of RNA is increased (Gates and Bonner, 1959). More recent work indicated only minor changes in leaf RNA after brief exposure to drought of corn (*Zea mays*) or bean plants but protein synthesis declined much more. A separation of polysomes into subunits occurred (Genkel *et al.,* 1967).

Some of the enzymes that have been studied undoubtedly arise from lysosomal particles or vacuoles since the usual procedures for extraction probably release many of these degradative enzymes. Ribonuclease in leaves of several species and protease in tobacco (*Nicotiana tabacum*) leaves increased with dehydration treatment (see Table I). One interpretation is that some of the particles do not rupture during extraction but may do so as a consequence of desiccation. Phosphomonesterase in cowpea (*Vigna sinensis*) leaves increased by nearly 50% as a result of drought whereas phosphatase activity in detached, rapidly desiccated wheat (*Triticum aestivum*) leaves decreased (Table I). It is entirely possible that decreased levels of enzymes found are the result first of denaturation followed by hydrolysis by proteinases. For example, α-amylase was not attacked by bacterial protease until this enzyme was slightly denatured (Okunuki, 1961).

Conflicting data can be found on effects of drought on enzyme activity (see catalase and peroxidase in Table I). Many of the studies are old and usually the methods and plant materials used differ. Water is measured in many different ways (e.g., leaf water content, soil water content, relative water content etc.) which make comparison difficult if not impossible. Figure 1 shows leaf protein content along with relative enzyme activities of peroxidase in wheat and nitrate reductase in corn. With increase in water deficit nitrate reductase activity disappears at a much greater rate than soluble protein, whereas peroxidase decreases at a lower rate than protein (Fig. 1). Thus there is preferential disappearance of cer-

TABLE I

CHANGES IN ENZYME LEVEL AS A CONSEQUENCE OF DEHYDRATION OF LEAVES OR SHOOTS

Enzyme	Plant source	Intact plant or detached leaf	Changes in enzyme level with progressive drought	Conditions and remarks	Reference
Hydrolases					
α-Amylase	Cowpeas (*Vigna sinensis*)	Attached	40% increase 10% decrease	41% of initial soil moisture 13% of initial soil moisture	Takaoki (1968)
β-Amylase	Cowpeas	Attached	80% increase 30% decrease	60% of initial soil moisture 13% of initial soil moisture	Takaoki (1968)
Amylase	Wheat (*Triticum aestivum*)	Attached	Decrease		Popova (1941)
Amylase	Sunflower (*Helianthus annuus*)	Detached	Increase	Followed starch content of leaves	Spoehr and Milner (1939)
Amylase	Tobacco (*Nicotiana tabacum*)	Detached	Increase	Followed starch content of leaves	Spoehr and Milner (1939)
Invertase	Wheat	Detached	Decrease		Sisakyan (1937)
Invertase	Wheat	Detached	50% decrease	At 50% RWC[a]	Todd and Yoo (1964)
Peptidase	Wheat	Detached	No change	To 25% RWC, slight decrease below 10% RWC	Todd and Yoo (1964)
Proteinase	Wheat	Detached	56% decrease	At 47% RWC	Todd and Yoo (1964)
Protease	Tobacco	Detached (curing)	Increase	Not found after curing completed; large molecular weight proteins disappeared; smaller ones accumulated	Kawashima *et al.* (1967, 1968)
Protease	Wheat	Attached	Increase		Sisakyan and Kobyakova (1938)

Enzyme	Species	State	Effect	Conditions	Reference
Ribonuclease	Tomato (*Lycopersicon esculentum*)	Attached	100% increase	After 6 days' drought	Kessler (1961)
Ribonuclease	Tomato	Attached	50–100% increase	At 50–35% RWC	Dove (1967)
Ribonuclease	Bean	Attached	Increase	Mild water stress	Genkel *et al.* (1967)
Ribonuclease	Corn (*Zea mays*)	Attached	Increase	Mild water stress	Genkel *et al.* (1967)
Ribonuclease	Wheat	Attached	Increase	Mild water stress	Yi and Todd (1970)
Ribonuclease	Apple (*Malus sylvestris*)	Detached	200% increase	At 30% RWC increased rapidly as tissue RWC decreased below 80%	Kessler (1961)
Phosphatase (monoesterase)	Cowpeas	Attached	230% increase	13% of initial soil moisture	Takaoki (1968)
Phosphatase	Wheat	Detached	50% decrease	At 50% RWC	Todd and Yoo (1964)
Oxidoreductases					
Catalase	Wheat	Attached	No change		Popova (1941)
Catalase	Wheat		Increase		Lukicheva (1969)
Catalase	Cowpeas	Attached	100% increase	28% of initial soil moisture	Takaoki (1968)
			25% increase	13% of initial soil moisture	
Catalase	Tobacco	Detached (curing)	Decrease		Albo (1968)
Peroxidase	Maize (*Zea mays*)	Attached	Increase	Up to permanent wilting point magnitude of increase variable depending upon stage of development	Petinov and Malysheva (1960)
Peroxidase	Tobacco	Detached (curing)	Increase then decreased	Increased with two days curing then decreased to low level at 6 days	Weston (1968)
Peroxidase	Cowpeas	Attached	No change or slight decrease	90–13% of initial soil moisture	Sheen and Calvert (1969) Takaoki (1968)
Peroxidase	Wheat		Decrease		Lukicheva (1969)

TABLE I (*Continued*)

Enzyme	Plant source	Intact plant or detached leaf	Changes in enzyme level with progressive drought	Conditions and remarks	Reference
Peroxidase	Wheat	Detached	No change	Mild stress	Todd and Yoo (1964)
Peroxidase	Wheat	Attached	Decrease	Severe stress	
			Decrease	Marked changes in slightly wilted (65% RWC); changes in isoenzymes	Stutte and Todd (1969)
Indoleacetic acid oxidase	Pea (*Pisum sativum*)	Attached	16% increase	−10 atm mannitol	Darbyshire (1971)
Indoleacetic acid oxidase	Wheat	Attached	20% decrease	At 80% RWC	Mills and Todd (1971)
			50% decrease	At 30% RWC	
Ascorbic acid oxidase	Sugar beet	Intact	170% increase	At 35% soil moisture	K. A. Zholkevich et al., (1958)
Ascorbic acid oxidase	Corn	Intact	Increase		Petinov and Malysheva (1960)
Polyphenolase	Grape (*Vitis vinifera*)		Increase	With loss of 20% leaf weight	Shou-ju (1962)
Polyphenolase	Sugar beet	Attached	160% increase	35% soil moisture	K. A. Zholkevich et al. (1958)
Polyphenoloxidase	Corn	Attached	Increase	To permanent wilting point, up to 2-fold increase depending upon developmental stage	Petinov and Malysheva (1960)
Phenoloxidase	Tobacco	Detached	Increase, followed by decrease	50% increase after 2 days	Weston (1968)
Phenoloxidase	Tobacco	Detached	Decrease	Linear with curing	Sheen and Calvert (1969)
Nitrate reductase	Corn	Attached	84% decrease	At wilting	Mattas and Pauli (1965)
Nitrate reductase	Barley (*Hordeum vulgare*)	Attached	50% decrease	After 3 days' drought	Huffaker et al. (1970)

Enzyme	Plant		Change	Condition	Reference
Nitrite reductase	Barley	Attached	28% decrease	After 3 days' drought	Huffaker et al. (1970)
Succinic (?) dehydrogenase	Wheat	Detached	20% increase 40% decrease	At 40% RWC At 20% RWC	Todd and Yoo (1964)
Lactic dehydrogenase	Wheat	Attached	Slight change Marked decrease	At 65% RWC At 35% RWC changes in isozyme	Stutte and Todd (1969)
Glucose 6-phosphate dehydrogenase	Wheat	Attached	Decrease	Wilted (30% RWC)	Tsai and Todd (1969)
Malic dehydrogenase	Wheat	Attached	Decrease	Wilted (30% RWC[a])	Tsai and Todd (1969)
Miscellaneous					
Phosphorylase	Cowpeas	Attached	100% increase	13% of initial soil moisture	Takaoki (1968)
Amino acid-activating enzymes	Wheat	Attached	Increase	Wilted compared to control	Khokhlova et al. (1969)
Pentose phosphate shunt	Corn and fodder beans	Attached	100% increase		Abrarov (1969)
Glycolysis	Corn and fodder beans	Attached	Decrease	Almost completely suppressed	Abrarov (1969)
Cellulose synthesis	Oat (*Avena sativa*)	Detached	Decrease	Osmotic stress	Ordin (1960)
	Sunflower	Attached	Decrease		Plaut and Ordin (1964)
Phosphoenol pyruvate carboxylase	Barley	Attached	21% decrease	After 3 days' drought	Huffaker et al. (1970)
Phosphoribulokinase	Barley	Attached	11% decrease	After 3 days' drought	Huffaker et al. (1970)
Ribulose-1,5-diP carboxylase	Barley	Attached	No change	After 3 days' drought	Huffaker et al. (1970)

[a] RWC, relative water content.

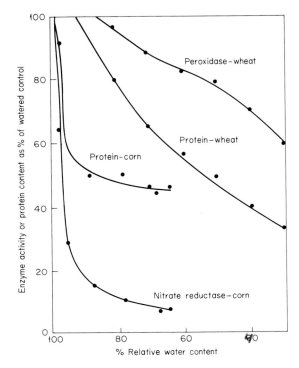

Fig. 1. Protein content and activity of two enzymes from leaves of wheat (Stutte and Todd, 1969) and corn (Mattas and Pauli, 1965) subjected to water stress. Note that methods for determining relative water content (RWC) were different. The corn leaves showed wilting at 88% RWC whereas wheat leaves did not wilt until a RWC of less than 60% was reached. Redrawn by permission.

tain enzymes. As nitrate reductase is an adaptive enzyme it might be expected to disappear first. In addition, accumulation of higher levels of amino compounds might be harmful and a control mechanism may shut off synthesis of the enzyme. It should be noted that relative water content (RWC) in these two studies was obtained by different methods. Values of RWC for corn were obtained by vapor equilibration, whereas those in wheat leaves by a method in which leaves were resaturated by floating them on water.

In a great many studies cited in Table I, control plants were compared to ones subjected to fairly severe water stress. More complete data are necessary before it can be surmised whether the loss of enzyme was gradual or sudden at some particular water deficit. For example, Lukicheva (1969) reported a decrease in peroxidase in wheat leaves with water stress. An examination of a wide range of water deficits with detached wheat leaves revealed little or no change until severe water stress was reached, whereas activity decreased quite precipitously when RWC fell to less than 60%. The marked decrease (with increasing moisture stress) of protein in the soluble fraction obtained with the usual methods of homogenization of leaves (Fig. 1) would suggest that amount of "soluble" enzymes would decrease with increasing water deficits if there were no new synthesis. Declines in lactic, malic, and glucose 6-phosphate dehydrogenase activities with progressive drought (Table I) coincide with sharp losses in soluble proteins resulting from desiccation. These particular enzymes are thought to be normally part of the soluble cytoplasm. It is recognized that a sizable portion of the proteins found in the soluble fraction after homogenization and centrifugation was probably originally parts of subcellular organelles such as chloroplasts (Smillie, 1963) or mitochondria.

The possibility of degradation of enzymes during normal dehydration cannot be disregarded as a possible cause of death of plants. During drying of *Larrea* leaves that are very resistant to drought there was little change in protein content (Duisberg, 1952). Similar small decreases in soluble protein were obtained with *Selaginella lepidophylla* (Murphy and Todd, 1965) in contrast to the losses observed in wheat leaves (Todd and Basler, 1965). Rapid drying of leaves does not bring about the same changes as does slow drying. During rapid drying of rye-grass leaves (*Lolium* spp.) (2.5 hr to 62% dry matter), protein breakdown was negligible whereas slow drying (24 hr to 40% dry matter) caused a loss of more than 16% of the protein (MacPherson, 1952).

Many of the enzymes are present in the cell in more than one form. Limited studies indicate that during desiccation of tissues these isozymes may change at different rates (Stutte and Todd, 1969). From Fig. 2 it is evident that succinic and glucose 6-phosphate dehydrogenases change quantitatively whereas malic dehydrogenase appears to change both quantitatively and qualitatively after the most severe water stress. Peroxidase isozymes, total, and iron-containing proteins show marked changes after desiccation of the leaves. These isozymes could arise either from changes in configuration of enzyme molecules or from different locations within the cell.

Glycolytic activity was suppressed almost completely in droughted leaves of corn and fodder bean, while at the same time the rate of oxidation via the pentose phosphate pathway doubled (Abrarov, 1969). This shift was a consequence of synthesis of new enzymes as the increase in pentose phosphate pathway metabolism was not observed when azauracil was applied during drought (which presumably would prevent RNA synthesis). This induced shift to the pentose pathway is especially interesting as stress induced by a leaf disease caused by a rust fungus induced a similar change

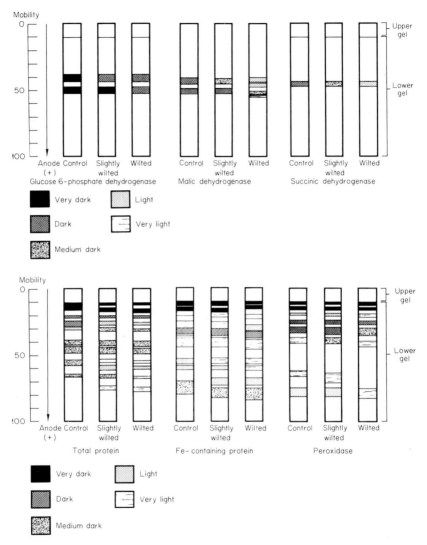

Fig. 2. Diagrammatic representations of polyacrylamide gel electrophoresis patterns obtained from soluble wheat leaf proteins (variety Ponca). Slightly wilted plants (RWC about 65%) deprived of water for 4 days; wilted plants (RWC about 35%) deprived of water for 7 days. Previously unpubli hed data of Tsai and Todd (1969) based on methods given in Stutte and Todd (1969).

(Shaw and Samborski, 1957). A shift in respiratory metabolism also occurred in corn leaves in which the respiratory quotient was lowered by drought (Petinov and Malysheva, 1960).

Careful study of Table I, which summarizes the effect of desiccation on levels of various enzymes, reveals a few overall tendencies:

1. Severe water deficits generally cause an overall decrease in enzyme level.

2. Levels of enzymes involving hydrolysis or degradation usually either remain the same or increase but they do not decrease until fairly severe desiccation has taken place.

3. Levels of some enzymes involved in synthesis are decreased and levels of others increase as a result of water deficits.

4. More work is needed to elucidate all aspects of changes in enzyme levels, especially with more comprehensive studies being performed with a given species.

III. EFFECT OF WATER DEFICITS ON ORGANELLE PROCESSES OR MULTIPLE ENZYME SYSTEMS

Many of the basic metabolic functions are carried on within organelles where enzyme systems are integrated within their structure. The process of drying on such systems could be different for each organelle. The specific effect would depend on how much the internal structure is organized around water and how dehydration affects the membrane, since the membrane may be extremely critical for proper functioning of the organelle. Mitchell's hypothesis (1965) for the coupling of high-energy phosphate production with proton separation at the membrane depends on maintenance of proper membrane structure which must include hydration state. As will be pointed out, high-energy phosphate production in both mitochondria and chloroplasts is reduced during dehydration.

A. MITOCHONDRIA

Mitochondria isolated from wilting pea (*Pisum sativum*) shoots had a reduced ratio of molecules of high-energy phosphate produced per atoms of oxygen taken up (P/O ratio) and with 24 days' drought (47% water content) there was no phosphorylative activity. With mild water stress the P/O ratio actually was increased (V. N. Zholkevich and Rogacheva, 1968).

Nir *et al.* (1970) prepared homogenates from maize root tips and assayed both mitochondrial and 20,000*g* supernatant fractions for cytochrome oxidase activity. The supernatant from untreated root tips gave no activity whereas roots dehydrated over NaCl (water loss of 78%) had 4.5 times as much cytochrome oxidase in the mitochondrial fraction as untreated samples and, in addition, the supernatant fraction gave considerable activity. This indicated that drying enhanced the activity of the enzyme, perhaps liberating it from controls imposed on the enzyme when it was present in the highly organized state in control mitochondria. The release of a tightly bound enzyme such as cytochrome oxidase indicates extensive disruption of the mitochondria.

Nir *et al.* (1970) also measured O_2 uptake by dehydrated root tips and found that a loss of about 50% of their water content reduced O_2 uptake by more than half after they were rehydrated. Root tips that had lost more than 70% of their initial water content regained turgor upon rehydration but no longer absorbed O_2.

The decrease in O_2 uptake with increase in water stress often observed with intact organs has been found to occur in mitochondria isolated from droughted tissues. V. N. Zholkevich and Rogacheva (1968) noted a reduction of 80% in O_2 uptake in mitochondria from water-stressed pea leaves at a water content of 72.5% or less (compared with 87% water content in control leaves). The P/O ratio actually increased during the same time that O_2 consumption was decreasing. The P/O ratio was nil at 47% water content although O_2 uptake still occurred. As a change in P/O ratio with rapidly desiccated cucumber cotyledons was not observed, V. N. Zholkevich and Rogacheva (1968) suggested that some poisonous metabolic intermediate accumulated during drying and in a short desiccation period did not have sufficient time to cause damage. Structural damage to tissue was noted on prolonged drying, however. This difference in extent of damage to tissue during rapid and slow drying may be very important as Flowers and Hanson (1969) found a much smaller decrease in soybean (*Glycine max*) respiration when KCl or sucrose was used to produce a lower water potential. State III mitochondrial O_2 uptake rates were greatly reduced by lower osmotic potentials whereas state IV respiration was affected very little. Differences were found between KCl and sucrose as the osmoticum. Flowers and Hanson (1969) suggested that salts may play a role in mitochondrial changes when dehydration occurs although there are complications in comparing water stress induced by osmotic gradients with that produced by other means such as evaporation, especially since the membranes usually are not perfect in their differentially permeable properties. Mitochondria isolated from water-stressed corn shoots were found to show altered swelling and ion transport characteristics when compared with mitochondria from non-stressed plants (Miller *et al.,* 1971).

Vacuum drying of rat liver mitochondria at 20°C destroyed oxidative phosphorylation activity whereas activity was not lost by drying at 0° or −20°C (Greiff and Myers, 1964). The presence of 0.5 M sucrose before freezing and lyophilization of rat liver mitochondria, followed by thawing in 5% dextran (mol. wt. 80,000) or 3% polyvinylpyrrolidone, yielded mitochondria that retained about 65–70% of their phosphorylative and oxidative capacities (Greiff and Myers, 1963; Greiff and Rightsel, 1966). The data suggested that protection by these large-molecular-weight substances was brought about by regulating the rate at which water entered the system upon rehydration. The possibility was suggested of a long-chain fatty acid in the membrane that promotes mitochondrial deformation and uncouples oxidation and phosphorylation.

B. Chloroplasts

Reduction in the water content of isolated chloroplasts was followed by a decrease in photosynthetic phosphorylation and reductive activity. On rehydration most of the reductive capacity as well as part of the phosphorylative activity was restored. Further reduction in water content caused losses in activity that could not be restored. The data suggested that dehydration caused changes in structure of the integranal lamellae (Nir and Poljakoff-Mayber, 1967).

Dehydration of a complex system, such as the chloroplast, might be expected to affect various components differently—some effects may involve dehydration of certain enzymes. Chloroplasts isolated from wheat leaves subjected to moisture stress were still capable of carrying out reduction of 2,6-dichlorophenolindophenol, although the numbers of chloroplasts extracted from wilted plants was reduced (Todd and Basler, 1965). Chloroplasts obtained from pea or sunflower leaves that had been rapidly dried showed a reduced ability to produce O_2, sunflower chloroplasts being affected at water potentials lower than −8 bars and peas at lower than −12 bars (Boyer and Bowen, 1970).

When chloroplasts were completely dehydrated (over $CaCl_2$ in a vacuum at 2°C) both Hill reaction activity and photophosphorylation were lost (Santarius and Heber, 1967). Protection could be achieved if sucrose, peptone, or bovine serum albumin were added before dehydration. Salts abolished protection. More sugar was required to protect cyclic or noncyclic photophosphorylation than electron transport. The protection was attributed to the water-binding ability of the additives; e.g., sucrose was twice as effective as glucose on a molar basis and it bound twice as much water. It appeared that high concentrations of sugars and proteins in plants were associated with high resistance to desiccation. However, Heber

(1968) subsequently reported the discovery of a protein isolated from cytoplasm of several plant sources that was 20 to 50 times more effective than sucrose in protecting chloroplasts from freezing. This protein, with a molecular weight of 10,000, was not present in leaves that were not frost hardy. It would be of great interest to know if such a protein might be present in drought-hardy plant cells to protect chloroplasts from drought.

Whereas sucrose protects chloroplasts during dehydration, high concentrations in the medium during measurements are detrimental to some components in the system. Chloroplasts placed in 3 M sucrose (which they calculate would correspond to 90% loss of water of a tissue) lost 75–95% of their cyclic photophosphorylation capacity but ferricyanide reduction was unaffected (Santarius and Ernst, 1967). The decreased photophosphorylation only occurred when dehydration took place in the light. The loss of phosphorylation was reversible, possibly indicating a physical disconnection between electron transport and phosphorylation.

C. MEMBRANES

There is evidence that the enzymes contained within the mitochondria and chloroplasts may be remarkably stable toward desiccation as long as their structure is retained intact and that damage to organelles from drying may be the result of unfavorable changes within the membrane or even destruction of membrane structure.

There are other indications that maintenance of membrane structure to ensure separation of components may be of primary importance in desiccation resistance. Small cells have advantage over larger ones (Iljin, 1957) although the important factors in resistance may be the amount of shrinkage of cells or the nature of the membranes. For example, shrinkage (decrease in thickness) of wheat leaves (Ridley and Todd, 1966) is much less than in *Impatiens balsamina* (Sengupta and Todd, 1970). In the latter species, after moderate drought stress the thickness of leaves is reduced by about half. Such drastic change might be expected to have adverse effects on membranes and may be related to the fact that between 30 and 50% of the membrane structure is water (Hechter, 1965).

Additional evidence for involvement of membrane components in dehydration injury comes from experiments involving air drying of yeast cells (Harrison and Trevelyan, 1963). Phospholipids decreased by 10–20% upon drying and loss of viability was attributed to membrane damage caused by lecithinase C. Yeast cells grown on a low nitrogen medium were not damaged by drying, suggesting that these cells either lacked the enzyme or it was not activated during drying.

If we can extrapolate from experiments on freezing to the drying process it is especially interesting to note that lipoproteins are more sensitive to freezing and thawing than are other proteins and they have a preponderance of nonpolar residues (Lovelock, 1957).

Lea and Hawke (1952) found that freezing or freeze-drying of lipoprotein of egg yolk (lipovitellin) brought about the release of lipid from the complex. If water is needed to maintain the normal configuration (e.g., clathrate structures) then removal of water by freezing or desiccation might be expected to cause irreversible damage to membranes. Heber and Santarius (1964) suggested that freezing caused disruption of hydrogen bonding of the lipoprotein of membranes which in turn brought about a loss in activity such as photophosphorylation. They suggested that protection endowed by sugars was owing to their substitution for hydrogen bonding of water, thereby stabilizing membrane structure. This might prevent intra- and intermolecular bonding of macromolecules rather than loss of lipid. Levitt and Dear (1970), also in connection with studies of freeze-injury, present evidence that dehydration of the cell surface may induce a rigidity (presumably due to SS bond formation) which is not reversed upon rehydration of the cell.

Lysosomal enzymes are released during freezing and thawing, or with changes in osmoticum or salt concentration (Tappel, 1966). These enzymes rapidly attack mitochondrial membranes and uncouple oxidative phosphorylation (Tappel, 1966). Insufficient research has been conducted on lysosomes from plant cells but these appear to be prime candidates for involvement in desiccation injury and offer another possibility for membrane disruption.

Thus several possibilities exist for explaining damage to cell membranes on drying:

1. Enzymatic destruction of membranes by enzymes released from lysosomes, vacuoles or elsewhere in the cell.

2. Dehydration-induced instability and conformational changes caused by disruption of hydrogen bonds.

3. Shrinkage due to loss of water.

4. Rigidity induced by SS bond formation which might be the result of reason 2 or 3 above.

It is entirely possible that any or all of these mechanisms are operative in addition to others that may be shown by future research.

D. Messenger RNA and Protein Synthesis

When plants are subjected to desiccation, recovery of cellular activity such as respiration and photosynthesis upon rewatering may be of little value unless the cells are capable of renewed protein synthesis. Limited but very promising studies have been made of such synthetic systems.

Webb and Walker (1968) studied the effect of dehydration on RNA content of F2 phage and β-galactosidase RNA from *Escherichia coli* and found that 1-inositol partially protected the RNA. However, they believed that damage to existing RNA was not involved in loss of cell viability but death occurred because of damage to the RNA-forming mechanism. They suggested that removal of water from DNA caused dimerization and that perhaps inositol replaced bound water on the DNA. DNA synthesis in developing *Vicia faba* seed ceased when water content dropped to 75% of cotyledon fresh weight (Brunori, 1967).

A study of this overall process in germinating wheat embryos has yielded some very interesting results (Chen *et al.*, 1968a,b). It had been observed some time ago that wheat seeds could be germinated for periods up to 48 hours and then air-dried without apparent injury upon germination. However, when the seeds were germinated for more than 48 hours before drying, death occurred following rehydration. Dehydration at either stage inactivated the RNA and arrested protein synthesis, although seeds germinated for less than 48 hours retained RNA in a latent state (Chen *et al.*, 1968a). Rehydration of the seeds germinated for 72 hours and subsequently dried resulted in transcription of RNA which was false and was inactive in producing the normal enzyme complement (Chen *et al.*, 1968b). It was suggested that this may have been due to breakdown of DNA as well as false messages which might arise from lack of proper initiation points, or a frame transition on the broken template. Extraction of DNA from the drought-damaged embryos revealed a disappearance of the highly polymerized components and appearance of smaller units. It was noted that DNA duplication and transcription were active in normal 72-hr germinating seedlings but were relatively inactive at earlier stages (none for 24 hr) (Chen *et al.*, 1968a). Ribosomes were not damaged by drying as they were still capable of incorporating phenylalanine in the presence of polyuridylic acid.

These results complement those obtained by Genkel's group (Genkel *et al.*, 1967). They found that drought caused disappearance of polysomes (4–10 ribosomes) in corn and bean leaves and appearance of free ribosomes and dimers (relative protein content decreased at the same time). Ribonuclease attacks mRNA more readily than ribosomal RNA. Since ribonuclease activity was found to increase in droughted plants, the breakdown of polysomes could easily be explained (Genkel *et al.*, 1967). Hsiao (1970) also noted rapid conversion of polyribosomes to the monomeric form with mild water stress but does not believe the change is due to increased ribonuclease activity. Ivanova (1969) found mild drought to cause a reversible decrease in nucleic acids in wheat leaves (mostly RNA) and accumulation of intermediate products of nucleic acid metabolism. More severe drought caused irreversible decomposition, a finding which agrees with other work with wheat (Todd and Basler, 1965; Stutte and Todd, 1968).

During drought hardening there are very likely to be changes in the enzyme complement. This would in turn open the possibility for changes in base ratios of nucleic acids. Kessler and Frank-Tishel (1962) reported an increased $(G + C)/(A + U)$ base ratio when olive (*Olea europaea*) leaves were subjected to water stress, whereas the ratio in *Ligustrum sinensis,* which is drought sensitive, did not change. An increased base ratio also was recorded in wheat leaves that were subjected to moderate water stress (Stutte and Todd, 1968). However, changes in the base ratio also were found when the plants were grown under different temperature regimes, showing that changes in base ratio can occur as a result of changes in more than one environmental variable.

For some time it has been apparent that water stress probably leads to reduced protein synthesis together with increased accumulation of amino acids (Petrie and Wood, 1938). Ribosomes isolated from water-deficient corn leaves incorporated amino acids at a slower-than-normal rate (Ramgopal and Hsiao, 1970). Substantial accumulations of certain amino acids, especially proline, occur concomitantly with water stress in many plants (Barnett and Naylor, 1966; Routley, 1966; Kemble and MacPhersen, 1954; Thompson *et al.,* 1966; Saunier *et al.,* 1968; Savitskaya, 1967). This accumulation is at the expense of carbohydrates as added sugars increased the amounts of amino acids formed, and when plants are placed in N_2 no accumulation occurs (Routley, 1966). The accumulation of amino acids probably relates to decreased protein synthesis inasmuch as turnover of proline is very slow (Barnett and Naylor, 1966). Turnip leaves under moisture stress formed considerable radioactive proline when discs had been supplied with either labeled glutamic or *N*-acetylglutamic acids (Morris *et al.,* 1969). Amino acids differ in behavior as tryptophan synthesis declined with reduced water supply (Prusakova, 1962). It has been suggested that the increase in amino acids may be a mechanism for preventing the buildup of NH_4^+ (Henckel, 1964), although accumulation of proline may be in itself a protective agent. Tyankova (1967) found that application of D,L-proline helped wheat plants to recover from drought.

IV. ENZYME ACTION AT VERY LOW MOISTURE CONTENTS

As the cytoplasm is dehydrated, activity of some enzymes is reduced before that of others. Such a situation would undoubtedly cause a metabolic upset which might contribute to lethal effects of drying. Much research on action of enzymes at low water contents has been an outgrowth of interest in food deterioration, especially storage of dried or lyophilized food products. Under otherwise similar conditions of storage, enzymatic

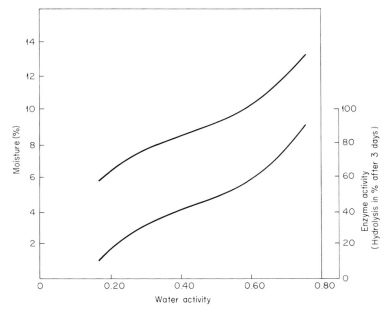

Fig. 3. Lipase activity at 25°C in defatted ground oats (substrate, 4% monoolein) as a function of water activity (lower curve) and the sorption isotherm (upper curve) of the same material. Acker (1969), reprinted by permission.

activity of ground food materials is greater than when such material is left intact. Cereal grains can be stored for years at a water content of 13% without deterioration whereas, after they are ground, deterioration occurs within a few weeks (Acker, 1969).

Enzymes such as lecithinase and urease are not active until sufficient adsorption of water has taken place to form a monomolecular layer. Water added beyond this point is apparently sufficient to allow diffusion of substrate to the active site on the enzyme. Activity of the water present is of greater importance than is the water content. Urease becomes active when kept a few hours at a relative humidity greater than 60% (20°C) (Skujins and McLaren, 1967). The calculated amount of water sorbed by urease when it becomes active is only 1.3 moles per mole of side-chain polar groups. Therefore, it was suggested that the only water that can be utilized in hydrolysis is that in excess of the stoichiometric minimum of 1 molecule of water per polar site. Other enzymes such as polyphenolase and lecithinase appear to function in a similar manner (Acker, 1969). A lipase was shown to function according to the liquidity of the fatty substrates supplied; splitting of tricaprin was much higher at 35° than 25°C.

This substrate is solid at 25°C but becomes liquid at 35°C. The enzyme was obtained from ground defatted oat seeds that had been in equilibrium with an atmosphere having a relative humidity of 15% (Acker, 1969).

Lipase activity of defatted ground oats parallels the activity of the water present (Fig. 3). At a water activity less than about 0.25 the water is probably held as a monomolecular layer; from a water activity between 0.25 and 0.65 it is probably held in bi- or multilayers, and at higher water activity in capillaries (Acker, 1969).

The capacity of some enzymes to operate at extreme conditions points up the limitations of making homogenates and extracting enzymes from water-stressed plants and attempting to characterize the activities taking place in the intact cell. Perhaps membrane-destroying systems are active long after many other systems have ceased to function inasmuch as they may operate very well at low moisture contents, if the experiments on seed lipases are any indication.

V. STATE OF ENZYMES IN ORGANS OR ORGANISMS THAT CAN BE AIR DRIED OR SUBJECTED TO VERY HIGH WATER DEFICITS

A. SEEDS

The seeds of a majority of higher plants are usually capable of surviving air-drying treatment, although there are certain exceptions such as *Acer* seeds which are killed when the moisture content drops below 30–34% (Jones, 1920). In fact, dry storage of seeds is usually conducive to extended viability. The embryonic cells in these seeds are capable of surviving desiccation to an extent far in excess of that which is lethal to most vegetative plant parts. There are several possibilities as to why enzymes, proteins, and nucleic acids are not damaged in the embryo during this severe desiccation:

1. Enzymes sensitive to drying are not produced during final stages of embryo formation or are lost during drying. In either case the early stages of germination would of necessity have to proceed without the use of such enzymes.
2. Sensitive enzymes are protected during the drying phase by special coatings or by compartmentation. The enzymes might be inactive in a stored state and require removal of the protective agent or "activation" by partial protein degradation.
3. The cellular environment is radically different in the embryo than in vegetative tissues owing to the presence of high levels of protective

agents or exclusion of agents that might be detrimental to dry storage, e.g., careful regulation of type and concentrations of electrolytes.

There is some evidence for operation of all these mechanisms, especially the first two, although much of the available evidence was collected with the objective of understanding the germination process. Some enzymes are not found during early stages of germination (e.g., α-amylase). In fact the germination process in wheat seeds can be halted after 2 days of germination by air drying of the seed with few if any observable detrimental effects. Enzymes such as phytases, transaminases, phosphatases, and amylases either appear or greatly increase 2 days after germination (Mayer and Poljakoff-Mayber, 1963) and if such seeds have germinated for 3 days, air drying is lethal (Chen et al., 1968b). Some enzymes such as glutamic acid–alanine transaminase and glutamic acid decarboxylase are present and active in wheat seeds at water contents lower than those required for germination to proceed (Linko and Milner, 1959). Wilson (1970) obtained incorporation of ^{32}P into ATP and glucose phosphate in crested wheatgrass (*Agropyron desertorum*) seed at water potentials as low as -130 atm. A number of hydrolytic enzymes such as lipases, proteases, and β-amylase are also present in dry seeds, some in a latent, inactive form (Mayer and Poljakoff-Mayber, 1963).

There generally is some lag in oxygen uptake, indicating absence of some systems until germination has proceeded to a stage at which the necessary enzymes have been activated, or more likely, produced *de novo*. A marked shift in respiratory pathways is indicated as glucose 6-phosphate and phosphogluconate dehydrogenases were present in early germination of mung beans and had nearly disappeared after 4 days of germination (Mayer and Poljakoff-Mayber, 1963). Mitochondrial oxidation in *Lupinus,* on the other hand, was very low at 12 hours and increased manyfold during the following several days of germination.

Shain and Mayer (1968) isolated zymogenlike granules from pea seeds containing amylopectin 1,6-glucosidase which was released as an active enzyme when the granules were treated with trypsin.

Lipids may play a role in protecting certain enzymes in dry seeds. Glutamate dyhydrogenase extracted from wheat embryos could be dried *in vacuo* without loss of activity but when lipids were removed with *n*-butanol followed by acetone, all activity was lost upon drying (Nations, 1966). Only 8.1% of the dry weight of enzyme was removed in the defatting process but this component was obviously very important in protecting the enzyme against drying. Nations (1966) also suggested that stability of the enzyme may depend on dissociation into subunits and that these were somehow dependent on the lipoidal component for protection.

B. VEGETATIVE TISSUES

Air drying is apparently lethal to leaves, roots, and other similar vegetative parts of most higher plants. By contrast, vegetative tissues of lower plants, including some mosses, ferns, lichens, algae, and microorganisms, can withstand such treatment (Levitt, 1956). The club moss *Selaginella lepidophylla* is especially interesting because it is large and can be air dried for periods of several years and remain viable. The air-dry plant can be fully hydrated within a few hours and apparently becomes fully functional within an hour or two. Dry plants fold up into a closed "ball"; turgid plants open into a flat configuration. Plants having relative water contents in excess of 50% and which were open carried on photosynthesis at essentially the same rate as fully turgid plants on a dry weight basis (Lamar and Todd, 1968). Uphof (1920) observed that in contrast to cells of mesophytic species of *Selaginella,* those of xerophytic *Selaginella* species contained large amounts of oil. He also noted that when xerophytic types were grown in the greenhouse they produced fewer oils, thus leading him to suggest that the oils in some way protected the cytoplasm from damage during dehydration.

The lichens *Cladonia* and *Parmelia,* which had been stored in a desiccator over $CaCl_2$ at room temperature for 1 year, were capable of fixing CO_2 photosynthetically at a rate equal to that obtained before the drying (Nifontova, 1968). Parker (1968) cited research on a number of mosses, liverworts, and ferns that are remarkably desiccation-resistant although apparently little research has been done on enzyme systems in conjunction with dehydration. Parker (1968) indicated that absence of plasmodesmata in mosses may protect them against dehydration. Genkel and Pronina (1968) found this to be true in the moss, *Neckera,* which in addition to the absence of plasmodesmata and vacuoles has a thick membrane that gives a characteristic stain with ruthenium red upon drying. The organic phosphorus fraction in this moss remains unchanged with changes in water content, whereas, in less resistant plants, it decreases by at least half (Genkel and Pronina, 1968). The flowering plant, *Myrothamnus flabellifolia,* which lacks plasmodesmata and has only small vacuoles, is also capable of withstanding air drying (Genkel and Pronina, 1969).

A fern, *Ceterach officinarum,* can withstand drying over $CaCl_2$. When dry, the cells show a high birefringence. This led Oppenheimer and Halevy (1962) to suggest that survival of *Ceterach* was related to formation of crystalline proteins. If this is true the time required to dissolve the protein crystals must be very short, inasmuch as they noted an increase in respiration within 10 min after rehydration.

VI. DENATURATION OF ENZYMES AT
LOW WATER CONTENTS

A. DEHYDRATION PHENOMENA

Probably most enzymes are inactivated when either pure or partially purified preparations are subjected to drying at room temperature. When intact tissues are dried in this way, degradative enzymes may destroy many enzymes and others may be denatured or inactivated because of elevated salt concentrations.

Freezing of intact tissue resembles drying as intracellular water may leave the cells to become part of ice nuclei outside the cells. This leads to dehydration as well as to increased salt concentration within cells. There was little apparent general success in drying enzymes until the freeze-drying process was devised and there seem to be relatively few reports on the effects of drying of enzyme preparations that were not first frozen. Air drying of thin layers obtained by spraying (without freezing) was used successfully on pepsin solutions (Bullock and Lightbown, 1942). Milk powder obtained in the same manner retained peroxidase activity. Spray drying was patented in 1872 so it is likely that crude preparations of enzymes were made long before 1942. Brosteaux and Eriksson-Quensel (1935) reported that when hemoglobin, serum albumin and globulin, edestin, or phycoerythrin were dried *in vacuo* at room temperature, the proteins were more or less "denatured" as judged by their inability to return to their former state of dispersion.

Dehydration causes a number of changes in the cytoplasm that could have profound effects on structure of proteins or other large polymers such as DNA or RNA which depend on water for their tertiary or quarternary structure. Examples of such changes are:

1. Loss of water between molecules allowing for increased hydrophilic or hydrophobic bonding.

2. Greater possibilities for binding with cations or anions as their concentration increases with dehydration.

3. Greater possibilities for binding with other molecules such as sugars.

4. Changes in structure due to pH changes occurring as a result of dehydration.

Each type of molecule would be affected somewhat differently, by any or all of these changes; thus we would expect some enzymes to be protected by conditions that would cause denaturation of others.

B. Effects of Sugars and Other Compounds

The addition (in decreasing order of effectiveness) of sugars, NaCl, alanine, glycine, or gelatin before drying protected hemoglobin, serum albumin and globulin, edestin, and phycoerythrin to some extent (Brosteaux and Eriksson-Quensel, 1935). Lactose was most effective of the compounds tested and the minimum quantity to achieve protection was that necessary to form a monomolecular layer around the protein molecules. This observation strongly indicates that coagulation or inactivation caused by desiccation is a consequence of intra- or intermolecular bonding when water is removed. Lactose would be expected to associate with hydrophilic groups, thereby preventing close spatial juxtaposition of hydrophilic or hydrophobic bonds of the same or different macromolecules.

Somewhat similar evidence has been obtained for human plasma β-lipoprotein. It can be freeze-dried in the presence of sucrose without denaturation (Lovelock, 1957). Lovelock (1957) also froze β-lipoprotein in different salt solutions and found that denaturation in the absence of sucrose coincided with removal of all the water from the solution.

Removal of water by freeze-drying prevents enzyme molecules from coming into contact with one another and thereby protects them against possible denaturation. This also can be accomplished by removal of water by displacement with organic solvents inasmuch as such preparations subsequently can be dried; examples are the preparation of trypsin and lipase from pancreas with ethylene dichloride (Levin, 1950) or the many enzyme extractions involving use of acetone powders (Umbreit et al., 1949).

The three-dimensional structure of proteins is determined by the amino acid sequence and, in the case of water-soluble proteins, the protein configuration is determined by the nature of the internal bonding possibilities and by the bonds exposed to the water. The configuration of proteins which are found within membrane structures or other nonaqueous systems is determined by whatever the phase external to the protein happens to be.

Removal of water from around the protein allows for new interactions of chemical groups either within the same protein molecule (but perhaps different chains) or between different molecules. Webb (1965) summarized the pertinent literature on drying of intact bacteria. He suggested that death of cells is due to removal of hydrogen-bonded water from such groups as —N, =N—H or —OH. This water is removed first and then water is removed from =C=O or =P=O groups, allowing for lethal interactions to occur. It has been suggested that, in freeze-drying, the bound water layer remains intact; when removed, loss of bacterial viability occurs

(Scott, 1958). Webb (1965) found that bacterial cells dried in the presence of a 10% solution of glucose were somewhat protected. All protective action was lost when α-methyl glucoside was used instead of glucose. Similarly, glucosamine was protective whereas N-acetylglucosamine was not. Webb (1965) concluded that "the importance of structural and steric configuration is due, in all probability, to the need for the solute not only to fit into the macromolecules, but to combine with it in such a manner as to preserve its natural configuration." He found that compounds which provided protection could form hydrogen bonds.

Several common sugars prevented the formation of a BSA-α-globulin complex resulting from heating (Hardt et al., 1946). However, only L-arabinose, D-ribose, digitoxose, and L-ascorbic acid prevented the heat coagulation of BSA, whereas several sugars, including galactose, fructose, sucrose, lactose, raffinose, and glucosamine, were ineffective (Fischer, 1947). Protection by maltose against heat coagulation was not equally effective against different proteins; e.g., 0.3 M maltose almost completely prevented coagulation of catalase that occurred in distilled water (10 min at 58°C) whereas coagulation of the enzyme "dehydrase" was only partially reversed in its presence (Christophersen and Precht, 1952).

Differences in amount of protection by different sugars for different proteins may be related to molecular configuration of the two molecules and probably involve hydrogen bonding. Giles and McKay (1962) employed refractometry to measure the interaction of gelatin, BSA, or casein with various sugars. D-glucose, D-galactose, D-mannose, L-sorbose, and mannitol formed complexes with at least one of these proteins, whereas L-glucose, L-galactose, D-fructose, and 4 disaccharides did not react.

Lysozyme was protected from inactivation by ultrasonic waves by BSA and a number of different amino acids, glutathione and methionine being the most effective (Dietrich, 1962). D-Glucose was ineffective.

C. Effects of Salts

Either cations or anions can change the water structure surrounding proteins. In depolymerization of F-actin the anion is of great importance, the effectiveness being $I^- > Br^- > Cl^-$ while the cation (Na, K, Cs, Rb) made little difference (Jencks, 1965). From studies based on the action of various ions, Jencks (1965) concluded that denaturation of proteins and nucleic acids involving hydrophobic bonds probably were accompanied by changes in structure of the solvent. Lithium salts appear to have a more profound influence in denaturing proteins than do other cations of the alkali series (Jencks, 1965; Chilson et al., 1965). Cations inhibit the action of α-amylase in the following decreasing order: $Li^+ > Na^+ > NH_4^+ >$

$Rb^+ > Cs^+ > K^+$, and anions in order of $I^- > F^- > NO_3^- > SO_4^{2-} > Br^- > Cl^-$ (Whitaker and Tappel, 1962). Inhibition of enzymatic activity was proportional to the square root of ionic strength and there appeared to be a relationship between the degree of inhibition and hydration of the ion.

The presence of electrolytes in the medium can greatly influence the capacity of protein to bind water. In the absence of electrolyte, 1 mole of egg albumin binds about 740 moles of water (Bull and Breese, 1968). Figure 4 shows that addition of chloride salts decreases water binding. The decrease in water binding for various alkali metals follows the lyotropic series $Li^+ > Na^+ > K^+ > Rb^+ > Cs^+$. Bull and Breese (1970) also found that LiCl, RbCl, and CsCl were bound to the extent of about 5.7 moles/ mole of protein, whereas NaCl and KCl were not bound. NaBr reduces water binding even more than NaCl (Fig. 4). Sodium sulfate caused a marked increase in water binding to 1350 moles per mole of egg albumin (assuming no salt binding) whereas NaI, NaCNS, and $CaCl_2$ were bound to egg albumin in preference to water. If the extent of water binding is important in preventing denaturation by drying then it is readily apparent that the amounts and kinds of salts present will greatly influence the results. This aspect of desiccation resistance needs to be studied further.

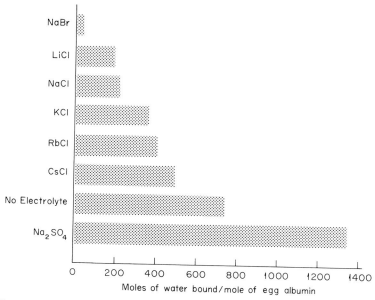

Fig. 4. Binding of water by egg albumin in the presence of various electrolytes (relative humidity 92%; 25°C). Used with permission, adopted from Bull and Breese (1970).

The chloride salts of sodium, potassium, magnesium, and calcium at a concentration of about 300 meq/l were found to reduce heat coagulation of thermophilic bacteria (20 min at 65°C) to almost nil (Ljunger, 1970). Thermostability may involve certain ions to maintain protein structure. A heat-stable α-amylase from *Bacillus stearothermophilus* contained 2 gm atoms of calcium after extensive dialysis. In addition to this tightly bound calcium, about 0.9 mg/ml of $CaCl_2$ were required for maximum activity which could not be duplicated by chlorides of Na, K, or Mg (Manning and Campbell, 1961).

D. Changes in pH

As desiccation proceeds and concentration of ions increases, certain salts become insoluble, thereby changing the composition and pH of the remaining solution. A thorough study has been made of the effect of freezing on the composition and pH of KH_2PO_4–Na_2HPO_4–H_2O solutions (van den Berg and Rose, 1959) and of the effect of addition of neutral salts (such as KCl) to the mixture (van den Berg, 1959). Both increases and decreases in pH were observed, although decreases in pH appear more likely in the cell inasmuch as freezing of cabbage juice caused reduction in pH (Harvey, 1918). Presumably similar types of changes might occur on desiccation of tissue and decreases in pH might be anticipated.

It is difficult to measure experimentally internal cell pH under conditions of cell desiccation. However, it is relatively simple to make measurements of pH as one rehydrates lyophilized plant material. This has the advantage of rapid dehydration but, of course, is a system in which all cell structure is lost. Figure 5 shows that when small amounts of water are added to such dry, ground, lyophilized material the pH is substantially lower than when the material is mixed with approximately the amount of water the tissue originally contained. Most of the lowering of pH was at a water content of less than 4 ml water/gm dry material, which corresponds to a relative water content of about 40%. Normally in wheat leaves, water content must be reduced to a value below 40% before death is probable (Salim *et al.,* 1969) which is also the point where pH was rapidly lowering. Such an extrapolation from dry, ground material to intact leaves probably is not justified but it does suggest that a lowering of pH with desiccation should be considered a possibility.

E. Sulfhydryl Effects

Wilting of plants was found to cause an increase in oxidation state of the tissue (Subbotina, 1959). Levitt (1962) suggested that protoplasmic

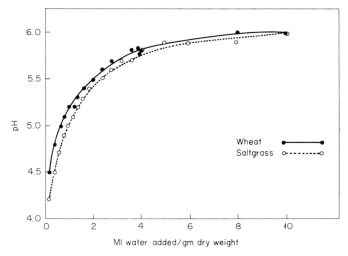

Fig. 5. The pH profile as water was added to lyophilized, ground leaves of wheat (*Triticum aestivum* L.) and saltgrass (*Distichlis stricta* Torr., Rydb.). From Todd (1969).

damage from drought or freezing may be due to formation of inter- or intramolecular SS bonds. Freezing of solutions of Thiogel markedly enhanced intermolecular SS bond formation as indicated by the rise in melting point of the gel (Levitt, 1965). The fact that dehydration is involved in freeze damage is emphasized by additional studies on bovine serum albumin (BSA). When BSA solutions were frozen in a dialysis sac, native BSA was dehydrated 90% whereas reduced BSA was dehydrated 100% (Goodin and Levitt, 1970). Only slight aggregation of native BSA occurred whereas nearly 80% of the mercaptoethanol-reduced BSA was aggregated. This was interpreted as the result of increased possibilities for SS bond formation during drying. Goodin and Levitt (1970) also suggested that hydrophobic bonding occurs during thawing.

Gaff (1966) measured changes in protein SH bonds during desiccation of cabbage leaves over $CaCl_2$ for periods up to 24 hr. He found a decrease of 50% in reactive SH protein bonds with drying. About 60% of the original SH groups that disappeared could be accounted for as SS bonds, apparently involving linkages of nonprotein to protein. After desiccation of the leaves to a water potential of -40 atm there was an increase in reactive SH groups on the structural protein. The total SH and SS content did not change during drying until a water potential of -94 atm was reached, and then considerable conversion of SH to SS occurred. Gaff (1966) indicated that these bonds were intermolecular protein–protein bonds and bond formation coincided with loss of viability of the tissue.

Protection should be afforded by adding agents to maintain SH bonds in the reduced state. Paricha and Levitt (1967) applied thiols to red cabbage (*Brassica oleracea*) leaves in an attempt to protect lipoprotein SH groups which they suggested were most sensitive. β-Mercaptoethanol at 10^{-2} M protected the plants from drought injury (as measured by plasmolysis-deplasmolysis) although glutathione was ineffective. They suggested the lack of effectiveness of glutathione was related to solubility at the proper site. It might be noted that, with equal amounts of oxidized and reduced glutathione in solution, for a negatively charged protein, the ratio of protein SH to protein SS is lowered with increasing ionic strength (Danielli and Davies, 1951). Desiccation would lead to increased solute concentration, and therefore an agent like glutathione might be expected to become less effective in maintaining proteins in a reduced state as desiccation increases.

On the basis of structural characteristics of a number of enzymes, Levitt (1966) concludes that heat stability depends upon a high proportion of strong hydrophobic bonds, intramolecular SS bonds, and absence of SH groups. It is probable that similar characteristics would improve desiccation resistance.

F. OTHER FACTORS IN CONFORMATIONAL CHANGES

The conformation of enzyme proteins in solution is determined by the nature of covalent bonds and noncovalent interactions of side-chain groups. Of the latter groups hydrophobic bonds may be quite important. There is considerable free energy change when such nonpolar groups interact. Such interaction is accompanied by a decrease in interactions with water (Némethy and Scheraga, 1962). Such nonpolar interactions should not be adversely affected by decreased water content although new possibilities for such interactions between different molecules might occur.

The stabilization of proteins by hydrogen bonding in solution is apparently marginal and can be destroyed by heating, with the extent of destruction depending on the protein structure (Joly, 1965). It was recognized many years ago that increasing the hydration of proteins increases their instability toward heat. Ewart (1897) cites work of Lewith (1890) who observed the coagulation of albumin; at 18% water it coagulated at 80°–90°C, whereas with 0% water, coagulation did not take place until a temperature of 160°–170°C was reached.

A more detailed study was made of heat stability of egg albumin over a wide range of water contents. The hydrated protein was heated for 10 min to either 70°, 76.5°, or 80°C. Figure 6 shows that increased hydration increased the amount of albumin that became water-insoluble (Bull and

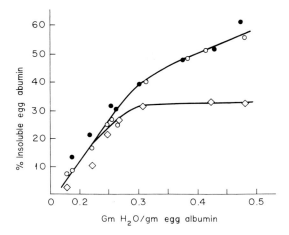

Fig. 6. Solubility of egg albumin as a function of water content during heating. Protein samples were hydrated to a known level, then exposed for 10 min to either 70°C (lower line, open squares), 76.5°C (upper line, open circles), or 80°C (upper line, closed circles). Used with permission, Bull and Breese (1968).

Breese, 1968). This figure also shows that the higher temperatures caused greater denaturation at a given level of hydration but only when a minimum level of hydration was exceeded. They performed similar experiments on lysozyme. In this case they were able to resolubilize the insoluble enzyme by recrystallization after it had been heat coagulated. It has been noted that adsorption of enzymes reduces their mobility and therefore reduces the ability to undergo lethal conformational changes when subjected to heat (James and Augenstein, 1966). Silman and Katchalski (1966) discuss the formation of water-insoluble derivatives of enzymes with cellulose polymers and *p*-aminophenylalanine leucine and other copolymers. These increased stability of trypsin, chymotrypsin, papain, and urease when stored in water at 4°C for several months. One trypsin derivative lost activity upon lyophilization although a maleic anhydride ethylene copolymer derivative of trypsin was active following lyophilization. As there are many macromolecules within cells that could act as surfaces for enzyme adsorption, such adsorption could provide a mechanism for successful storage of enzymes in a cell in the dehydrated state (e.g., seeds).

There are indications of conformational changes in soluble proteins extracted from leaves of droughted plants. Dielectric measurements indicated increased bound water which was thought to be due to exposure of hidden groups which could bind water (Sedykh and Khokhlova, 1967). This increase in binding was lost when the plants were rewatered before extraction of the protein.

Klotz (1965) suggested that hydrophobic interactions cause an ordering of water around nonpolar groups into giant water crystals or clathrate structures. Water ordered in this manner would be icelike and would increase protein stability; increased temperatures would break such structures. Klotz (1965) studied the stability against heat coagulation of a polymer, polyvinylmethyloxazolidinone, which cannot form hydrogen bonds. The polymer is water soluble up to a temperature of 40°C, but it becomes insoluble at higher temperatures. He concludes that the influence of small molecules cannot be on hydrogen bonding within the polymer but they are effective by modifying the solvent structure which indirectly changes the conformation of the macromolecule. Tanford (1970) concluded that effects of electrolytes on denaturation are due to their affinity for exposed peptide groups and not on their reorganization of the ordering of water.

VII. HYDRATION OF MACROMOLECULES

Hydration of proteins has been thought for some time to be related to desiccation resistance, with resistant species being able to retain more bound water than susceptible ones (Newton and Martin, 1930). The amount of tightly bound water should be related to the number of free polar groups on the protein. Fisher (1965) devised the following formula for predicting the amount of hydration of a protein based on the volume of external polar groups:

$$H_t = 0.725 \times 10^{-24} V_e$$

where $H_t =$ grams of water bound per gram of protein and $V_e =$ volume of the external polar layer of the protein. V_e can be calculated from the amino acid composition of the protein. The calculated values agreed fairly well with experimentally determined values. Calculated hydration for 34 common water-soluble proteins ranged from 0.221 to 0.330 gm of water per gram of protein and was almost independent of molecular weight. The narrow range for the amounts bound by these different proteins is remarkable.

Bull and Breese (1968) measured water binding of proteins using an isopiestic method and then derived an equation for water binding based on the nature of polar residues. The derived regression line giving the best fit with experimental values was:

$$y = -0.97 \times 10^{-3} + 6.77\, X_1 - 7.63\, X_2$$

where $y =$ moles of water bound per gram of protein at 25°C and 92% relative humidity; $X_1 =$ sum of moles of hydroxyls, carboxyls, and basic groups per gram of protein; and $X_2 =$ moles of amide per gram of protein. Measurements of hydration of 10 different proteins showed a poor correlation with polar groups unless amide residues were excluded. When amides were excluded an excellent correlation of 6 moles of water complexed per mole of polar residues was obtained.

Bull and Breese (1968) suggested that amide groups do not bind water themselves but inhibit neighboring polar groups from doing so, perhaps by hydrogen bonding with them. Bovine serum albumin did not fit the pattern for the other proteins but it is known that a number of carboxyl groups in the native protein are masked. Similar results for water binding of proteins were obtained by nuclear magnetic resonance (NMR) signals taken at $-35°C$ (Kuntz et al., 1969). This bound water did not freeze and was not in a highly ordered state.

In considering how important a role bound water might play in protecting proteins against drying, the distribution and location of the water would be very important, especially if the protection involves reducing the possibility of intermolecular hydrophobic bonding. Bull and Breese (1968) made calculations for cytochrome c to obtain some idea of this distribution. They assumed that the molecule was spherical, that all of the polar groups were on the surface, and that each polar group occupied 25 $Å^2$ units. Under these assumptions they concluded that only about one half of the surface is occupied by hydrophilic groups. For soluble proteins there should then be a direct relation between protection against dehydration coagulation and water-binding capacity. Before this approach can be tested more needs to be known of particular protein structure as water binding could not be predicted from calculations based on numbers of polar groups for silk fibroin, zein, or collagen (Bull and Breese, 1968). Some of the deviations from expected water-binding behavior may be a function of swelling or contraction of macromolecules. Dry agar swells during hydration, thereby continually exposing new surfaces (Masuzawa and Sterling, 1968); thus the capacity to bind water changes. Nucleic acids bind even more water than protein. Denatured tRNA binds 3–5 times more water than proteins as determined by NMR at $-35°C$ (Kuntz et al., 1969).

VIII. DRYING OF PURIFIED AND CRYSTALLINE ENZYMES

Informative literature on effects of drying on isolated enzymes is difficult to locate because it is widely scattered and details of preparation of enzymes usually are confined to brief statements in sections on methodology. The information is largely of interest only to those who plan to prepare enzymes in purified states. Many enzymes can be lyophilized in a purified state as shown by the fact that over 40 different enzymes are available in various states of purification from two commercial sources

TABLE II

Some Enzymes That Can Be Lyophilized following Crystallization

Enzyme	Source	Reference
Alcohol dehydrogenase	Yeast or horse liver	[a]
ATP-creatine transphosphorylase	Rabbit muscle	Noda et al. (1955)
ATP-AMP transphosphorylase	Rabbit muscle	Noda and Kuby (1963)
Chymotrypsin A and B	Beef pancreas	[a]
Chymotrypsinogen A and B	Beef pancreas	[a]
Deoxyribonuclease I	Beef pancreas	[a]
Lysozyme	Egg white	[a]
Pepsin	Hog stomach mucosa	[a]
Pepsinogen	Hog stomach mucosa	[a]
Ribonuclease	Beef pancreas	Fruchter and Crestfield (1965)
Trypsin	Beef or hog pancreas	[a]
Urease	Jack bean	[a]

[a] Commercially available from such sources as Nutritional Biochemicals Corp., 26201 Miles Road, Cleveland, Ohio, or Worthington Biochemical Corp., Freehold, New Jersey.

(Worthington Biochemicals and Nutritional Biochemicals). Over 20 additional enzymes can be obtained commercially as dry lyophilized powders in a less purified state.

Of 48 crystalline enzymes available from these commercial sources only 9 are supplied as lyophilized powders (Table II). The other 39 are supplied usually in saturated aqueous solutions of ammonium sulfate. Whether the ones not supplied in the lyophilized state are actually inactivated by lyophilization or whether they are supplied in this form for convenience could not be ascertained. The high salt concentration also acts as a preservative as well (Schwimmer and Pardee, 1953).

A search of the literature revealed very few reports of enzymes that are inactivated by lyophilization (see Table III). Further work may reveal

TABLE III

Enzymes Reported To Be Inactivated by Lyophilization

Enzyme	Source	Reference
α-Amylase	Hog pancreas	Caldwell et al. (1952)
D-Amino acid oxidase	Hog or sheep kidney	Burton (1955)
Carboxypeptidase A	Beef pancreas	Nutritional Biochemicals Co. (1969)
Phosphodiesterase	Lamb brain	Healy et al. (1963)
Phosphomonoesterase	Cow's milk	Morton (1951)
D-Xylulokinase	Calf liver	Ashwell (1961)

special conditions that would make lyophilization successful in those instances where it was not satisfactory. For example D-amino oxidase can be protected by addition of substrate or coenzyme during lyophilization, whereas without such addition activity is lost (Burton, 1955).

Even if more information were gathered on effects of drying highly purified or crystalline enzymes this would not necessarily provide answers to questions of protection or damage in cells. Cells contain a wide diversity of enzymes many of which are located in specific combinations with other enzymes, proteins, lipids, carbohydrates, etc., plus a great variety of smaller molecules and inorganic salts.

IX. CONCLUSIONS

1. A wide variety of enzymes can be dried by lyophilization probably because freezing prevents conformational changes during the drying process.

2. A number of agents can offer protection to enzymes undergoing dehydration, notably sugars and proteins, or the prior combination of the enzyme with other large molecules. These probably protect the enzyme by preventing undesirable interactions.

3. Inactivation of enzymes during dehydration can be at least partially explained by formation of intra- or intermolecular SS bond formation or by hydrophobic interactions.

4. Probably very few enzymes can be air dried from solution at room temperature without subsequent loss of activity. Such drying would maximize the opportunity for interaction of intra- and intermolecular bonds.

5. Water deficits of intact cells lead to release or activation of degradative enzymes. Because of a reduced rate of synthesis a number of crucial metabolic activities may cease.

6. Water deficits in intact cells may cause direct damage to enzymes through interactions with other large molecules, conformational changes, or interactions with the increasing amounts of salts present.

7. Under certain circumstances it is possible that cells can synthesize enzymes having a similar activity but which are less susceptible to damage during dehydration.

8. Enzymes present within highly organized subcellular structures may be rather resistant to changes in water deficits because of their somewhat fixed structural relationship to the other molecules within the organelle.

9. Maintenance of membrane structure may be especially important in intact cells, as they may be directly involved in energy production as well as maintaining separate compartments that prevent undesirable interactions of the contents of the various compartments.

REFERENCES

Abrarov, A. A. (1969). The relation of the respiratory systems of plants during drought. *Biol. Abstr.* **49**, 74523.

Acker, L. W. (1969). Water activity and enzyme activity. *Food Technol.* **23**, 27.

Albo, J. (1968). The evolution of oxidizing enzymes in tobacco from harvesting to fermentation. *Biol. Abstr.* **49**, 21040.

Ashwell, G. (1961). D-Xylulokinase from calf liver. *Methods Enzymol.* **5**, 208.

Barnett, N. M., and Naylor, A. W. (1966). Amino acid and protein metabolism in bermuda grass during water stress. *Plant Physiol.* **41**, 1222.

Boyer, J. S., and Bowen, B. L. (1970). Inhibition of oxygen evolution in chloroplasts isolated from leaves with low water potentials. *Plant Physiol.* **45**, 612.

Brosteaux, J., and Eriksson-Quensel, I. (1935). Etude sur la dessication des protéines. *Arch. Phys. Biol.* **12**, 209.

Brunori, A. (1967). Relationship between DNA synthesis and water content during ripening of *Vicia faba* seed. *Caryologia* **20**, 333.

Bull, H. B., and Breese, K. (1968). Protein hydration. I. Binding sites. *Arch. Biochem. Biophys.* **128**, 488.

Bull, H. B., and Breese, K. (1970). Water and solute binding by proteins. I. Electrolytes. *Arch. Biochem. Biophys.* **137**, 299.

Bullock, K., and Lightbown, J. W. (1942). Spray drying of pharmaceutical products. I. Infusions and extracts. *Quart. J. Pharm. Pharmacol.* **15**, 228.

Burton, K. (1955). D-amino acid oxidase from kidney. *Methods Enzymol.* **2**, 199.

Caldwell, M. L., Adams, M., Kung, J. T., and Toralballa, G. C. (1952) Crystalline pancreatic amylase. II. Improved method for its preparation from hog pancreas glands and additional studies of its properties. *J. Amer. Chem. Soc.* **74**, 4033.

Chen, D., Sarid, S., and Katchalski, E. (1968a). Studies on the nature of mRNA in germinating wheat embryos. *Proc. Nat. Acad. Sci. U. S.* **60**, 902.

Chen, D., Sarid, S., and Katchalski, E. (1968b). The role of water stress in the inactivation of messenger RNA of germinating wheat embryos. *Proc. Nat. Acad. Sci. U. S.* **61**, 1378.

Chilson, O. P., Costello, L. A., and Kaplan, N. O. (1965). Effect of freezing on enzymes. *Fed. Proc., Fed. Amer. Soc. Exp. Biol.* **24**, S-55.

Christophersen, J., and Precht, H. (1952). Untersuchungen zum Problem der Hitzeresistenz. II. Untersuchungen an Hefenzellen. *Biol. Zentralbl.* **71**, 585.

Crafts, A. S. (1968). Water deficits and physiological processes. *In* "Water Deficits and Plant Growth" (T. T. Kozlowski, ed.). Vol. 2, p. 85. Academic Press, New York.

Danielli, J. F., and Davies, F. T. (1951). Reactions at interfaces in relation to biological problems. *Advan. Enzymol.* **11**, 35.

Darbyshire, B. (1971). The effect of water stress on indoleacetic acid oxidase in pea plants. *Plant Physiol.* **47**, 65.

Dietrich, F. M. (1962). Inactivation of egg-white lysozyme by ultrasonic waves and protective effect of amino acids. *Nature (London)* **195**, 146.

Dove, L. D. (1967). Ribonuclease activity of stressed tomato leaflets. *Plant Physiol.* **42**, 1176.

Duisberg, P. C. (1952). Some relationships between xerophytism and the content of resin nordihydroguaiaretic acid and protein of *Larrea divaricata Cav. Plant Physiol.* **27**, 769.

Ewart, A. J. (1897). On the power of withstanding desiccation in plants. *Liverpool Biol. Soc. Proc. Trans.* **11**, 151.

Fischer, R. (1947). Verhütüng der Hitzekoagulation von Serumproteinen durch Zucker. *Experientia* **3**, 29.

Fisher, H. F. (1965). An upper limit to the amount of hydration of a protein molecule. A corollary to the "limiting law on protein structure." *Biochim. Biophys. Acta* **109**, 544.

Flowers, T. J., and Hanson, J. B. (1969). The effect of reduced water potential on soybean mitochondria. *Plant Physiol.* **44**, 939.

Fruchter, R. G., and Crestfield, A. M. (1965). Preparation and properties of two active forms of ribonuclease dimer. *J. Biol. Chem.* **240**, 3868.

Gaff, D. F. (1966). The sulfhydryl-disulfide hypothesis in relation to desiccation injury of cabbage leaves. *Aust. J. Biol. Sci.* **19**, 291.

Gates, C. T., and Bonner, J. (1959). Response of the young tomato plant to a brief period of water shortage. IV. Effects of water stress on the RNA metabolism of tomato leaves. *Plant Physiol.* **34**, 49.

Genkel, P. A., and Pronina, N. D. (1968). Factors underlying dehydration resistance of Poikiloxerophytes. *Fiziol. Rast.* **15**, 68.

Genkel, P. A., and Pronina, N. D. (1969). Anabiosis with desiccation of the Poikiloxerophytic flowering plant *Myrothamnus flabellifolia*. *Fiziol. Rast.* **16**, 745.

Genkel, P. A., Satarova, N. A., and Tvorus, E. K. (1967). Effect of drought on protein synthesis and the state of ribosomes in plants. *Fiziol. Rast.* **14**, 754.

Giles, C. H., and McKay, R. B. (1962). Studies in hydrogen bond formation. XI. Reactions between a variety of carbohydrates and proteins in aqueous solutions. *J. Biol. Chem.* **237**, 3388.

Goodin, R., and Levitt, J. (1970). The cryoaggregation of bovine serum albumin. *Cryobiology* **6**, 333.

Greiff, D., and Myers, M. (1963). Oxidative phosphorylation by suspensions of mitochondria following freezing and drying by sublimation in *vacuo*. *Biochim. Biophys. Acta* **78**, 45.

Greiff, D., and Myers, M. (1964). Enzymatic activities of mitochondria following freezing and freeze-drying. In "Aspects théoriques et industriels de la lyophilisation" (L. Rey, ed.), p. 351. Hermann, Paris.

Greiff, D., and Rightsel, W. (1966). Recent research in freezing, drying, and kinetics of thermal degradation of viruses. In "Advances in Freeze-Drying" (L. Rey, ed.), p. 103. Hermann, Paris.

Harrison, J. S., and Trevelyan, W. E. (1963). Phospholipid breakdown in baker's yeast during drying. *Nature (London)* **200**, 1189.

Hardt, C. R., Huddleson, I. F., and Ball, C. D. (1946). An electrophoretic analysis of changes produced in blood serum and plasma proteins by heat in the presence of sugars. *J. Biol. Chem.* **163**, 211.

Harvey, R. B. (1918). Hardening processes in plants and developments from frost injury. *J. Agr. Res.* **15**, 83.

Healy, J. W., Stollar, D., and Levine, L. (1963). Preparation of lamb brain phosphodiesterase. *Methods Enzymol.* **6**, 49.

Heber, U. (1968). Freezing injury in relation to loss of enzyme activities and protection against freezing. *Cryobiology* **5**, 188.

Heber, U., and Santarius, K. A. (1964). Loss of adenosine triphosphate synthesis caused by freezing and its relationship to frost hardiness problems. *Plant Physiol.* **39**, 712.

Hechter, O. (1965). Role of water structure in the molecular organization of cell membranes. *Fed. Proc., Fed. Amer. Soc. Exp. Biol.* **24**, S-91.

Henckel, P. A. (1964). Physiology of plants under drought. *Annu. Rev. Plant Physiol.* **15**, 363.

Hsiao, T. C. (1970). Rapid changes in levels of polyribosomes in *Zea mays* in response to water stress. *Plant Physiol.* **46**, 281.

Huffaker, R. C., Radin, T., Kleinkopf, G. E., and Cox, E. L. (1970). Effects of mild water stress on enzymes of nitrate assimilation and of the carboxylative phase of photosynthesis in barley. *Crop Sci.* **10**, 471.

Iljin, W. S. (1957). Drought resistance in plants and physiological processes. *Annu. Rev. Plant Physiol.* **8**, 257.

Ivanova, A. P. (1969). The effect of drought on the available water and nucleic acid composition of wheat leaves. *Biol. Abstr.* **49**, 64004.

James, L. K., and Augenstein, L. G. (1966). Adsorption of enzymes at interfaces: Film formation and the effect on activity. *Advan. Enzymol.* **28**, 1.

Jencks, W. P. (1965). Water structure and protein denaturation. *Fed. Proc., Fed. Amer. Soc. Exp. Biol.* **24**, S-50.

Joly, M. (1965). "A Physico-chemical Approach to the Denaturation of Proteins." Academic Press, New York.

Jones, H. A. (1920). Physiological study of maple seeds. *Bot. Gaz.* **69**, 127.

Kawashima, N., Imai, A., and Tamaki, E. (1967). Studies on protein metabolism in higher plants. IV. Changes in protein components in detached leaves of tobacco. *Plant Cell Physiol.* **8**, 595.

Kawashima, N., Fukushima, A., Imai, A., and Tamaki, E. (1968). Studies on protein metabolism in higher plants. V. Some properties of a tobacco leaf protease increasing during curing. *Agr. Biol. Chem.* **32**, 1141.

Kemble, A. R., and MacPherson, H. T. (1954). Liberation of amino acids in perennial rye grass during wilting. *Biochem. J.* **58**, 46.

Kessler, B. (1961). Nucleic acids as factors in drought resistance of higher plants. *Recent Advan. Bot.* **2**, 1153.

Kessler, B., and Frank-Tishel, J. (1962). Dehydration-induced synthesis of nucleic acids and changing of composition of ribonucleic acid: A possible protective reaction in drought-resistant plants. *Nature (London)* **196**, 542.

Khokhlova, L. P., Alekseeva, V. Y., and Murav'eva, A. S. (1969). Influence of weak dehydration of the properties of amino acid-activating enzymes of wheat leaves. *Fiziol. Rast.* **16**, 54.

Klotz, I. M. (1965). Role of water structure in macromolecules. *Fed. Proc., Fed. Amer. Soc. Exp. Biol.* **24**, S-24.

Kuntz, I. D., Jr., Brassfield, T. S., Law, G. D., and Purcell, G. V. (1969). Hydration of macromolecules. *Science* **163**, 1329.

Lamar, E. W., and Todd, G. W. (1968). Unpublished observations.

Lea, C. H., and Hawke, J. C. (1952). Lipovitellin, 2. The influence of water on the stability of lipovitellin and the effects of freezing and of drying. *Biochem. J.* **52**, 105.

Levin, E. (1950). Dried defatted enzymatic material. *Chem. Abstr.* **44**, 5413i.

Levitt, J. (1956). "The Hardiness of Plants." Academic Press, New York.

Levitt, J. (1962). A sulfhydryl-disulfide hypothesis of frost injury and resistance in plants. *J. Theor. Biol.* **3**, 355.

Levitt, J. (1965). Thiogel—a model system for demonstrating intermolecular disulfide bond formation on freezing. *Cryobiology* **1**, 312.

Levitt, J. (1966). Cryochemistry of plant tissue. *Cryobiology* 3, 243.

Levitt, J., and Dear, J. (1970). The role of membrane proteins in freezing injury and resistance. *Frozen Cell, Ciba Found. Symp.* p. 149.

Linko, P., and Milner, M. (1959). Enzyme activation in wheat grains in relation to water content. Glutamic acid-alanine transaminase and glutamic acid decarboxylase. *Plant Physiol.* 34, 392.

Ljunger, C. (1970). On the nature of the heat resistance of thermophilic bacteria. *Physiol. Plant.* 23, 351.

Lovelock, J. E. (1957). The denaturation of lipid-protein complexes as a cause of damage by freezing. *Proc. Roy. Soc., Ser. B* 147, 427.

Lukicheva, E. L. (1969). The changes in some oxidation-reduction enzymes of spring wheat in drought. *Biol. Abstr.* 50, 83490.

MacPherson, H. T. (1952). Changes in nitrogen distribution in crop conservation. II. Protein breakdown during wilting. *J. Sci. Food Agr.* 3, 365.

Manning, G. B., and Campbell, L. L. (1961). Thermostable α-amylase of *Bacillus sterothermophilus*. *J. Biol. Chem.* 236, 2952.

Masuzawa, M., and Sterling, C. (1968). Gel-water relationships in hydrophilic polymers: Thermodynamics of sorption of water vapor. *J. Appl. Polym. Sci.* 12, 2023.

Mattas, R. E., and Pauli, A. W. (1965). Trends in nitrate reduction and nitrogen fractions in young corn plants during heat and moisture stress. *Crop Sci.* 5, 181.

Mayer, A. M., and Poljakoff-Mayber, A. (1963). "The Germination of Seeds." Macmillan, New York.

Miller, R. J., Bell, D. T., and Koeppe, D. E. (1971). The effects of water stress on some membrane characteristics of corn mitochondria. *Plant Physiol.* 48, 229.

Mills, V. M., and Todd, G. W. (1971). Unpublished data.

Mitchell, P. (1965). Chemiosmotic coupling in oxidative and photosynthetic phosphorylation. *Biol. Rev.* 41, 445.

Morris, C. J., Thompson, J. F., and Johnson, C. M. (1969). Metabolism of glutamic acid and N-acetylglutamic acid in leaf disc and cell-free extracts of higher plants. *Plant Physiol.* 44, 1023.

Morton, R. K. (1951). Phosphomonoesterase of milk. *Methods Enzymol.* 2, 533.

Murphy, K., and Todd, G. W. (1965). Unpublished observations.

Nations, C. (1966). An enzyme garrison in wheat embryos. Ph.D. Thesis, Oklahoma State University, Stillwater.

Némethy, G., and Scheraga, H. A. (1962). The structure of water and hydrophobic bonding in proteins. III. The thermodynamic properties of hydrophobic bonds in proteins. *J. Phys. Chem.* 66, 1773.

Newton, R., and Martin, W. M. (1930). Physico-chemical studies on the nature of drought resistance in crop plants. *Can. J. Res.* 3, 336.

Nifontova, M. G. (1968). Aftereffect of dehydration and high temperatures of photosynthesis of lichens. *Chem. Abstr.* 68, 102563g.

Nir, I., and Poljakoff-Mayber, A. (1967). Effect of water stress on the photochemical activity of chloroplasts. *Nature (London)* 213, 418.

Nir, I., Poljakoff-Mayber, A., and Klein, S. (1970). The effect of water stress on mitochondria of root cells. *Plant Physiol.* 45, 173.

Noda, L., and Kuby, S. A. (1963). Myokinase, ATP-AMP transphosphorylase. *Methods Enzymol.* 6, 223.

Noda, L., Kuby, S. A., and Lardy, H. (1955). ATP-creatine transphosphorylase. *Methods Enzymol.* **2**, 605.

Nutritional Biochemicals Co. (1969). Technical Specifications for NBC Enzymes. Cleveland, Ohio. 64 pp.

Okunuki, K. (1961). Denaturation and inactivation of enzyme proteins. *Advan. Enzymol.* **23**, 29.

Oppenheimer, H. R., and Halevy, A. H. (1962). Anabiosis of *Ceterach officinarum* Lan. et DC. *Bull. Res. Counc. Isr. Sect. D* **11**, 127.

Ordin, L. (1960). Effect of water stress on cell metabolism of Avena coleoptile tissue. *Plant Physiol.* **35**, 443.

Paricha, P. C., and Levitt, J. (1967). Enhancement of drought tolerance by applied thiols. *Physiol. Plant.* **20**, 83.

Parker, J. (1968). Drought-resistance mechanisms. *In* "Water Deficits and Plant Growth" (T. T. Kozlowski, ed.), Vol. I, p. 195. Academic Press, New York.

Petinov, N. S., and Malysheva, K. M. (1960). Effect of drought on efficiency of respiration in corn leaves. *Fiziol. Rast.* **7**, 455.

Petrie, A. H. K., and Wood, J. G. (1938). Studies on the nitrogen metabolism of plants. I. The relation between the content of proteins, amino acids, and water in the leaves. *Ann. Bot. (London)* [N. S.] **2**, 33.

Plaut, Z., and Ordin, L. (1964). The effect of moisture tension and nitrogen supply on cell wall metabolism of sunflower leaves. *Physiol. Plant.* **17**, 279.

Popova, Z. N. (1941). Enzyme activity in spring wheat in relation to soil drought. *Chem. Abstr.* **35**, 5538-2.

Prusakova, L. D. (1962). Tryptophane synthesis in wheat leaves in relation to stage of cell development and conditions of water supply. *Fiziol. Rast.* **9**, 353.

Ramgopal, S., and Hsiao, T. C. (1970). Isolation of polyribosomes from green corn leaves: Effect of water stress. *Plant Physiol.* **46S**, 4 (Abstract).

Ridley, E. J., and Todd, G. W. (1966). Anatomical variations in the wheat leaf following internal water stress. *Bot. Gaz.* **127**, 235.

Routley, D. G. (1966). Proline accumulation in wilted ladino clover leaves. *Crop Sci.* **6**, 358.

Salim, M. H., Todd, G. W., and Stutte, C. A. (1969). Evaluation of techniques for measuring drought resistance in cereal seedlings. *Agron. J.* **61**, 182.

Santarius, K. A., and Ernst, R. (1967). Das Verhalten von Hill-Reaktion und Photophosphorylierung isolierter Chloroplasten in Abhängigkeit vom Wassergehalt. I. Wasserentzug mittels Kozentrierter Lösungen. *Planta* **73**, 91.

Santarius, K. A., and Heber, U. (1967). Das Verhalten von Hill-Reaktion und Photophosphorylierung isolierter Chloroplasten in Abhängigkeit vom Wassergehalt. II. Wasserentzug über CaCl₂. *Planta* **73**, 109.

Saunier, R. E., Hull, H. M., and Ehrenreich, J. H. (1968). Aspects of the drought tolerance in creosotebush (*Larrea divaricata*). *Plant Physiol.* **43**, 401.

Savitskaya, N. N. (1967). Accumulation of free proline in barley plants under conditions of water deficiency in the soil. *Fiziol. Rast.* **14**, 737.

Schwimmer, S., and Pardee, A. B. (1953). Principles and procedures in the isolation of enzymes. *Advan. Enzymol.* **14**, 375.

Scott, W. J. (1958). The effect of residual water on the survival of dried bacteria during storage. *Jour. Gen. Microbiol.* **19**, 624.

Sedykh, N. V., and Khokhlova, L. P. (1967). Investigation of the influence of drought on the cytoplasmic protein structure of plant leaves by the dielcometric method of ultra high frequencies. *Fiziol. Rast.* **14**, 429.

Sengupta, S. P., and Todd, G. W. (1970). Unpublished observations.

Shain, Y., and Mayer, A. M. (1968). Activation of enzymes during germination: Amylopectin-1,6-glucosidase in peas. *Physiol. Plant.* **21**, 765.

Shaw, M., and Samborski, D. J. (1957). The physiology of host-parasite relations. III. The pattern of respiration in rusted and mildewed ceral leaves. *Can. J. Bot.* **35**, 389.

Sheen, S. J., and Calvert, J. (1969). Studies on polyphenol content, activities and isozymes of polyphenol oxidase and peroxidase during air-curing in three tobacco types. *Plant Physiol.* **44**, 199.

Shou-ju, O. (1962). A physiological characteristic of drought resistance of certain grape varieties. *Fiziol. Rast.* **9**, 564.

Silman, I. H., and Katchalski, E. (1966). Water-insoluble derivatives of enzymes, antigens, and antibodies. *Annu. Rev. Biochem.* **35**, 873.

Sisakian, N. M. (1937). Prevailing direction of enzymic action as an index of drought resistance in cultivated plants. I. The prevailing direction in drought-resistant and nonresistant strains of wheat. *Biokhimiya* **2**, 687.

Sisakian, N. M., and Kobyakova, A. (1938). Prevailing direction of enzymic action as an index of drought resistance in cultivated plants. II. Prevailing direction of protease action in drought-resistant and non-resistant strains of wheat. *Biokhimiya* **3**, 796.

Skujins, J. J., and McLaren, A. D. (1967). Enzyme reaction rates at limited water activities. *Science* **158**, 1569.

Smillie, R. M. (1963). Formation and function of soluble proteins in chloroplasts. *Can. J. Bot.* **41**, 123.

Spoehr, H. A., and Milner, H. W. (1939). Starch dissolution and amylolytic activity in leaves. *Proc. Amer. Phil. Soc.* **81**, 37.

Stutte, C. A., and Todd, G. W. (1968). Ribonucleotide compositional changes in wheat leaves caused by water stress. *Crop Sci.* **8**, 319.

Stutte, C. A., and Todd, G. W. (1969). Some enzyme and protein changes associated with water stress in wheat leaves. *Crop Sci.* **9**, 510.

Subbotina, N. V. (1959). Influence of wilting on redox conditions in plants. *Fiziol. Rast.* **6**, 39.

Takaoki, T. (1968). Relation between drought tolerance and aging in higher plants. II. Some enzyme activities. *Bot. Mag.* **81**, 297.

Tanford, C. (1970). Protein denaturation. Part C. Theoretical models for the mechanism of denaturation. *Advan. Protein Chem.* **24**, 1.

Tappel, A. L. (1966). Effects of low temperatures and freezing on enzymes and enzyme systems. *In* "Cryobiology" (H. T. Meryman, ed.), p. 163. Academic Press, New York.

Thompson, J. F., Stewart, C. R., and Morris, C. J. (1966). Changes in amino acid content of excised leaves during incubation. I. Effect of water content of leaves and atmospheric oxygen level. *Plant Physiol.* **41**, 1578.

Todd, G. W. (1969). Unpublished study.

Todd, G. W., and Basler, E. (1965). Fate of various protoplasmic constituents in droughted wheat plants. *Phyton (Buenos Aires)* **22**, 79.

Todd, G. W., and Yoo, B. Y. (1964). Enzymatic changes in detached wheat leaves as affected by water stress. *Phyton (Buenos Aires)* **21**, 61.

Tsai, S., and Todd, G. W. (1969). Unpublished data.

Tyankova, L. A. (1967). Influence of proline on the sensitivity of wheat plants to drought. *Chem. Abstr.* **66**, 10090z.

Umbreit, W. W., Burris, R. H., and Stauffer, J. F. (1949). "Manometric Techniques and Tissue Metabolism." Burgess, Minneapolis, Minnesota.

Uphof, J. G. T. (1920). Physiological anatomy of xerophytic Selaginellas. *New Phytol.* **19**, 101.

van den Berg, L. (1959). The effect of additions of sodium and potassium chloride to the reciprocal system: KH_2PO_4—Na_2HPO_4—H_2O on pH and composition during freezing. *Arch. Biochem. Biophys.* **84**, 305.

van den Berg, L., and Rose, D. (1959). Effect of freezing on the pH and composition of sodium and potassium phosphate solutions: The reciprocal system KH_2PO_4—$Na_2HPO_4 \cdot H_2O$. *Arch. Biochem. Biophys.* **81**, 319.

Webb, S. J. (1965). "Bound Water in Biological Integrity." Thomas, Springfield, Illinois.

Webb, S. J., and Walker, J. L. (1968). The influence of cell water content on the inactivation of RNA by partial desiccation and ultraviolet light. *Can. J. Microbiol.* **14**, 565.

Weston, T. J. (1968). Biochemical characteristics of tobacco leaves during flue-curing. *Phytochemistry* **7**, 921.

Whitaker, J. R., and Tappel, A. L. (1962). Modification of enzyme activity. II. Effect of salts on α-amylase, alcohol dehydrogenase, peroxidase and hematin catalysis. *Biochim. Biophys. Acta* **62**, 310.

Wilson, A. M. (1970). Incorporation of ^{32}P in seeds at low water potentials. *Plant Physiol.* **45**, 524.

Yi, C., and Todd, G. W. (1970). Unpublished observations.

Zholkevich, K. A., Prusakova, L. D., and Lizandr, A. A. (1958). Translocation of assimilants and respiration of conducting pathways in relation to soil moisture. *Fiziol. Rast.* **5**, 333.

Zholkevich, V. N., and Rogacheva, A. Y. (1968). P/O ratio in mitochondria isolated from wilting plant tissues. *Fiziol. Rast.* **15**, 450.

CHAPTER 6

WATER DEFICITS AND NUTRIENT AVAILABILITY*

Frank G. Viets, Jr.

SOIL AND WATER CONSERVATION RESEARCH DIVISION
AGRICULTURAL RESEARCH SERVICE, U.S.D.A., FORT COLLINS, COLORADO

I. INTRODUCTION

Although nutrient and water absorption are independent processes in the root, the necessity for available water in both the plant and soil for growth and nutrient transport makes them intimately related. This close relationship makes it difficult to clearly define the effects of drought on nutrition.

Nutrients are the 13 mineral elements generally accepted as essential for plant growth and reproduction. They are generally absorbed as ions, and water is the transporting medium. When they are present in the plant in sufficient amounts and their absorption rate is sufficient to meet the

* Contribution from the Northern Plains Branch, Soil and Water Conservation Research Division, Agricultural Research Service, USDA, Fort Collins, Colorado.
† Soil Scientist and Soil Fertility Investigations Leader, USDA, Fort Collins, Colorado.

needs for plant growth, we do not pay much attention to them, but when a deficiency limits growth or nutritional value, or an excess is toxic to the plant or to the animals that eat them, then we do become concerned. The International Soil Science Society and the Soil Science Society of America define nutrient availability as the actual plant uptake (plant content). However, various indexes of nutrient availability, such as the amount of an element extractable from soil or even the amount of fertilizer applied, are often confused with nutrient availability.

Drought is a general term meaning a sustained period of significantly subnormal water or soil moisture supply. It has different meanings to agriculturists, meteorologists, and hydrologists as discussed by Hoffman and Rantz (1968). To the hydrologist it means a period of either short or long duration having subnormal streamflow. To the ecologist it may mean a period of either long or short duration that determines the ability of a plant species to survive competition and reproduce itself in a community of other plant competitors.

To the farmer it may mean a short period of water deficiency when his newly seeded crop germinates but fails to survive, or a period of hot, dry weather when his corn is tasseling and fails to set kernels. To the plant physiologist a drought may be a chronic water deficit alleviated by only brief rains, which allows the perennial plant to accomplish its important functions and then survive in a state of inactivity. Drought may mean a short period of acute water deficit and loss of turgor resulting in permanent damage to leaves and death. Another kind of drought affecting root activity can be distinguished in which the plant is getting its water from one part of the soil deficient in nutrients, but the soil's main supply of nutrients is locked up in dry soil lacking active roots. Hence, drought is a somewhat general term, the specific meaning depending on the circumstances.

II. THE IMPORTANCE OF WATER TO NUTRITION

In soils, water in the suction range of about 0.1–10 bars is essential for every process promoting nutrient availability, from the reactions affecting their concentration in the soil solution through their transport by diffusion and mass flow to the root surface, to their absorption by the root. The root's ability to absorb them, in turn, is affected by its capacity for absorption, its ability to translocate them from the root to the leaves, and its capacity to extend its root system to more distant points of supply. Only some of the more pertinent points are discussed here. The reader is referred to comprehensive reviews (Viets, 1967; S. R. Olsen and Kemper, 1968; Barber, 1962; Barley, 1970).

In drought, we are concerned only with the drier end of the available water scale when the plant is adapting successfully to water stress or suffering the consequences. Nutrient availability for most nonhydrophytes is highest when soil water suction is near field capacity. Field capacity depends on the site situation, but generally is defined as the water the soil will hold against percolation after several days of drainage. Field capacity may range from 0.1 to 0.3 bar of suction. Soil water content near field capacity allows for the best combination of sufficient air space for oxygen diffusion, the greatest amount of nutrient in soluble form, the greatest cross-sectional area for diffusion of ions and mass flow of water, and favorable conditions for root extension. Lesser amounts of water result in progressively poorer conditions for nutrient availability. Because of the experimental difficulties of maintaining constant water suction at suctions much above $\frac{1}{3}$ bar with transpiring plants, definitive information is lacking on how much nutrient availability is impaired by progressively drier soil conditions. Whether the reduced nutrient uptake is of significance is another important question that is not settled, but some significant observations will be discussed.

III. WATER DEFICIT AND NUTRIENT TRANSPORT TO THE ROOT

The quantity of water in the soil affects not only the amount (concentration times volume) of nutrient in the soil solution, but the rate of movement to the root by diffusion and flow in the water (mass flow) as water is absorbed by the root. The water content of a soil expressed on a volumetric basis can fluctuate through an eightfold range for sands and nearly a twofold range for clays between the wilting percentage (15 bars suction) and saturation. Between the wilting percentage and field capacity, the water content doubles for many soils. These changes depend on the water-retention characteristics of the soil and the stratification of the soil horizons. Xerophytes and other plants in situations where they can get most of their water from deeper depths can reduce water content to below wilting percentage in part of the soil. Furr and Reeve (1945) define the first wilting point as the lower limit of water available for vegetative growth, and the ultimate wilting point as the lower limit for maintaining life. These limits define the wilting range. Haise et al. (1955) define a "minimum point" to which the upper portion of the soil profile may dry when the crop is getting its water from lower depths, as occurs in the U. S. Great Plains in rainless periods when wheat is maturing. They show that the minimum point corresponds to 26 bars of water suction. So little water is held between the permanent wilting point and either the ultimate wilting point or the minimum point that it is of little significance for nutrient availability, although it may be sufficient to help plants survive a drought or help to mature a grain crop.

Very little information exists on the changes in concentration of essential nutrient ions in the soil solution as the soil dries from field capacity to the wilting range by evaporation or water extraction by roots. We are considering now the soil solution as a strictly chemical system, ignoring ion absorption by roots and the effects of water on microorganisms that may change the concentration of ions in the solution. Nutrients that occur mostly or entirely in the soil solution are ones not absorbed or only very weakly absorbed by the mineral and organic fractions of the soil. They are NO_3^-, Cl^-, and SO_4^{2-} (in soils lacking gypsum). Theoretically, the concentration of these ions should double as the volume percent of water in the soil declines by half. Most of the other ions, such as K, Ca, and phosphate species, are in equilibrium adsorption with solid-phase minerals and organic matter, so that the solution concentration may not change much during soil drying over the range of water contents of significance to plants. Reitemeier and Richards (1944) showed that the concentrations of K, Mg, Na, and Cl in pressure membrane extracts of a soil, extracted at water contents of 2.85, 1.78, and 1.24 times the 15-bar percentage, increased with a decrease in the initial water content.

Whether the change in concentration per se of the various ions in the soil solution on drying is really of much significance to nutrient uptake is not known. The root extracts nutrients from the solution and from the labile pool of nutrients readily exchangeable with the ions in solution. Direct exchange of a nutrient between the solid phase and root surface has been postulated and supported with evidence (Jenny and Overstreet, 1939), only to be questioned by others (Dean and Rubins, 1945; Lagerwerff, 1958; R. A. Olsen and Peech, 1960).

The water content of the soil in relation to the amount of nutrient in solution may be of most significance for ions such as nitrate, and possibly sulfate, whose source in the absence of applied fertilizers is microbial decomposition of organic matter. Several workers (Robinson, 1957; Miller and Johnson, 1964; Reichman et al., 1966) have shown that ammonification of organic matter and nitrification decrease proportionally to decreases in soil water content over the range from field capacity to the wilting point. Figures 1 and 2 illustrate the effect of constant soil moisture content on the microbial release of total soluble inorganic N (ammonia plus nitrate) and subsequent nitrification from the soil organic matter of two Northern Great Plains soils. For the Parshall soil, a Chestnut, 13.2% of saturation represents 100 bars water suction and 52.3% represents 0.2 bar. For the Gardena, a Chernozem, 13.2% of saturation is 100 bars and 48.1% is 0.2 bar. The increase in rate of organic matter mineralization and sub-

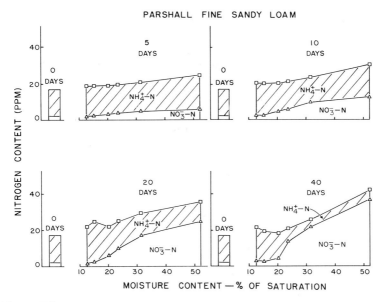

Fig. 1. Effect of constant soil water content on microbial release of mineral N from organic matter of a Parshall soil. From Reichman *et al.* (1966).

sequent nitrification of the released ammonia with increase in water content up to 0.2 bar is apparent. Dommergues (1961) reported that organic matter mineralization to NH_4^+ and NO_3^- started at pF 4.8 to 5.2 (pF is negative log of suction in centimeters of H_2O) in tropical black soils, but that little mineralization occurred in ferrolitic red soils at such low water contents.

Water content of soil affects nutrient transport to the root surface by affecting the rate of diffusion and the mass flow of water to the root. In the latter, ions in solution are simply swept along in the water flow. Ions such as nitrate that are entirely soluble can be completely extracted from a soil by mass flow. Barber *et al.* (1963) state that mass flow can account for most of the transport of Ca, Mg, and N, but is inadequate to account for much of the transport of P and K. Whether transport of nutrients by mass flow is reduced by increasing water deficit depends on the effect of the increasing water suction on the rate of water absorption. Diffusion rate is strongly water-suction dependent because drying diminishes the thickness of water films and the effective water-filled cross-sectional area of the soil, and increases the tortuosity of the path. S. R. Olsen *et al.* (1965) showed that the diffusion coefficient for phosphate decreased about eightfold for a twofold decrease in the volumetric water content for two soils (see Fig.

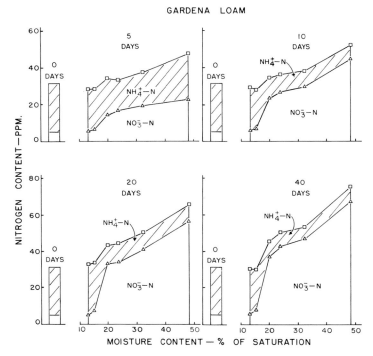

Fig. 2. Same kind of data as in Fig. 1 for a Gardena soil containing 2.2 times more total N than the Parshall. From Reichman *et al.* (1966).

3). The apparent diffusion coefficients of Rb (Place and Barber, 1964) and molybdate (Lavy and Barber, 1964) depended on the water content of the soil. Rb is used as a tracer for K in many ion absorption studies on plant roots because of the plant's inability to distinguish between them.

Water content appears to be very important in nutrient transport to the root, but probably just as important is its effect on rate of root extension. Rates of root extension may be 5 cm per day, but the distances of ion diffusion are only 1–5 mm, even over much longer periods. Rate of root elongation is highly dependent on available water because root growth is in part a hydration process (Gingrich and Russell, 1956, 1957). Peters (1957) showed that elongation of corn roots was favored by low soil water suction and high water content at a given suction (see Fig. 4). Increase in soil water suction also increases the shearing strength of soils and may increase the mechanical impedance to root extension, particularly in dense soils (Barley, 1963).

Through the combination of a number of effects, the most important

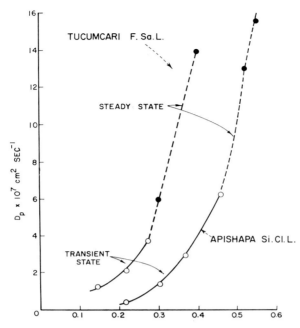

Fig. 3. Porous system self-diffusion coefficient for P in two soils as affected by volumetric water content. From S. R. Olsen *et al.* (1965).

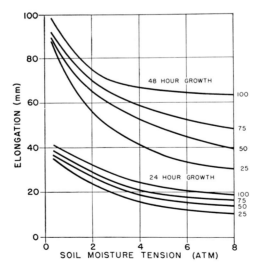

Fig. 4. Elongation of corn roots as a function of soil water suction. The numbers on the right are the percentages of clay soil in a soil–sand mixture. From Peters (1957).

being the effects of water deficit on transport to the root and on root extension, the plant in dry soil or with a part of its roots in dry soil should have more difficulty absorbing nutrients than one growing in a better watered soil. Most of the experimental data support this conclusion. One of the problems of getting information on nutrient uptake in relation to water availability is the problem of maintaining constant soil water suction at suctions of $\frac{1}{3}$ bar or more. A number of novel approaches have been used that involved either nontranspiring seedlings or split-root or split-medium techniques.

IV. NUTRIENT ABSORPTION AT CONTROLLED SOIL WATER SUCTION

Danielson and Russell (1957) studied the 24-hr uptake of ^{86}Rb by corn seedlings in closed boxes over a soil water-suction range of $\frac{1}{3}$ to 12 bars. Uptake of the isotope and water content of the seedling decreased almost logarithmically as suction increased, with uptake decreasing sharply at suctions above 1 bar. Uptake of Rb was proportional to water content of the soil over the whole suction range. S. R. Olsen et al. (1961), using the same technique, showed that ^{31}P uptake by corn was proportional to water content in each of four soils (see Figs. 5 and 6). Other investigators have used split-medium techniques in which part of the root system was in

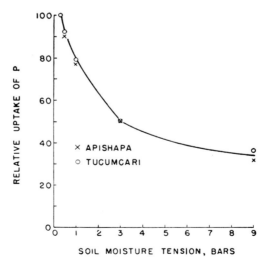

Fig. 5. Relative uptake of P by corn seedlings in relation to soil moisture tension in two soils. From S. R. Olsen et al. (1961).

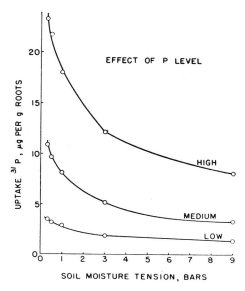

Fig. 6. Effect of P fertility level and soil moisture tension on P uptake by corn seedlings. From S. R. Olsen *et al.* (1961).

soil or sand with high water content to supply most of the plant's water needs, and the remainder of the root system was in soil containing nutrients and water at a definite suction. Thus, Mederski and Wilson (1960) found that corn grown for 25 days took most of its water from the upper sand layer and but little water from the soil below separated from the sand by an impermeable barrier. They found that the dry weight of tops and roots increased linearly with increasing soil water content as did the uptake and plant percentage of P, K, and Mg. For example, when the soil initially contained 12.1% water by weight (15 bar percentage was 6.4%), top dry weight of plants was about 2.2 gm and they contained 0.13% P and 2.6% K. When the plants were grown in soil with 20.8% water initially (⅓ bar percentage of the Wooster silt loam was 25%), top dry weight was 2.9 gm containing 0.21% P and 3.8% K. These data are for plants grown at normal air humidity that averaged about 50%, but fluctuated between 35 and 85%. During the experiment the soil with initial 12.1% water dropped to 8.6% and the 20.8% treatment dropped to 14.6%. This experimental technique does not ensure constant water suction or content, but certainly minimizes the changes during a 25-day period. Hobbs and Bertramson (1950) divided the root systems of tomatoes longitudinally and put half in surface soil and half in quartz sand or subsoil. Little uptake of native or fertilizer B occurred from the surface soil if it was kept dry and the plant got its water from the subsoil or the sand. Boron-deficiency symptoms were induced by keeping the surface soil dry, but no deficiency symptoms occurred if the surface soil was watered.

Meyer and Gingrich (1966) found that osmotic stress obtained by use of colloidal suspensions of Carbowax applied to half the root system of wheat reduced the rate of P and N uptake, resulting in declines in N and P contents of the whole plant within 24 hr. Dry matter production was not decreased under their conditions of low evapotranspiration.

Most studies show that root penetration and nutrient uptake are minimal from soils below the wilting point. In such experiments there is always the probability of water transfer from wet soil through the root to dry soil surrounding the root, as pointed out by Breazeale (1930) and confirmed by Volk (1947) and Hunter and Kelley (1946). Under such conditions, uptake of K (Breazeale, 1930) and of K and N (Volk, 1947) from soils initially below the wilting percentage were shown. Hunter and Kelley (1946), Volk (1947), and Boatwright *et al.* (1964) obtained no significant uptake of P from dry soils. Hunter and Kelley (1946) showed that guayule (*Parthenium argentatum*) could absorb ^{32}P from moist soil, 122 cm from the surface, and translocate it to the top through 122 cm of soil that was at or below the wilting percentage.

V. NUTRIENT AVAILABILITY AS AFFECTED BY THE SOIL WATER REGIME

A plant growing in the field is usually subject to fluctuations in water availability that may range from soil saturation to extreme drought. Continued removal of water from the soil by evapotranspiration produces a time-linear decrease in water content and a logarithmic increase in soil water suction until rain or irrigation restores part or all of the root zone to suctions above field capacity. Thus, the plant lives in a water suction regime in which the water availability fluctuates between 100% and some fraction of it. Complicating the picture is that various parts of the root zone usually have different suctions and water contents. Because constant suction or water availability cannot be maintained at suctions much below field capacity because of the slow movement of water in soil, much of our information on nutrient availability in relation to water must come from water regime experiments in which watering frequency and amount are varied so that one degree of depletion can be compared to another; e.g., 25 vs 75% depletion or 1 vs 16 bars suction at some depth. Such experiments are difficult to interpret because much of the nutrient uptake may occur soon after the soil is rewetted. In this brief period, the plant may be able to meet all of its nutrient requirements, provided the soil nutrient pool is adequate. No one has successfully measured the rate of nutrient uptake as a continuous function of the soil water availability. Nevertheless, data from comparisons of nutrient uptake in different soil water regimes are useful.

At this point, I should comment on two measures of nutrient availability often used. One is the total uptake of the element, e.g., K or N. If drought or water stress decreases growth and total nutrient content also decrease, then by definition drought also decreases nutrient availability. The other measure is the percentage composition. Thus, an increase in percentage composition of the plant with increasing stress may be considered to be nutrient absorption at a faster rate than increase in dry weight. Water stress is inhibiting growth more than nutrient uptake. A decrease in percentage composition means that net assimilation and growth have not been affected as much as nutrient availability. Both concentration and total uptake data are needed in making complete interpretations. Moisture regime experiments have produced much conflicting evidence on the effect of water stress on nutrient availability. The major elements (N, P, and K) and secondary elements (Ca, Mg, and S) have received most attention. Richards and Wadleigh (1952) summarized the existing data on nutrient availability in relation to soil water availability and concluded that decreasing water supply produced a definite increase in N concentration, a definite decrease in K concentration, and variable effects on the P, Ca, and Mg concentrations in the plant.

Some of the better work on water supply in relation to nutrient availability has been done with fertilizers tagged with radioisotopes. One precaution in interpretation is that if the wetter water regime favors more growth, then a greater proportion of fertilizer nutrient as compared to native soil nutrient is absorbed, so that the effect of moisture regime on nutrient availability may be overemphasized (Grunes, 1959). Haddock (1952) found with ^{32}P-tagged phosphate placed 10 cm deep, and 15 cm to the side of the sugar beet row at thinning, that less than 24% of the P in the leaves came from the fertilizer on a wet regime but more than 40% came from the fertilizer on a dry one. Presumably, the availability of the soil P was reduced in the drier soil and the plants took more of their P from the zone of higher concentration of fertilizer P when it became wet at each irrigation. If the fertilizer was broadcast, the water regime did not affect the ratio of soil P to fertilizer P absorbed by the beet. Power et al. (1961) found that the soil water regime had a considerable effect on the ratio of fertilizer P to soil P taken up by spring wheat, as shown in Fig. 7. The four water regimes were two amounts of soil water in the profile at seeding in factorial combination with two amounts of growing season water as

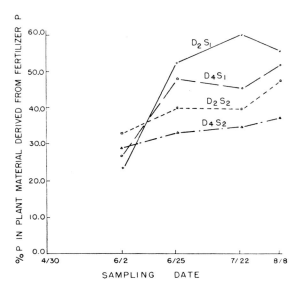

Fig. 7. Percent P at various stages of growth in spring wheat plants derived from 11 lb of P in concentrated superphosphate tagged with ^{32}P banded with the seed under four moisture conditions in the field. D_2, upper 20 in. of soil moist; D_1, upper 48 in. of soil moist; S_1, 3.34 in. of growing season precipitation; S_2, 6.80 in. of growing season precipitation and supplemental irrigation. From Power *et al.* (1961).

natural rainfall with and without supplemental irrigation. Water regime did not affect the absolute amount of P taken up from the ^{32}P-tagged fertilizer, but there was more soil P taken up the wetter the regime (Fig. 8). Lipps *et al.* (1957) investigated the uptake of ^{32}P placed at different depths in the soil by subirrigated alfalfa. The water table was about 240 cm deep. Up to late June, when the whole profile contained available water maintained by frequent rains, ^{32}P-uptake from the upper 46 cm of profile was very active and there was practically no uptake from deeper depths. In July and August, when available water in the upper layers became depleted, uptake of ^{32}P from the deeper, wetter layers became very active. Jenne *et al.* (1958) compared the growth and nutrient uptake of corn that was irrigated and corn that was depleting the soil moisture supply in western Nebraska. All plots were near field capacity to a depth of 78 in. at planting time, as shown in Fig. 9. Differences in dry matter accumulation and total accumulation of N, P, and Mg became apparent by July 28 when the irrigated plots were tasseling. In the next week, accumulation of K and Ca on the dry, unirrigated soils declined relative to the irrigated

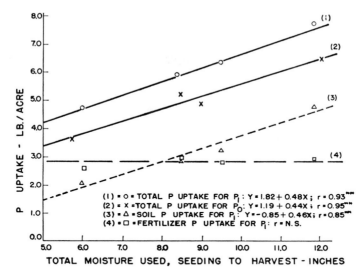

Fig. 8. Influence of available water on total, soil, and fertilizer P uptake by spring wheat. From Power *et al.* (1961).

ones. From August 4 to September 27, when corn was using water from below the 75-cm depth in the dry plots, relative dry matter production was decreased less than the relative N and P accumulation. Magnesium accumulation was affected less than N and P accumulation, whereas Ca accumulation was affected very little.

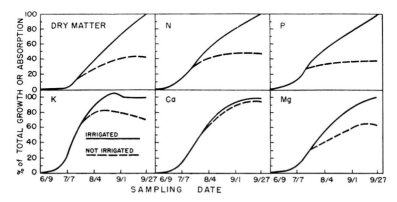

Fig. 9. Relative dry weight and mineral accumulation in tops of corn as affected by water availability. Values obtained on September 27 for corn with adequate water considered to be 100%. From Jenne *et al.* (1958).

VI. THE CHANGING WATER AND NUTRIENT AVAILABILITY PATTERNS IN THE FIELD

Some native vegetation and agricultural crops are grown in wet climates or in site situations in drier climates where water availability is always high. But for most plants, the soil water status and active root distribution change constantly with depth and time. Most soils go through wetting and drying cycles, with the upper layers going through more frequent and wider fluctuations than lower layers. The upper layers are nearest the source of water infiltration and also subject to the most rapid water withdrawal by surface evaporation.

When the whole root zone is near field capacity, the moisture extraction pattern with depth is generally regarded as about proportional to the density of absorbing roots (Gardner, 1966). As water is extracted from the soil surface layers, the plant gets more water from the lower zones, which have lower water suction and higher water availability. Figure 10 illustrates changes in water content of a soil with depth and time after watering of a sorghum crop. Figure 11 shows the calculated and observed average soil suction from the same experiment reported by Gardner (1964). Thus, an established sugarbeet crop in midsummer in the first week after irrigation may get 50% of its water from the first 30 cm, 35%

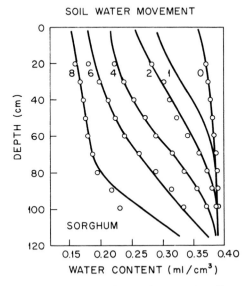

Fig. 10. Observed (circles) and calculated (curves) soil water content of soil under transpiring sorghum after irrigation. Numbers on curves are days after irrigation. From Gardner (1964).

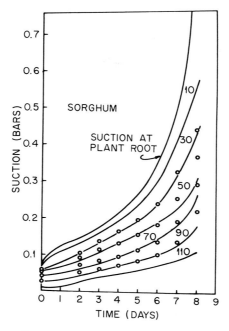

Fig. 11. Average soil suction as a function of time for the same data given in Fig. 10. From Gardner (1964).

from the second, 10% from the third, and only 5% from the fourth 30 cm. However, in the fourth week it may get only 8% from the surface 30 cm, 17% from the second, 30% from the third, and 45% from the fourth 30 cm. The soil water suction may increase to about 12 bars at the 15-cm depth, but to only 5 bars at 150 cm. At the end of 4 weeks, growth may be diminished (Taylor and Haddock, 1956).

The important point is that water availability fluctuates most in the surface soil that usually contains highest concentration of most of the soil and fertilizer nutrients. The concentrations of extractable and exchangeable nutrients are generally higher in the surface soil because of the accumulation of organic matter, and the cycling of nutrients from the subsoil to the surface by plants and earth-dwelling fauna. Fertilizers are generally applied to the surface or incorporated in the plow depth by tillage and remain there by adsorption on clays and organic matter. Only nitrate, sulfate, and chloride are readily leachable. Tables I and II show the distribution with depth of some extractable nutrients for two unfertilized, cultivated soils in former grasslands. For the Keiser clay loam, only the surface horizon contains sufficient N, P, and K to produce much plant growth. For the Pullman soil, only the surface horizon contains enough extractable P to support

TABLE I
PLANT NUTRIENT ESTIMATES OF VARIOUS HORIZONS OF KEISER
CLAY LOAM FROM MONTANA[a]

Depth (in.)	Horizon	NaHCO₃– soluble P₂O₅ (ppm)	Total N (%)	NO₃–N prod. (ppm)[b]	Exchangeable K (mEq/100 gm)
0–8	A$_p$	11.0	0.114	68.3	0.84
8–13	B$_3$	2.2	0.060	12.6	0.48
13–24	C$_{Ca-1}$	2.0	0.049	5.9	0.50
24–36	C$_{Ca-2}$	1.7	0.038	2.1	0.52
36–60	C	2.2	0.032	1.2	0.56

[a] From Reuss and Campbell (1961).
[b] Nitrate N produced by incubation of a soil–vermiculite mixture for 2 weeks.

much plant growth even though the total P in the profile does not show much change with increasing depth. The exceptions to the rule that extractable nutrients decrease with depth occur in desert soils where little vegetation grows and mineral weathering is minimal, and in some highly leached tropical soils in which nutrient availability is very low throughout the profile unless the forest or "bush" has been burned recently.

Since in a drying cycle the plant gets its water from progressively deeper depths where nutrient supply is lower, it can be postulated that plants have lower mineral absorption during dry periods. However, data on simultaneous uptake of water and nutrients for short periods in a drying cycle are not available.

That nutrient uptake is lower when the surface soil is dry is generally substantiated by the better crop response to fertilizers and animal manures

TABLE II
PLANT NUTRIENT ESTIMATES OF VARIOUS HORIZONS OF A PULLMAN
SILTY LOAM FROM TEXAS[a]

Depth (in.)	Horizon	NaHCO₃– soluble P (ppm)	Total P (ppm)	Total N (%)	Exchangeable K (ppm)
0–5	A$_{1p}$	20.5	450	0.106	475
5–8	B$_{21}$	3.9	374	0.105	440
8–13	B$_{22}$	2.0	328	0.082	425
13–19	B$_{22a}$	1.0	344	0.062	450
19–28	B$_{2Ca}$	2.0	322	0.053	475
28–38	B$_{2b1}$	4.2	327	0.049	425
38–56	B$_{2b2}$	—	—	0.035	425
56–65	C$_{Cab}$	—	—	0.021	210

[a] From Eck et al. (1965).

that are placed deep (ca 20 to 25 cm) than shallow or broadcast applications in dry years. In wet years, depth of placement may have little or no effect (see Black, 1966). Patrick *et al.* (1959) found that deep tillage and deep fertilization of four soils with "traffic pans" in Louisiana permitted deeper root growth and use of subsoil moisture during dry years. Very deep or subsoil placement of inorganic fertilizers, however, has often produced yields no better than shallower placements (Jamison and Thornton, 1960; Larson *et al.*, 1960). Fertilizer application to correct nutrient deficiencies promotes deeper root development and more extraction of water from lower depths (Viets, 1962), but under most circumstances, deep placement even of immobile elements like K or P is not essential. In the southeastern United States where subsoils with high acidity, deficiencies of Ca, and toxic concentrations of aluminum restrict root growth, correction of acidity with deep-placed lime would promote root growth, if a way could be found to lime the subsoil economically. Roots are able to penetrate infertile soils, providing they are not acidic and low in Ca, by translocating the needed N, P, and other elements from the tops or surface roots. Calcium appears to be the exception in that it is not translocated to root tips (Pearson, 1966; Rios and Pearson, 1964). McClure and Jackson (1968) claim that a nutrient translocated to a growing root tip may sometimes be less effective in root growth than one adsorbed from the external medium near the root tip.

The conclusion then appears to be justified that most plants do not need a constant rate of nutrient uptake, but do have at least limited capacity to accumulate nutrients from surface layers when they are wet and store them for future use. Boatwright and Viets (1966) found that wheat and intermediate wheat grass grown in solution cultures with adequate P for 5 weeks (up to heading for wheat) and then transferred to P-free solutions produced as much dry matter or grain as plants kept in P-containing solutions until maturity. These species, probably like most others, have the capacity to store P for later use. Whether the plant growing in wet soil has sufficient capability to absorb sufficient P and store it for use during periods when P availability in surface soil is low depends on the concentration of P in the soil solution and the duration and frequency of the wet periods.

VII. THE SIGNIFICANCE OF LOWER NUTRIENT AVAILABILITY ASSOCIATED WITH DROUGHT

The preponderance of evidence indicates that drought decreases nutrient availability to plants as measured by total nutrient uptake and sometimes in reduced concentration. If the nutrient concentration in the plant grown with different water supplies remains constant, but drought restricts growth, then the total nutrient uptake is less and so nutrient availability by definition was less. If nutrient concentration declines, then nutrient availability has been inhibited more than growth. This situation can occur when most of the nutrients are in the surface soil that becomes dry while the plant is getting its water from deeper depths. The accumulation of nitrate in some species under drought conditions, which are generally accompanied with high air temperatures, is an interference with the reductase enzyme system, discussed in Chapter 5. As Slatyer (1969) points out: "The effect of water stress on mineral nutrition is difficult to resolve clearly . . . the key point is whether or not reduced nutrient uptake retards growth and development in a plant under stress."

The classic example of drought-induced deficiency symptoms is that of B in crops, particularly alfalfa, grown on soils with inadequate B supply. The usual explanation is that the available B is contained in the organic matter of the surface soil and its availability is low when the surface soil becomes dry (Hobbs and Bertramson, 1950). Decreases in P concentration associated with dry soils have been noted for white clover (Low and Piper, 1960), coffee (Warden, 1961), and sugarbeets (Haddock, 1952). This behavior in dry soils may be especially significant to forage quality, if the forage is low in P under the best of conditions. Will (1961) reported that Mg-deficiency symptoms in pine seedlings grown in nurseries in two locations in New Zealand were intensified by drought.

Another aspect of the nutrient–water availability complex of great significance, both ecologically and agriculturally, is the broad relationship that exists between the amount of precipitation and the need for and response to fertilizers. In humid areas there is a general need for N, P, and K, and sometimes other elements in agriculture. Native vegetation often responds to fertilization. Some fertilizers are now even being used on forests and on native grasslands. This response is generally due to the greater production of vegetation because of water sufficiency and the high intensity of leaching that removes soluble nutrients from the soil and from dead vegetation. The kind of land use, and even of the natural vegetation, may be controlled by the amount of nutrients available and whether they can be profitably added.

In drier areas of the virgin grasslands of the Great Plains, the response to fertilizers is generally low. Response of crops like wheat has been low but has been increasing as soil organic matter has become depleted and varieties with higher yield potentials have become available. In dry years, N fertilization can decrease small grain yields by increasing the rate of water use so that the plant lacks sufficient available soil water to carry it through periods of stress. Such instances of reduced yields are not common, but they do occur (Viets, 1962). Fertilization of native grasslands in the Great Plains frequently results in significant yield responses and shifts in species composition, but nitrogen fertilization can result in loss of stands during years of drought (Kipple and Retzer, 1959). The long-term effect of N fertilization on the climax vegetation of grasslands is not known, since the experiments have not been carried through enough long-term weather cycles.

In semideserts and deserts, little is known about the response of native vegetation to fertilizers. However, if desert or grassland soils are put under irrigation so that drought does not occur, an immediate need for N fertilizers occurs followed soon by need for P and sometimes other elements. The elements needed will depend on the kind of soil and, to some extent, on the kinds of crops.

There can be no doubt that removal of drought by irrigation or more efficient water management soon (or eventually) results in the need for fertilizers.

Whether the kind of native vegetation in desert and semidesert regions subject to chronic or acute periods of drought is ever controlled by nutrient availability is unknown. Moisture availability and temperature may completely dominate what will survive over a long period of time. Most books on plant ecology mention the fertility of the soil but do not give it a dominant role in controlling what will survive in a dry climate.

The principal effect of salinity in a soil is reduction in water availability. Specific effects of one ion in high concentration on the absorption of others, and toxicities of B, Li, Cl, and Na do occur. However, these effects are not due to reduced water availability. Salty subsoil may restrict root distribution and water availability so that there is little uptake of nutrients from the subsoil. As Bernstein (1962) says, "It is, perhaps, superfluous to point out that fertilization of an extremely saline soil will not be rewarding."

VIII. CONCLUSIONS

"Water deficits generally reduce plant growth. The reasons for such reductions are still not completely known. It is quite clear that much of the reduction in growth is associated with reduction in turgor and cell wall development. However, it appears unlikely that turgor can be considered the key to the general response." Little evidence or reasons justify amending this quotation from Vaadia and Waisel (1967) with the claim that reduced mineral nutrient availability contributes to poor growth under conditions of decreasing water availability in the soil.

Decreasing water availability culminating in plant distress symptoms, popularly associated with drought, results in reduced total nutrient uptake and frequently in reduced concentrations of mineral nutrients in plant tissues. Boron-deficiency symptoms are the only ones generally associated with drought, and then only on soils borderline in B supply.

Available water near field capacity favors microbial and solution reactions in the soil, transport of ions by diffusion, and mass flow to the root, and root development into unexploited soil. All of these processes are inhibited as the soil dries toward the wilting range. The plant can probably absorb and store when soil moisture availability is high all of the minerals it will need during the dry phase of the soil-water cycle when growth is being arrested concurrently.

Although ion and water absorption by the root from solutions are independent, in soil they are intimately and complexly linked because of the dominance of water availability on all microbial, physical, and physiological processes.

REFERENCES

Barber, S. A. (1962). A diffusion and mass-flow concept of soil nutrient availability. *Soil Sci.* 93, 39–49.

Barber, S. A., Walker, J. M., and Vasey, E. H. (1963). Mechanisms for the movement of plant nutrients from the soil and fertilizer to the plant root. *J. Agr. Food Chem.* 11, 204–207.

Barley, K. P. (1963). Influence of soil strength on growth of roots. *Soil Sci.* 96, 175–180.

Barley, K. P. (1970). The configuration of the root system in relation to nutrient uptake. *Advan. Agron.* 22, 159–201.

Bernstein, L. (1962). Salt affected soils and plants. *Arid Zone Res.* 18, 139–174.

Black, C. A. (1966). Crop yields in relation to water supply and soil fertility. *In* "Plant Environment and Efficient Water Use" (W. H. Pierre *et al.*, eds.), pp. 177–206. Amer. Soc. Agron., Madison, Wisconsin.

Boatwright, G. O., and Viets, F. G., Jr. (1966). Phosphorus absorption during various growth stages of spring wheat and intermediate wheat grass. *Agron. J.* 58, 185–189.

Boatwright, G. O., Ferguson, H., and Brown, P. L. (1964). Availability of P from superphosphate to spring wheat as affected by growth stage and surface soil moisture. *Soil Sci. Soc. Amer., Proc.* 28, 403–405.

Breazeale, J. F. (1930). Maintenance of moisture equilibrium and nutrition of plants at and below the wilting percentage. *Ariz., Agr. Exp. Sta., Tech. Bull.* 29.

Danielson, R. E., and Russell, M. B. (1957). Ion absorption by corn roots as influenced by moisture and aeration. *Soil Sci. Soc. Amer., Proc.* 21, 3–6.

Dean, L. A., and Rubins, E. J. (1945). Absorption by plants of phosphorus from a clay-water system: Methods of ensuing observations. *Soil Sci.* 59, 437–448.

Dommergues, Y. (1961). Nitrogen mineralization at low moisture contents. *Trans. Int. Congr. Soil Sci., 7th, 1960* Vol. 2, pp. 672–678.

Eck, H. V., Hauser, V. L., and Ford, R. H. (1965). Fertilizer needs for restoring productivity on Pullman silty clay loam after various degrees of soil removal. *Soil Sci. Soc. Amer., Proc.* 29, 209–213.

Furr, J. R., and Reeve, J. O. (1945). Range of soil-moisture percentages through which plants undergo permanent wilting in some soils from semiarid irrigated areas. *J. Agr. Res.* **71**, 149–170.

Gardner, W. R. (1964). Relation of root distribution to water uptake and availability. *Agron. J.* **56**, 41–45.

Gardner, W. R. (1966). Soil water movement and root absorption. *In* "Plant Environment and Efficient Water Use" (W. H. Pierre *et al.*, eds.), pp. 127–149. Amer. Soc. Agron., Madison, Wisconsin.

Gingrich, J. R., and Russell, M. B. (1956). Effect of soil moisture tension and oxygen concentration on the growth of corn roots. *Agron. J.* **48**, 517–520.

Gingrich, J. R., and Russell, M. B. (1957). A comparison of effects of soil moisture tension and osmotic stress on root growth. *Soil Sci.* **84**, 185–194.

Grunes, D. L. (1959). Effect of nitrogen on the availability of soil and fertilizer phosphorus to plants. *Advan. Agron.* **11**, 369–396.

Haddock, J. L. (1952). The influence of soil moisture condition on the uptake of phosphorus from calcareous soils by sugarbeets. *Soil Sci. Soc. Amer., Proc.* **16**, 235–238.

Haise, H. R., Haas, H. J., and Jensen, L. R. (1955). Soil moisture studies of some Great Plains soils. II. Field capacity as related to ⅓ atmosphere percentage, and "minimum point" as related to 15- and 26-atmosphere percentages. *Soil Sci. Soc. Amer., Proc.* **19**, 20–25.

Hobbs, J. A., and Bertramson, B. R. (1950). Boron uptake by plants as influenced by soil moisture. *Soil Sci. Soc. Amer., Proc.* **14**, 257–261.

Hoffman, W., and Rantz, S. E. (1968). What is drought? *J. Soil Water Conserv.* **23**, 105–106.

Hunter, A. S., and Kelley, O. J. (1946). The extension of plant roots into dry soil. *Plant Physiol.* **21**, 445–451.

Jamison, V. C., and Thornton, J. F. (1960). Results of deep fertilization and subsoiling on a claypan soil. *Agron. J.* **52**, 193–195.

Jenne, E. A., Rhoades, H. F., Yien, C. H., and Howe, O. W. (1958). Change in nutrient element accumulation by corn with depletion of soil moisture. *Agron. J.* **50**, 71–74.

Jenny, H., and Overstreet, R. (1939). Cation interchange between plant roots and soil colloids. *Soil Sci.* **47**, 257–272.

Kipple, G. E., and Retzer, J. L. (1959). Response of native vegetation of the Central Great Plains to applications of corral manure and commercial fertilizer. *J. Range Manage.* **12**, 239–243.

Lagerwerff, J. V. (1958). Comparable effects of adsorbed and dissolved cations on plant growth. *Soil Sci.* **86**, 63–69.

Larson, W. E., Lovely, W. G., Pesek, J. T., and Burwell, R. F. (1960). Effect of subsoiling and deep fertilizer placement on yields of corn in Iowa and Illinois. *Agron. J.* **52**, 185–189.

Lavy, T. L., and Barber, S. A. (1964). Movement of molybdenum in the soil and its effect on availability to the plant. *Soil Sci. Soc. Amer., Proc.* **28**, 93–97.

Lipps, R. C., Fox, R. L., and Koehler, F. E. (1957). Characterizing root activity of alfalfa by radioactive tracer techniques. *Soil Sci.* **84**, 195–204.

Low, A. J., and Piper, F. J. (1960). The influence of water supply on the growth and phosphorus uptake of Italian ryegrass and white clover in pot culture. *Plant Soil* **13**, 242–252.

McClure, G. W., and Jackson, W. A. (1968). Nutrient distribution in root zones. II. The concept of external vs. internal nutrient supply. *Agrochimica* **12**, 353–364.

Mederski, H. J., and Wilson, J. H. (1960). Relation of soil moisture to ion absorption by corn plants. *Soil Sci. Soc. Amer., Proc.* **24**, 149–152.

Meyer, R. E., and Gingrich, J. R. (1966). Osmotic stress effects on wheat using a split root solution culture system. *Agron. J.* **58**, 377–381.

Miller, R. D., and Johnson, D. D. (1964). The effect of soil moisture tension on carbon dioxide evolution, nitrification, and nitrogen mineralization. *Soil Sci. Soc. Amer., Proc.* **28**, 644–647.

Olsen, R. A., and Peech, M. (1960). The significance of the suspension effect in the uptake of cations by plants from soil-water systems. *Soil Sci. Soc. Amer., Proc.* **24**, 257–261.

Olsen, S. R., and Kemper, W. D. (1968). Movement of nutrients to plant roots. *Advan. Agron.* **20**, 91–151.

Olsen, S. R., Watanabe, F. S., and Danielson, R. E. (1961). Phosphorus absorption by corn roots as affected by moisture and phosphorus concentration. *Soil Sci. Soc. Amer., Proc.* **25**, 289–294.

Olsen, S. R., Kemper, W. D., and van Schaik, J. C. (1965). Self-diffusion coefficients of phosphorus in soil measured by transient and steady-state methods. *Soil Sci. Soc. Amer., Proc.* **29**, 154–158.

Patrick, W. H., Jr., Sloane, L. W., and Phillips, S. A. (1959). Response of cotton and corn to deep placement of fertilizer and deep tillage. *Soil Sci. Soc. Amer., Proc.* **23**, 307–310.

Pearson, R. W. (1966). Soil environment and root development. *In* "Plant Environment and Efficient Water Use" (W. H. Pierre *et al.*, eds.), pp. 95–126. Amer. Soc. Agron., Madison, Wisconsin.

Peters, D. B. (1957). Water uptake of corn roots as influenced by soil moisture content and soil moisture tension. *Soil Sci. Soc. Amer., Proc.* **21**, 481–484.

Place, G. A., and Barber, S. A. (1964). The effect of soil moisture and Rb concentration on diffusion and uptake of Rb 86. *Soil Sci. Soc. Amer., Proc.* **28**, 239–243.

Power, J. F., Reichman, G. A., and Grunes, D. L. (1961). The influence of phosphorus fertilization and moisture on growth and nutrient absorption by spring wheat. II. Soil and fertilizer P uptake in plants. *Soil Sci. Soc. Amer., Proc.* **25**, 210–213.

Reichman, G. A., Grunes, D. L., and Viets, F. G., Jr. (1966). Effect of soil moisture on ammonification and nitrification in two Northern Plains soils. *Soil Sci. Soc. Amer., Proc.* **30**, 363–366.

Reitemeier, R. F., and Richards, L. A. (1944). Reliability of the pressure-membrane method for extraction of soil solution. *Soil Sci. Soc. Amer., Proc.* **25**, 119–135.

Reuss, J. O., and Campbell, R. E. (1961). Restoring productivity to leveled land. *Soil Sci. Soc. Amer., Proc.* **25**, 302–304.

Richards, L. A., and Wadleigh, C. H. (1952). Soil water and plant growth. *In* "Soil Physical Conditions and Plant Growth" (B. T. Shaw, ed.), pp. 73–251. Academic Press, New York.

Rios, M. A., and Pearson, R. W. (1964). The effect of some chemical environmental factors on cotton root behavior. *Soil Sci. Soc. Amer., Proc.* **28**, 232–235.

Robinson, J. B. D. (1957). The critical relationship between soil moisture content in the region of the wilting point and the mineralization of natural soil nitrogen. *J. Agr. Sci.* **49**, 100–105.

Slatyer, R. O. (1969). Physiological significance of internal water relations in crop yield. *In* "Physiological Aspects of Crop Yield" (J. D. Eastin *et al.,* eds.), pp. 53–79. Amer. Soc. Agron., Madison, Wisconsin.

Taylor, S. A., and Haddock, J. L. (1956). Soil moisture availability related to power required to remove water. *Soil Sci. Soc. Amer., Proc.* **20,** 284–288.

Vaadia, Y., and Waisel, Y. (1967). Physiological processes as affected by water balance. *In* "Irrigation of Agricultural Lands" (R. M. Hagan *et al.,* eds.), pp. 354–372. Amer. Soc. Agron., Madison, Wisconsin.

Viets, F. G., Jr. (1962). Fertilizers and the efficient use of water. *Advan. Agron.* **14,** 223–264.

Viets, F. G., Jr. (1967). Nutrient availability in relation to soil water. *In* "Irrigation of Agricultural Lands" (R. M. Hagan *et al.,* eds.), pp. 458–471. Amer. Soc. Agron., Madison, Wisconsin.

Volk, G. M. (1947). Significance of moisture translocation from soil zones of low moisture tension to zones of high moisture tension by plant roots. *J. Amer. Soc. Agron.* **39,** 93–106.

Warden, J. C. (1961). Experiments and observations on the contribution of a dry topsoil to mineral deficiencies in the coffee tree. *Coffee Res. Sta. Lyamunga Res. Rep., 1960* pp. 34–42.

Will, G. M. (1961). Magnesium deficiency in pine seedlings growing in pumice soil nurseries. *N. Z. J. Agr. Res.* **4,** 151–160.

CHAPTER 7

WATER DEFICITS AND NITROGEN METABOLISM

Aubrey W. Naylor

DEPARTMENT OF BOTANY, DUKE UNIVERSITY, DURHAM, NORTH CAROLINA

I. INTRODUCTION

Characteristically, with incipient wilting there is a slowing of growth rate. A balance between water supply and available inorganic and organic nutrients may well be crucial at such a stage. Critical data to back up this hypothesis are not abundant but a body of circumstantial evidence has accumulated.

After it was observed that citrus trees are limited in their growth before the transpiration rate is affected by the moisture content of the root absorbing zone, that is, while the top 3–4 ft of soil were above the wilting point, it was suspected that a balance between water supply and nutrients is crucial in the growth process (Furr and Taylor, 1939; Oppenheimer and Elze, 1941). Added to this picture is the fact that the stomata of citrus trees close when the water deficit of the soil is only about −3.5 bars (Mendel, 1945). Thus with only a minor water deficit, lack of carbohydrate in citrus could lead to a carbohydrate deficit and consequently an amino acid deficiency even though reduced nitrogen might be available in the plant. The inescapable consequence would be slow growth or growth at the expense of other parts of the plant.

Research on biochemical changes accompanying water stress in higher plants was overshadowed until the 1950's and 1960's by attention given to gross physiological and biophysical changes in cells and tissues undergoing desiccation. This is well illustrated in reviews such as those of Iljin (1957), Vaadia et al. (1961), Henckel (1964), and Philip (1966), which empha-

241

size water stress caused changes in protoplasmic viscosity and elasticity. Stocker (1960), however, devoted considerable attention in his review to the biochemical aspects of water-stressed plants.

Shortage of water in leaves has the obvious effect of causing them to wilt. Farm crops react to prolonged drought in a manner that is reminiscent of the development of nutrient-deficiency symptoms. Following slowing of the growth rate there is often loss of chlorophyll accompanied by yellowing first in the lower leaves and progressively upward to the top. The stem becomes somewhat woody and the lower leaves may die.

Many plants react to successive wilting regimes by shedding their lower leaves. Commonly, with drought there is a reduction of both stem and leaf tissue. In the case of moderate drought in cereal crops, leaf loss may not be accompanied by a drop in grain production but the crop is likely to mature early. With increasing severity of drought, the crop will come to maturity earlier and earlier and yield of grain falls more rapidly than that of straw until a condition is reached where the flowers abort (Widtsoe, 1912).

Where water is not the limiting factor, nitrogen and phosphorus probably are the two plant nutrients that most limit dry weight accumulation. Nitrogen readily becomes limiting when rainfall is high because the inorganic forms are only lightly held by soil particles. Conversely, as drought conditions develop, nitrogen may accumulate. Diminishing soil moisture, if it occurs rapidly, should result in the soil solution becoming concentrated. This could be reflected in an accumulation of mineral nutrients in plants limited in growth by moisture supply. Wadleigh and Richards (1951) have reviewed the literature on this point for several vegetable crops and conclude that most experimental evidence shows that for a given level of fertility, decreasing moisture supply is associated with a definite increase in total nitrogen content of the plant tissue, a definite decrease in potassium content, and a variable effect upon the content of phosphorus, calcium, and magnesium. Henrici (1953) reported an increase in nitrate. Using a radiotracer technique, Dove (1969) showed that absorption of ^{32}P by roots of soybean and maize was severely inhibited by water potentials below -12 to -15 bars. He concluded that exposure of roots to air reduced total ^{32}P-uptake by decreasing active absorption.

Unfortunately nothing is known about the effect of drought on uptake of salts of iron, molybdenum, copper, and boron. It is not known either if nitrate accumulates or if nitrate reduction is affected in the soil. One might expect that as drought conditions develop nitrogen-fixing bacteria would slow their rates of fixation. They and the ammonifying and nitrifying bacteria along with soil bacteria and fungi might be expected to become increasingly competitive with seed plants for the nitrogen made available from degradation of plant and animal residues. Thus, while water content of the soil will affect the ecology of the soil microflora, we should also expect to find microfloral composition changes adjacent to the roots to be induced by changes in composition of the root exudate. This conjecture has been borne out by Rovira's data (1959, 1965). He pointed out that the total amount and relative quantities of different amino acids found in root exudate were markedly affected by environmental conditions.

While small leaves are typical of dry habitats, it has long been noted that leaves of many plants growing in bogs and other wet habitats have xeromorphic structures. Usually such plants are said to be subjected to physiological drought. Caughey (1945) questioned that assumption; and others (Mothes, 1932; Albrecht, 1940; Stocker, 1960) have expressed the view that the xeromorphic form is an expression of nitrogen deficiency or a combination nutrient deficiency.

Even short-term water deficits have a marked effect on metabolism as reflected in dry weight determinations. For example, Iljin (1923) reported that wheat harvested on a particularly dry day weighed as much as 30% less than that harvested the day before when the humidity was high. Similar losses in dry weight were observed when *Impatiens parviflora* leaves were harvested on successive days of wilting (Maximov and Krasnoselsky-Maximov, 1924). Since wilting is accompanied by degradative activity by hydrolytic enzymes (Molisch, 1921; Mothes, 1931), perhaps there is rapid transport of solubilized material out of the leaf.

The relationship of water stress to protein hydrolysis in leaves is not fully clarified. Certainly, proteolysis is common in cut plants that are allowed to wilt (MacPherson, 1952; Kemble and MacPherson, 1954). Mothes (1931) was among the claimants that protein hydrolysis is a result of water stress. But as Petrie and Wood (1938) pointed out, Mothes' measurements were confounded by concurrent starvation of his plants. Mothes' claim was supported to some degree, however, by the finding that some hydrolysis of protein appears to occur in tobacco leaves during drought (Petrie and Arthur, 1943). Gates (1964) believes that many claims of proteolysis resulting from water stress are subject to reinterpretation for the following reasons: (1) Protein synthesis may be interrupted in stressed plants but not in the controls, thereby giving an impression of hydrolysis when none has occurred; (2) pooling of samples of old and young leaves is common practice. The normal loss of protein in physiologically old leaves may overshadow the much smaller amount actually in and retained by the young tissue. The second point of Gates' hypothesis was put to test by Dove (1964), who found that old tomato leaves lost protein during water stress whereas young ones did not.

Until the mid-1960's the question of what happens to the soluble nitrogen fraction during water stress was not determined. Is it translocated or does it remain *in situ?* Translocation of nitrogen away from a leaf can result in irreparable damage yielding symptoms resembling those associated with nitrogen deficiency. In investigations with tomato, Dove (1968) has shown that translocation of soluble nitrogen occurs to upper leaves of intact plants in dry air and lower leaves when the plants are topped, i.e., have the upper shoot removed. Under the latter conditions there is failure to show typical yellowing of lower leaves in water-deficient plants. Thus it is reasonable to hypothesize that nitrogen in droughted plants is mobile and moves along concentration gradients.

II. WATER STRESS AND PROTEIN METABOLISM

During the 1960's the idea grew that water-stress injury has a metabolic base that is concerned with damage to the protein-synthesizing mechanism. Isotope-labeling studies have made it clear that turnover of proteins is occurring continuously in cells (Hellebust and Bidwell, 1963). Obviously, replacement or repair of the loss must occur all of the time. In addition, growing cells have a net gain in newly formed protein. Thus new growth is dependent on protein synthesis.

While it has been known for some time that leaf protein undergoes accelerated hydrolysis as water stress develops (Mothes, 1928, 1931, 1956; Petrie and Wood, 1938; MacPherson, 1952; Kemble and MacPherson, 1954; Zholkevich and Koretskaya, 1959; Dove, 1968), it is only since radioisotope techniques became available that the question of synthesis during water stress could be asked with a reasonable expectation that a quantitative answer could be obtained. And it is still more recent that the answer could be determined whether the new protein is replacement protein or enzymatic proteins—some possibly leading to further destruction of cell constituents.

We shall now consider the little that is known of the impact of water stress on the machinery of protein synthesis including the several classes of RNA used in synthesis, the incorporation of amino acids into protein, and the effect of water stress on the behavior of specific enzymatic proteins.

RNA synthesis does not seem to change appreciably in the early stages of water deficit of older tissue but ribonuclease activity may rise sharply as water stress increases (Dove, 1967; Kessler, 1961; Kessler and Monselise, 1959). Surprisingly small water deficits can, however, have marked effects on chlorophyll accumulation, ribonucleic acid synthesis, (Bourque and Naylor, 1971) and nitrate reductase synthesis (Huffaker *et al.,* 1970) in young tissue. Soluble nitrogen levels rise to higher levels than can be accounted for by the degradation of nucleic acids. The unspecified nitrogen is evidently from degraded protein or newly synthesized amino acids that cannot be incorporated into protein. The ribonuclease activity observed in the water-stressed plants has been found (Kessler and Engelberg, 1962) to be associated with the difficult-to-sediment fraction of the leaf homogenate and can be rated highly stable.

Attempts have been made to determine if the degree of drought resistance can be correlated with the speed with which ribonuclease appears in the tissues subjected to desiccation. Kessler (1961) proposed that the speed with which an adenine–RNA complex is broken up is correlated with the degree of drought resistance. He further suggested that adenine is unstable in water stress-sensitive strains of plants. Whatever the validity of this hypothesis, it is clear that there are definite ribonucleotide compositional changes in corn (West, 1962), olive leaves (Kessler and Frank-Tishel, 1962), and wheat (Stutte and Todd, 1968) that are traceable to water deficits. Since the base composition of the extracted RNA is also affected by the temperature regime under which the plants are grown, interpretation of the water stress effects alone is difficult. In wheat and olive leaves the guanidine + cytidine/adenine + uridine ratio increases rapidly as water stress increases. Nevertheless, Stutte and Todd (1968) finally concluded that the use of the nucleotide composition of wheat RNA alone does not appear to be a suitable method of determining a plant's potential for drought resistance.

Because of the action of ribonucleases on the amino acid and protein-synthesizing machinery, one should expect prolonged drought to affect the size and composition of the free amino acid pools and the kinds of proteins synthesized in plants and plant parts undergoing desiccation. Obviously, the lack of one or more key enzymes could cause slow resumption of normal metabolic activity after relief from drought conditions. Mothes (1956) is of the opinion that despite ultimate recovery of turgidity, irreversible damage to the protein-making machinery may occur which can lead to death. Sissakian (1940) and Sissakian and Kobjakova (1939) reported reduced protein-synthesizing activity as desiccation occurred in drought-resistant plants while drought-sensitive species showed no net synthesis.

Water stress is accompanied by definite changes in the levels of free amino acids. Is uniform hydrolysis indicated? No. The number and amounts of the amino acids do not reflect hydrolysis of the "average" protein of the cell. There is an especially marked accumulation of free proline (Chen et al., 1964; Kemble and MacPherson, 1954; Prusakova, 1960; Barnett and Naylor, 1966; Routley, 1966; Stewart et al., 1966) and amides (Chen et al., 1964; Mothes, 1956). The accumulation of amides is thought to be the result of incorporation of free ammonia released by deamination of amino acids which arise through proteolysis during water stress (Mothes, 1956).

The effects of water stress on amino acid levels and turnover of both free and protein-bound amino acids have been studied by means of ^{14}C-labeling techniques by Barnett and Naylor (1966). Water stress levels of -15 bars (moderate stress) and -30 bars (severe stress) were used with two clonal strains of Bermuda grass. $^{14}CO_2$ was provided for 30 minutes and label in the free amino acids and protein was followed for the succeeding 77 hours. Initial incorporation of ^{14}C was greatest into, and the most label was retained in, the free amino acid fraction from moderately stressed shoots. But more label was retained in the free amino acids of the severely stressed plants than the controls (Fig. 1). Greater treatment differences, however, were found in the incorporation of label into protein. The watered controls accumulated label into protein continuously during the sampling period. Label in protein from moderately stressed shoots reached a maximum after 5 hours and declined thereafter. After 77 hours the amount of label in moderately stressed shoots was only 20% of that of the controls. In contrast, very little label was incorporated into protein in severely stressed shoots. The data support the conclusion that free amino acids are synthesized during water stress. Incorporation of the newly synthesized amino acids into protein, however, did not occur as readily in the stressed plants as in the unstressed controls.

Upon examining the kinds of amino acids accumulating in Bermuda grasses during desiccation, Barnett and Naylor (1966) found free proline

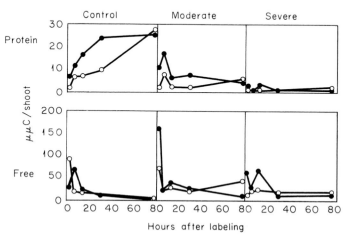

Fig. 1. Time course of change in radioactivity of the free amino acid fraction and soluble protein fraction in Bermuda grass with increasing water stress. ●, common; ○, coastal. From Barnett and Naylor (1966).

increased from 10 to 125 times the control value (Fig. 2). Valine tripled. In contrast, the amounts of glutamic acid and alanine decreased. Since a pulse-labeling technique was used, it was possible to tell if the newly formed free amino acids were limited in number and amount. Serine, glycine, and alanine were the free amino acids that became most highly labeled. Lesser amounts of label were incorporated into aspartate, glutamate, γ-aminobutyrate, and asparagine. All of the free amino acids found in the controls, except proline, became more highly labeled in the moderately stressed shoots than in any other treatment. Proline seemed to become labeled slowly. It also lost label slowly. The low specific activity of

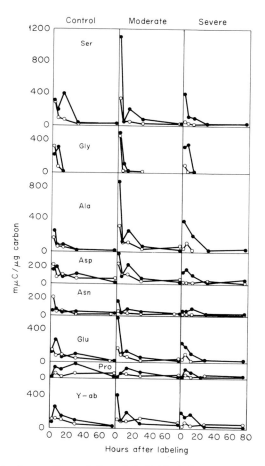

Fig. 2. Change in specific activity of individual free amino acids from $^{14}CO_2$-labeled Bermuda grass shoots. From Barnett and Naylor (1966).

proline tended to obscure the fact that the proline present in the stressed plants contained more than half the activity remaining in the free amino acid fraction after 77 hours. By varying the levels of carbohydrate and using inhibitors of glycolysis and the tricarboxylic acid cycle, Stewart *et al.* (1966) were able to provide further evidence with wilted turnip leaves that most of the proline formed was newly synthesized.

In answer to the question whether there is any indication of selectivity among water–soluble proteins that are degraded during water stress, Barnett and Naylor (1966) found indications there is selectivity. There was a 20–40% decrease in arginine-containing proteins. This is thought to have special significance since it is the nucleoproteins and ribosomal proteins that are especially rich in arginine.

The pulse-labeling experiments of Barnett and Naylor (1966) also make it possible to obtain data useful in answering the question of whether or not water stress affects the kinds of amino acids incorporated into pro-

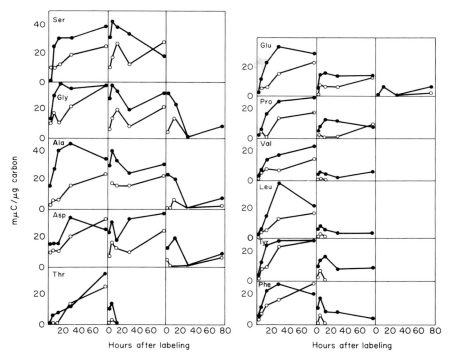

Fig. 3. Changes in specific activity of individual soluble protein amino acids from $^{14}CO_2$-labeled Bermuda grass shoots. Specific activities were calculated from average protein composition values for common Bermuda controls and the measured radioactivity figures. From Barnett and Naylor (1966).

tein. Perhaps not unexpectedly the answer is yes (see Fig. 3). The proteins of the moderately desiccated Bermuda grass shoots 20 hours after pulse labeling with CO_2 had no label in threonine, but otherwise there was a good complement of labeled amino acids. The severely water-stressed shoots, however, incorporated only labeled glycine, alanine, aspartate, and glutamate into protein. No labeled serine, threonine, proline, valine, leucine, tyrosine, or phenylalanine found its way into protein. The reason for this is not yet known but may be of special significance in understanding the nature of water deficit-induced injury. Presumably water stress leads to blockage of synthesis of these amino acids at one or more points in the metabolic pathways leading to their formation. If this is true, then protein synthesis would be brought to a halt as soon as the unlabeled pools of these essential amino acids are exhausted.

Stutte and Todd (1967) have used gel filtration and acrylamide gel electrophoresis techniques to study the effects of water stress on soluble protein components in wheat leaves. Two size classes were followed—a large-molecular-weight fraction (mol. wt. greater than 100,000) and a small-molecular-weight fraction (less than 100,000). Plants subjected to water stress showed a decrease in the large-molecular-weight fraction and an increase in the small-molecular-weight fraction. A correlation was found between the ability to withstand water stress and slowness of degradation of the large-molecular-weight proteins. Drought-resistant varieties maintained a higher percentage of the large-molecular-weight proteins under water stress conditions than the nonresistant varieties. These changes in wheat leaf composition are accompanied by changes in number and location of protein bands on the acrylamide gels.

Because of the marked changes in metabolism that occur when plants undergo water stress it should be expected that there would be appreciable changes in enzymatic complement in the cell. Several enzymes have been found to show increased activity as dehydration occurs; for example, amylase in wheat leaves (Spoehr and Millner, 1939), catalase and reductase in sunflower and tobacco leaves (Golovina, 1939), polyphenol and ascorbic acid oxidases in sugar beet (Zholkevich et al., 1958). Todd and Yoo (1964) found with wheat that saccharase activity rapidly decreased to a low level upon desiccation, whereas phosphatase decreased to a lesser extent and peptidase still less. Peroxidase decreased with drying and succinic dehydrogenase showed a sharp decline when the leaves had lost about 60% of their original water content. Upon subjecting wheat leaf homogenates to acrylamide gel electrophoresis Stutte and Todd (1969) found that with increasing water stress there were larger numbers of bands and quantities of iron-containing proteins. Under similar conditions the number of isozymes of lactic dehydrogenase declined. Certain peroxidase bands disappeared and new ones appeared with progressive water stress. Nitrate reductase activity is very susceptible to water-stress conditions (Mattas and Pauli, 1965; Younis et al., 1965; Huffaker et al., 1970). Up to half the nitrate reductase activity is lost in barley in a 4-day stress period. Nitrite reductase activity also declines to some extent. Huffaker et al. (1970) found that phosphoenol–pyruvate carboxylase activity decreased during water stress but only to about half the extent of nitrate reductase. Phosphoribulokinase and ribulose 1,5-diphosphocarboxylase, however, were little affected by water stress. Vieira-da-Silva (1970) has also given close scrutiny to variations in enzymatic activity accompanying changes in water potential. He found in stressed cotton plants that there was an increase in activity of catalase, acid phosphatase, RNase, invertase, and α- and β-amylase. The increase in activity was thought to reflect an increase in enzyme solubilization and an increase in total activity. Most of the solubilization occurred in the chloroplast fraction. The drought-resistant species, Gossypium anomalum, was slow to solubilize acid phosphatase. Chloramphenicol inhibited part of the total activity of acid phosphatase and RNase under osmotic stress conditions, but had no effect on invertase activity, which was inhibited by cycloheximide.

The basis for water-stress effects on enzymatic activity is obscure. Stocker (1960) has provided an incisive review of current ideas relevant to our understanding of the relationship of water to proteins in general. These concepts apply equally well to enzymatic, structural, and storage proteins. Unfortunately, very few purified enzymes have been studied intensively under conditions of varying water potential. While it is true that many enzymes are relatively unaffected by a wide range of osmotic conditions, the indirect evidence is clear that the effects are drastic with respect to some proteins. Invertase is one of these and is perhaps the most thoroughly studied of the purified enzymes.

By combining the data of Nelson and Schubert (1928) and Walter (1931) for sucrose hydrolysis in graded concentrations of sucrose solutions, one observes striking effects on the velocity of hydrolysis by β-fructofuranoxidase (Fig. 4). Dixon and Webb (1964) attribute this to lowered water activity. In addition, sucrose has been found to inhibit malate dehydrogenase (Flowers and Hanson, 1969) which catalyzes a reaction where water is not a reactant. The steady decline in activity with increasing sucrose concentration might be attributed to gradual changes of water structure surrounding the enzyme (Klotz, 1958). Because of differential hydrogen bonding of exposed hydroxyl, carboxyl, amide, sulfide, and methyl groups of enzymes, changes in the structure of the water shell surrounding enzymes can cause drastic changes in the properties of enzymes.

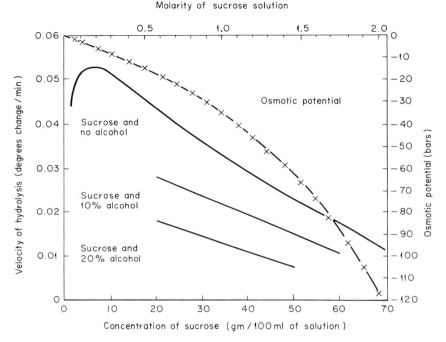

Fig. 4. The effect of water potential on the hydrolysis of sucrose by β-fructofuranosidase.

This in turn will be reflected in an alteration of the speed and even use of specific metabolic pathways.

REFERENCES

Albrecht, W. A. (1940). Calcium-potassium-phosphorus relation as a possible factor in ecological array of plants. *J. Amer. Soc. Agron.* **32**, 411–418.

Barnett, N. M., and Naylor, A. W. (1966). Amino acid and protein metabolism in Bermuda grass during water stress, *Plant Physiol.* **41**, 1222–1230.

Bourque, D. P., and Naylor, A. W. (1971). Large effects of small water deficits on chlorophyll accumulation and ribonucleic acid synthesis in etiolated leaves of jack bean (*Canavalia ensiformis* [L.] DC.). *Plant Physiol.* **47**, 591–594.

Caughey, M. G. (1945). Water relations of pocosins or bog shrubs. *Plant Physiol.* **20**, 671–689.

Chen, D., Kessler, B., and Monselise, S. P. (1964). Studies on water regime and nitrogen metabolism of citrus seedlings grown under water stress. *Plant Physiol.* **39**, 379–386.

Dixon, M., and Webb, E. C. (1964). "Enzymes," 2nd ed., pp. 73–75. Longmans, Green, New York.

Dove, L. D. (1964). Relationships between water stress and organic nitrogen constituents of tomato leaves. Ph.D. Dissertation, Duke University.

Dove, L. D. (1967). Ribonuclease activity of stressed tomato leaflets. *Plant Physiol.* **42**, 1176–1178.

Dove, L. D. (1968). Nitrogen distribution in tomato plants during drought. *Phyton (Buenos Aires)* **25**, 49–52.

Dove, L. D. (1969). Phosphate absorption by air-stressed root systems. *Planta* **86**, 1–9.

Flowers, T. J., and Hanson, J. B. (1969). The effect of reduced water potential on soybean mitochondria. *Plant Physiol.* **44**, 939–945.

Furr, J. R., and Taylor, C. A. (1939). Growth of lemon fruits in relation to moisture content of the soil. *U.S., Dep. Agr., Tech. Bull.* **640**.

Gates, C. T. (1964). The effect of water stress on plant growth. *J. Aust. Inst. Agr. Sci.* **30**, 3–22.

Golovina, A. S. (1939). [The effect of drought on the chemical processes of plants.] In Russian. *Nauch. Rab. Krasnodar. Stud., Gos. Pedagog. Inst.* pp. 51–58; *Chem. Abstr.* **36**, 4546-2 (1942).

Hellebust, J. A., and Bidwell, R. G. S. (1963). Protein turnover in wheat and snapdragon leaves. *Can. J. Bot.* **41**, 969–983.

Henckel, P. A. (1964). Physiology of plants under drought. *Annu. Rev. Plant Physiol.* **15**, 363–386.

Henrici, M. (1953). Further nutrition studies on Tribulus terrestris. *Union S. Afr., Dep. Agr., Forest. Sci. Bull.* **348**.

Huffaker, R. C., Radin, T., Kleinkopf, G. E., and Cox, E. L. (1970). Effects of mild water stress on enzymes of nitrate assimilation and of the carboxylative phase of photosynthesis in barley. *Crop Sci.* **10**, 471–474.

Iljin, V. S. (1923). L'influence de la secheresse sur la regulation des stomates et sur l'accrossement des plantes. *Preslia* **2**, 43–55 (cited in Maximov, 1929).

Iljin, V. S. (1957). Drought resistance in plants and physiological processes. *Annu. Rev. Plant Physiol.* **8**, 257–274.

Kemble, A. R., and MacPherson, H. T. (1954). Liberation of amino acids in perennial ryegrass during wilting. *Biochem. J.* **58**, 46–50.

Kessler, B. (1961). Nucleic acids as factors in drought resistance of higher plants. *Advan. Bot.* **2**, 1153–1159.

Kessler, B., and Engelberg, N. (1962). Ribonucleic acid and ribonuclease activity in developing leaves. *Biochim. Biophys. Acta* **55**, 70–82.

Kessler, B., and Frank-Tishel, J. (1962). Dehydration induced synthesis of nucleic acids and changing of composition of ribonucleic acid: A possible protective reaction in drought resistant plants. *Nature (London)* **196**, 542–543.

Kessler, B., and Monselise, S. P. (1959). Studies on ribonuclease, ribonucleic acid and protein synthesis in healthy and zinc-deficient leaves. *Physiol. Plant.* **12**, 1–7.

Klotz, I. M. (1958). Protein hydration and behavior. *Science* **128**, 815–822.

MacPherson, H. T. (1952). Changes in nitrogen distribution in crop conservation. II. Protein breakdown during wilting. *J. Sci. Food Agr.* **3**, 365–367.

Mattas, R. E., and Pauli, A. W. (1965). Trends in nitrate reduction and nitrogen fractions in young corn (*Zea mays* L.) plants during heat and moisture stress. *Crop Sci.* **5**, 181–184.

Maximov, N. A. (1929). "The Plant in Relation to Water." Allen & Unwin, London.

Maximov, N. A., and Krasnoselsky-Maximov, T. A. (1924). Wilting of plants in connection with drought resistance. *J. Ecol.* **12**, 95–110.

Mendel, K. (1945). Orange leaf transpiration under orchard conditions. Part 2. Soil moisture content decreasing. *Palestine J. Bot., Rehovot Ser.* **5**, 59–85.

Molisch, H. (1921). Über den Einfluss der Transpiration auf das Verschwinden der Stärke in den Blättern. *Ber. Deut. Bot. Ges.* **39,** 339–344.

Mothes, K. (1928). Die Wirkung des Wassermangels auf den Eiweissumsatz im höhern Pflanzen. *Ber. Deut. Bot. Ges.* **46,** 59–67.

Mothes, K. (1931). Zur Kenntnis des N-Stoffwechsels höherer Pflanzen. 3. Beitrag (unter besonderer Berücksichtigung des Blattalters und des Wasserhaushaltes). *Planta* **12,** 686–731.

Mothes, K. (1932). Ernährung, Struktur, und Transpiration. Ein Beitrag zur Kansalen Analyze der Xeromorphosen. *Biol. Zentralbl.* **52,** 193–223.

Mothes, K. (1956). Der Einfluss des Wasserzustandes auf Fermentprozesse und Stoffumsatz. *In* "Handbuch der Pflanzenphysiologie" (W. Ruhland, ed.), Vol. 3, pp. 656–664. Springer-Verlag, Berlin and New York.

Nelson, J. M., and Schubert, M. P. (1928). Water concentration and the rate of hydrolysis of sucrose by invertase. *J. Amer. Chem. Soc.* **50,** 2188–2193.

Oppenheimer, H. R., and Elze, D. L. (1941). Irrigation of citrus trees according to physiological indicators. *Rehovoth Agr. Res. Sta., Bull.* **31.**

Petrie, A. H. K., and Arthur, J. I. (1943). Physiological ontogeny in the tobacco plant. The effect of varying water supply on the drifts in dry weight and leaf area and on various components of the leaves. *Aust. J. Exp. Biol. Med. Sci.* **21,** 191–200.

Petrie, A. H. K., and Wood, J. G. (1938). Studies on the nitrogen metabolism of plants. III. On the effect of water content on the relationship between proteins and amino acids. *Ann. Bot. (London)* [N. S.] **2,** 887–898.

Philip, J. R. (1966). Plant water relations: Some physical aspects. *Annu. Rev. Plant Physiol.* **17,** 245–268.

Prusakova, L. D. (1960). Influence of water relations on tryptophan synthesis and leaf growth in wheat. *Fiziol. Rast.* **7,** 139–148.

Routley, D. G. (1966). Proline accumulation in wilted Ladino clover leaves. *Crop Sci.* **6,** 358–361.

Rovira, A. D. (1959). Root excretion in relation to the rhizosphere effect. IV. Influence of plant species, age of plant, light, temperature and calcium nutrition on exudation. *Plant Soil* **11,** 53–64.

Rovira, A. D. (1965). Plant root exudation and their influence upon soil-organisms. *In* "Ecology of Soil-Borne Plant Pathogens" (K. F. Baker and W. Snyder, eds.), pp. 170–184. Univ. of Calif. Press, Berkeley.

Sisakian, N. M. (1940). "The Biochemical Character of Drought-Resistant Plants." Publ. House Acad. Sci. U.S.S.R. Moskva; Leningrad (in Russian).

Sisakian, N. M., and Kobjakova, A. M. (1939). Direction of enzymatic action, as an indication of drought resistance in cultivated plants. II. The direction of protease action in drought-resistant and non-resistant strains of wheat. *Biokhimiya* **3,** 796–803.

Spoehr, H. A., and Milner, H. W. (1939). Starch dissolution and amylolytic activity in leaves. *Proc. Amer. Phil. Soc.* **81,** 37–78.

Stewart, C. R., Morris, C. L., and Thompson, J. F. (1966). Changes in amino acid content of excised leaves during incubation. II. Role of sugar in the accumulation of proline in wilted leaves. *Plant Physiol.* **41,** 1585–1590.

Stocker, O. (1960). Physiological and morphological changes in plants due to water deficiency. *In* "Plant-Water Relationships in Arid and Semi-Arid Conditions," Vol. 15, pp. 63–104. UNESCO, Paris.

Stutte, C. A., and Todd, G. W. (1967). Effects of water stress on soluble leaf proteins in *Triticum aestivum* L. *Phyton* (*Buenos Aires*) **24**, 67–75.

Stutte, C. A., and Todd, G. W. (1968). Ribonucleotide compositional changes in wheat leaves caused by water loss. *Crop Sci.* **8**, 319–321.

Stutte, C. A., and Todd, G. W. (1969). Some enzyme and protein changes associated with water stress in wheat leaves. *Crop Sci.* **9**, 510–512.

Todd, G. W., and Yoo, B. Y. (1964). Enzymatic changes in detached wheat leaves as affected by water stress. *Phyton* (*Buenos Aires*) **21**, 61–68.

Vaadia, Y., Raney, F. C., and Hagan, R. M. (1961). Plant water deficits and physiological processes. *Annu. Rev. Plant. Physiol.* **12**, 265–292.

Vieira-da-Silva, J. B. (1970). Contribution à l'étude de la resistance à la secheresse dans le genre *Gossypium*. II. La variation de quelques activités enzymatiques. *Physiol. Veg.* **8**, 413–447.

Wadleigh, C. H., and Richards, L. A. (1951). Soil moisture and the mineral nutrition of plants. *In* "Mineral Nutrition of Plants" (E. Truog, ed.), Chapter 17, pp. 437–440. Univ. of Wisconsin Press, Madison.

Walter, H. (1931). "Die Hydratur der Pflanze," p. 161. Fischer, Jena.

West, S. H. (1962). Protein, nucleotide and ribonucleic acid metabolism in corn during germination under water stress. *Plant Physiol.* **37**, 565–571.

Widtsoe, J. A. (1912). The production of dry matter with different quantities of water. *Utah, Agr. Exp. Sta., Bull.* **116**, 1–64.

Younis, M. A., Pauli, A. W., Mitchell, H. L., and Stickler, F. C. (1965). Temperature and its interaction with light and moisture in nitrogen metabolism of corn (*Zea mays* L.) seedlings. *Crop Sci.* **5**, 321–326.

Zholkevich, V. N., and Koretskaya, T. F. (1959). Metabolism of pumpkin roots during soil drought. *Fiziol. Rast.* **6**, 690–700.

Zholkevich, V. N., Prusakova, L. D., and Lizandr, A. A. (1958). Movement of assimilates and respiration of conductive paths in relation to the soil moisture. *Fiziol. Rast.* **5**, 337–344.

CHAPTER 8

WATER DEFICITS AND HORMONE RELATIONS

Avinoam Livne

THE NEGEV INSTITUTE FOR ARID ZONE RESEARCH, BEERSHEVA, ISRAEL

and

Yoash Vaadia

THE VOLCANI INSTITUTE OF AGRICULTURAL RESEARCH, BET DAGAN, ISRAEL

I. INTRODUCTION

The current and accepted approach in research of plant water relations considers the plant as an aqueduct between the soil and atmosphere. Water fluxes in the plant are determined and controlled by water potential differences within the system as well as by changes in water permeability. Some water deficits must occur in this system whenever transpiration rates exceed absorption rates. During the day, when transpiration rates may be high, water deficits can be observed even when soil water is ample. At night, on the other hand, such deficits are restored. Thus, even under controlled conditions under a light–dark cycle, plants may experience continuously changing levels of water deficits.

Changes which occur in water potentials or in permeability to water in plants are mediated at least in part via metabolic regulatory mechanisms.

Changes in water potential may be effected through electrolyte transport, biochemical transformations, cell wall elasticity, and other factors. Changes in tissue and membrane permeability can occur because of changes in membrane composition, spatial arrangement, changes in electrochemical potentials, and ionic composition around membranes. Thus, both water potential and membrane permeability may be modified and controlled through metabolic regulation. Therefore, the response of plants under conditions of water deficits can be metabolically regulated.

Plant growth in general is regulated through the intricate mechanisms of control which rest in the genome and respond to changes in environment. It is well known that various hormones play an important role in this regulation. The now common examples of photoperiodism, dormancy, flowering, and abscission all involve hormonal regulation and respond to environmental changes.

It appears as a reasonable hypothesis to assume that hormonal regulation is involved in control of water potential and membrane permeability and thus in the control of plant water deficits. In accordance with this view, the response of plants to environmental water stress is associated in a regulatory fashion with changes in levels and activity of various hormones in the plant. Such change in turn may provide a mechanism of adaptation of plants to varying environmental conditions.

The effects of plant hormones on water uptake and water permeability of plant cells have been reviewed by Pohl (1961) and Stadelmann (1969). This chapter will evaluate the possible role of some plant hormones in regulation of water deficits in plants.

II. EFFECT OF WATER DEFICITS ON HORMONAL DISTRIBUTION

A. WATER DEFICITS AND CYTOKININ ACTIVITY

When plants are subjected to various forms of stress, such as water deficits, salinity, or flooding, the leaves show symptoms of enhanced aging. In mature leaves of plants, under water stress, the levels of proteins and RNA decrease rapidly. Even in well-watered plants the protein contents of mature leaves decrease as the leaves age (Shah and Loomis, 1965). A decline in protein content and an elevated protein degradation in plants under water deficits have also been shown by other workers (Vaadia et al., 1961; Barnett and Naylor, 1966).

Itai and Vaadia (1965) suggested that the metabolic shifts and enhanced aging in shoots of stressed plants may be due to a reduced supply of cytokinins from the roots. This hypothesis was based on evidence that kinetin treatment is effective in retarding the aging process in detached leaves (Richmond and Lang, 1957; Mothes *et al.,* 1959; Osborne and McCalla, 1961; Leopold and Kawase, 1964; Gunning and Burkley, 1963). Furthermore, Kulaeva (1962) showed that senescence of detached leaves was delayed by a crude xylem exudate of tobacco plants, while Kende (1965) demonstrated the presence of cytokinins, capable of stimulating division of callus cells, in the xylem exudate of sunflower plants. The involvement of cytokinins, which apparently are synthesized in the roots, in the regulation of senescence of the shoot thus seems well established (Kende, 1971).

As predicted by their hypothesis, Itai and Vaadia (1965) and Itai *et al.* (1968) found a reduced cytokinin activity in the root exudate of plants subjected to a water stress. Furthermore, Shah and Loomis (1965) showed that the reduction of nucleic acids and proteins in water-stressed plants could be counteracted, at least in part, by treating the sugar beet leaves with benzyladenine.

The normal level of cytokinin in the xylem sap of sunflower plants was attained 2 days after removal of the water stress, while an overshoot was apparent after the first day of recovery from stress. The gradual addition of NaCl (up to 0.1 M) and of mannitol (up to 0.16 M) to the nutrient medium of sunflower plans resulted in a decreased level of cytokinin in the xylem exudate (Itai *et al.,* 1968; Vaadia and Itai, 1968). When applied to tobacco plants, these same treatments resulted also in a reduced incorporation of ^{14}C-leucine into leaf disc proteins. Kinetin pretreatment prior to incubation with labeled leucine partially restored the incorporation in leaf discs taken from stressed plants (Ben-Zioni *et al.,* 1967).

The cytokinin level is also affected by other forms of stress which affect normal growth. Burrows and Carr (1969) showed reduced concentration of cytokinins in the xylem exudate of flooded sunflower plants. Andreenko *et al.* (1964) reported that the levels of cytokinins were markedly reduced when the pH of the root medium was lowered from 7.0 to 4.0. Krikon *et al.* (1970) observed a marked decrease in the cytokinin content in tomato plants infected by *Verticillium dahliae.* The decline in cytokinin activity in the exudate was apparent before appearance of symptoms of the wilt disease, such as stunting and chlorosis.

These various forms of stress, although seemingly unrelated, all result in a reduction in cytokinin activity in treated plants. It is possible that water stress is common to all the above treatments, and thus availability of water and water balance may be of importance in regulation of cytokinin level in plants. However, it is also possible that changes in levels of cytokinins are induced because of other factors—such as ionic strength, aeration, etc.—and, in turn, affect the regulation of water balance.

Although the roots are apparently a major source of cytokinins, water stress could conceivably affect the level of cytokinins elsewhere. Itai and Vaadia (1971) examined the effect of water stress applied to tobacco shoots through enhanced evaporative demands. The plants were exposed for 30 min to an airstream which caused slight wilting of the leaves and then were allowed to recover turgor. Although the amount of exudate did not differ between treated and control plants, this short period of water stress nevertheless caused a reduction of about half in the cytokinin activity of the exudate as well as a smaller reduction in the leaves. Itai and Vaadia concluded that the biosynthesis in the root ceased when water tension in the leaf was enhanced. The signal to the site of synthesis may in this case be a change in the water potential which is transmitted through the plant.

The level of cytokinin in the shoot is apparently regulated not only by the rate of supply from the roots, but also by chemical transformations in the leaf itself. Detached tobacco leaves exposed to a dry atmosphere (over $CaCl_2$) for durations of 10–30 min had a reduced cytokinin activity. Cytokinin activity in these leaves was partly restored after an additional 18 hr in a humid chamber. Furthermore, when detached leaves were fed ^{14}C-kinetin and subjected to a dry atmosphere, the distribution pattern of the label, extracted from the wilting leaves, differed from that found in extracts of control leaves (Itai and Vaadia, 1971). A similar modified distribution of label from ^{14}C-kinetin was found in abscisic acid-treated leaves (Back et al., 1972).

B. WATER DEFICITS AND ABSCISIC ACID

The role of abscisic acid as a plant hormone has been intensively studied and amply reviewed (Addicott and Lyon, 1969; Milborrow, 1969; Wareing and Ryback, 1970). It has been suggested that abscisic acid is involved in regulation of physiological responses of plants under water stress (Wright, 1969; Wright and Hiron, 1969; Mizrahi et al., 1970). It is anticipated that this topic will gain much attention in the future.

The effect of wilting on the content of inhibitor β and its abscisic acid (ABA) equivalent in excised first leaves of wheat seedlings was studied by Wright (1969). He showed that levels of ABA increased in wilted leaves. The increase was related to the degree of wilting. In leaves which lost 6% of their fresh weight in 10 min, no change in ABA was observed. A severalfold increase in ABA was observed in these leaves after 2 hr. A fortyfold increase in ABA occurred in leaves which lost 9% of their fresh weight and were kept for 4 hr in this wilted condition. The marked increase of ABA levels in the extracts of wilting leaves was verified by optical rotatory dispersion measurement (Wright and Hiron, 1969), thus clearly demonstrating a dramatic change in the level of abscisic acid as a response to water deficit.

The increase in ABA is temperature dependent, and no change in the level of this hormone was apparent when the wilted leaves were maintained at 2°C. Wright concluded that the time lapse requirement and the temperature dependency of the formation of abscisic acid suggest an enzymatic conversion from a precursor, possibly from the glucopyranoside of ABA, which was isolated from plant tissues (Koshimizu et al., 1968).

Recently, Mizrahi and Richmond (1971) showed by bioassay and gas–liquid chromatography methods that the quantity of bound ABA decreased in leaves of tobacco plants soon after the addition of salt to the root medium. Conversely, the amount of free ABA increased in the same leaves.

Pegg and Selman (1959) noted an increase in inhibitor β in intact tomato plants infected with *Verticillium albo-atrum* and experiencing visible water deficits. Similarly, Steadman and Sequiera (1970) reported that an initial wilting of the upper leaves of tobacco plants inoculated with the wilt-inducing bacterium, *Pseudomonas solanacearum,* was correlated with an increase in an inhibitor (apparently ABA).

Mizrahi *et al.* (1970) subjected tobacco plants to osmotic stress (by addition of 0.1 M NaCl or addition of 0.17 M mannitol to the nutrient medium of the roots). They found that leaves of plants under either stress contained substantially greater amounts of ABA. More inhibitor was measured in the plants subjected to 48 hr of stress than in plants subjected to 4 hr of stress, even though the leaves regained turgor under the longer period of stress. They concluded that the increase in the ABA content of leaves is not necessarily dependent on reduced leaf turgor. In their opinion ABA may be involved in a more general aspect of water balance in plants, namely, the regulation of the adaptive response of plants to osmotic stress. This aspect will be dealt with in a later section of this chapter.

C. Water Deficits and Other Hormones

Flooding of root systems of plants is associated with a change of hormonal distribution in the plants. Phillips (1964) showed that auxin levels in shoots of sunflower plants with waterlogged root systems increased within 2 wk to threefold over the levels in control plants. The high auxin levels, however, subsequently decreased to those of controls. Reid *et al.* (1969) reported that in waterlogged tomato plants the level of gibberellins in the xylem sap is reduced. They suggested that this reduction may be responsible for the decreased growth in waterlogged plants.

In a recent paper, Darbyshire (1971) tested the activity of indoleacetic acid (IAA) oxidase in the shoots of stressed pea and tomato plants. Stress was applied by the addition of mannitol to the nutrient solution. In both pea and tomato plants, increased activity of IAA oxidase was observed with increasing water deficits.

At present available demonstrations of the involvement of IAA and gibberellins in water deficits are few. Yet the above reports suggest that these hormones are involved.

III. EFFECT OF HORMONES ON TRANSPIRATION

A. AUXINS

Since most of the water transpired by leaves diffuses through the stomata, attempts have been made to conserve water through the bio-chemical control of stomatal opening. A variety of chemical components, including metabolic regulators, plant growth regulators, and metal-chelating compounds, were shown by Zelitch (1961, 1963) to prevent stomatal opening at millimolar concentrations. Zelitch (1961) also presented evidence that, by chemical induction of stomatal closure in intact tobacco leaves, it was possible to reduce transpiration loss at high light intensities without appreciably diminishing photosynthetic CO_2 assimilation.

Of the plant growth regulators tested, IAA was found relatively ineffective in reducing stomatal opening in tobacco leaf discs even at a concentration of 10 mM. Other auxins, such as Na 2,4-dichlorophenoxyacetate and 2-naphthoxyacetic acid, caused appreciable stomatal closure at a 1 mM concentration (Zelitch, 1961).

Earlier work by Johansen (1954) also showed that IAA had virtually no effect on stomatal behavior when applied at 0.6 mM to leaves of *Sinapis alba*. On application to the soil of potted plants, a 3 mM concentration induced stomatal closure, but this effect apparently was indirect, since stomata reopened when leaves from treated plants were cut and dipped in water or 0.6 mM IAA.

Several investigators (Brown, 1946; Ferri and Lex, 1948; Ferri and Rachid, 1949; Player, 1950; Brandbury and Ennis, 1952) related stomatal closure in auxin-treated plants to a direct effect of the auxins on leaves. Other workers (Takaoki, 1962; Allerup, 1964; Kozinka, 1967, 1968, 1970) emphasized the effect of auxins on water uptake by the roots.

Allerup (1964) reported transpiration measurements made by means of a hygrometer with barley seedlings grown in water culture. Replacement of nutrient solutions by 1 mM solutions of 2-indoleacetic acid, 2-indolebutyric acid, or 1-naphthaleneacetic acid induced immediate transient transpiration increases. The increase lasted for less than 30 min and was followed by a gradual fall in transpiration. If the plants were exposed to the growth regulators for a sufficient length of time, the effect became irreversible and eventually greatly impeded passage of water through the root. Allerup explained the immediate rise in transpiration by a decomposition effect of the auxins, at relatively high concentrations, on plasma membranes in the roots cells. This explanation is supported by the observation of von Guttenberg and Beythien (1951) that $0.6 \times 10^{-3} M$ IAA induced a visible decomposition of the protoplasmic membranes within 10–15 min, while the entire protoplast dissolved within 20–30 min. 2,4-dichlorophenoxyacetic acid had no protoplasm-destroying effect, and indeed no transient increase of transpiration in barley plants was apparent.

Kozinka (1967) reported rapid inhibition of water uptake in intact pea plants caused by millimolar concentrations of several auxins (1-naphtaneleneacetic acid, 2-indoleacetic acid, 2,4-dichlorophenoxyacetic acid, and 2-methyl-4-chlorophenoxyacetic acid) applied to the roots. Kozinka (1970) further showed that inhibition of water uptake in one half of the coleus roots, caused by synthetic auxins, was accompanied by an increase of water uptake in the remaining roots.

Mansfield (1967) reexamined the effect of auxins on stomatal behavior. Several auxins caused stomatal closure within 1 hr or less when applied at a 1 mM concentration to *Xanthium* leaves through the cut petiole. In the case of 2-naphthoxyacetic acid and 1-naphthylacetic acid the closure was readily reversed in an atmosphere of CO_2-free air. On the other hand, the closure by 2,4-dichlorophenoxyacetic acid was not reversed by flushing the leaves with CO_2-free air, indicating that this regulator exerts a more direct effect on the guard cells.

In contrast to the short-term effects of auxins, in which reduced transpiration was observed, Tal and Imber (1971b) observed that prolonged treatment (8 sprays within 2 wk) of 3-wk-old tomato plants by 2,4-dichlorophexyacetic acid at relatively low concentration (about $4 \times 10^{-6} M$) resulted in an increased transpiration rate.

The studies concerning the effect of auxins on stomatal behavior and transpiration reviewed in this section show a potential use of auxins in affecting water balance in plants. However, the high concentrations of the auxins used in many of these studies are much above the physiological range. The role of endogenous auxins in regulating transpiration is yet to be clarified.

B. CYTOKININS

Stomates commonly fail to open in the presence of various chemical compounds and metabolic inhibitors. An interesting exception to this rule is *p*-chloromercuric benzoate. This compound increases stomatal opening in *Commelina* leaves even in the dark, presumably by inhibiting ATPase activity, which is responsible for efflux of potassium ions from the guard cells (Fujino, 1967).

Cytokinins and apparently also gibberellic acid are unique among the compounds which affect stomatal opening in that they simultaneously promote stomatal opening and other physiological processes. It was demonstrated (Livne and Vaadia, 1965) that kinetin enhances transpiration rates and the opening of stomata of excised mature barley leaves (Fig. 1). This stimulation is rapid. One or two hours are required to discern this stimulation, but an effect of kinetin within less than 30 min is not uncommon. A dose–response curve for the stimulation of transpiration has an optimum at $3 \times 10^{-6} M$ kinetin. Figure 1 shows that during the daily course of transpiration a noon dip was observed in all treatments. In the kinetin treatment, however, the noon dip was smaller, despite the increased transpiration. On the basis of the findings of Wright (1969), the noon dip could conceivably result from a transient increase in the content of ABA

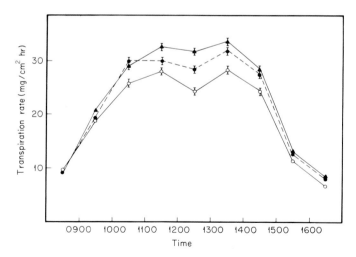

Fig. 1. Course of transpiration of recently mature, excised barley leaves with their bases dipping into kinetin solutions or water ($\bigcirc$, H₂O control; $\bullet$, $10^{-5} M$ kinetin; $\blacktriangle$, $10^{-6} M$ kinetin). Plants were 13 days old at the second-leaf stage. Measurements commenced immediately after excision and immersion of leaf bases into the test solution. From Livne and Vaadia (1965).

in the leaves because of temporary water deficit. The increase of ABA, in turn, reduces stomatal opening in the leaves. Thus, the reduced noon dip in the kinetin-treated leaves apparently reflects an outcome of the combined effects of both hormones on stomatal opening. The combined effect of both hormones on stomatal opening was demonstrated by Mittelheuser and Van Steveninck (1969) and by Mizrahi *et al.* (1970) in wheat and tobacco leaves, respectively.

The early observations that kinetin increases stomatal opening and stimulates transpiration rate in excised barley leaves have been confirmed in several laboratories and have been further extended to a variety of cytokinins and plant species (Table I). The range of concentrations and structural requirements for stimulation of transpiration by substituted purines were similar to those found for promotion of cell division. The effect thus seems to be of a general nature. However, Luke and Freeman (1968) reported that excised leaves or plants of several species did not exhibit increased transpiration when treated with various concentrations of kinetin. These plants include peas (*Vigna sinensis* (Torner) Savi), bean (*Phaseolus lunatus* L. and *P. vulgaris* L.), soybean (*Glycine max* (L.) Marr.), sunflower (*Helianthus annus* L.), pumpkin (*Cucurbita pepo* L.), maple (*Acer saccharinum* L.), and sweet gum (*Liquidambar styraciflua* L.).

Why do some species respond to exogenous cytokinins whereas others do not? The answer may be related to the endogenous level of cytokinin as well as other hormones and to the physiological status of the tissue tested. We have observed that the kinetin promotes the transpiration rate of recently matured, fully expanded barley leaves, but not of young leaves (Livne and Vaadia, 1965). Young leaves (e.g., primary leaves from 10-day-old plants) had to be excised and kept for 24 hr with the bases dipped in water before stimulation of transpiration by kinetin could be observed.

Using the sensitivity of the stomatal response of oat leaves to cytokinins, Luke and Freeman (1967) were able to develop a bioassay for cytokinins (Fig. 2). The relatively short assay time and the high sensitivity, among other advantages, qualify it as an attractive bioassay. In this oat assay system, gibberellic acid did not stimulate transpiration. However, Livne and Vaadia (1965) observed that the rate of transpiration of barley leaves was stimulated by gibberellic acid ($1.5 \times 10^{-4} M$). A possible explanation for the apparent discrepancy between the two systems may be related not only to the different plants employed, but also to the difference in the procedure. Livne and Vaadia observed (1968) that the response to gibberellic acid was limited to freshly excised barley leaves. When kept excised for 24 hr, stimulation of transpiration by gibberellic acid was no

TABLE I

STIMULATION OF TRANSPIRATION BY CYTOKININS

Plant	Species	Cytokinin tested	Concentrations[a]	Reference
Barley	Hordeum vulgare L.	Kinetin	3×10^{-6}	Livne and Vaadia (1965)
		Kinetin	(3×10^{-6})	Meidner (1967)
Wheat	Triticum vulgare L.	Kinetin	(10^{-6})	Luke and Freeman (1967)
		Kinetin	(10^{-6})	Luke and Freeman (1967)
		Kinetin	(2.5×10^{-5})	Mittelheuser and Van Steveninck (1969)
Oat	Avena sativa L.	Kinetin	$2.4 \times 10^{-8}\text{–}4 \times 10^{-5}$	Luke and Freeman (1968)
		Kinetin	(2.4×10^{-6})	Pallas and Box (1970)
		Zeatin	$5 \times 10^{-9}\text{–}5 \times 10^{-6}$	Luke and Freeman (1968)
		N^6-Benzylaminopurine	$2.4 \times 10^{-8}\text{–}4 \times 10^{-5}$	Luke and Freeman (1968)
		N^6-Phenylaminopurine	$>5 \times 10^{-5}\text{–}5 \times 10^{-5}$	Luke and Freeman (1968)
		N^6-(2-Butoxyethyl)aminopurine	$>2.4 \times 10^{-7}\text{–}2 \times 10^{-5}$	Luke and Freeman (1968)
		N^6-Hexylaminopurine	$5 \times 10^{-7}\text{–}5 \times 10^{-5}$	Luke and Freeman (1968)
		N^6-(2-Ethoxyethyl)aminopurine	$5 \times 10^{-6}\text{–}10^{-5}$	Luke and Freeman (1968)
Tobacco	Nicotiana rustica L.	Kinetin	(5×10^{-5})	Mizrahi et al. (1970)
Groundsel	Senecio odoris L.	Kinetin	5×10^{-6}	Livne (1966)
		Kinetin	5×10^{-6}	Livne (1966)
Tomato	Lycopersicon esculentum	N^6-Benzylaminopurine	$(10^{-5}\text{–}10^{-4})$	Tal and Imber (1971a)
Ryegrass	Lolium multiflorum Lam.	Kinetin		Luke and Freeman (1968)

[a] Optimal concentration or range of detectable concentrations are included when specified by the authors. If a single concentration was used, the figures are given in parentheses.

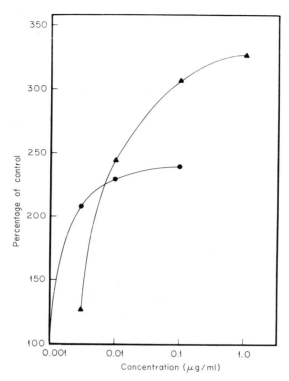

Fig. 2. Effect of concentration of cytokinin on the rate of transpiration of excised oat leaves. The basal ends of the leaves were immersed in the solution tested for 12 or 24 hr under continuous light and the leaves then transferred to distilled water for measurement of transpiration, which lasted an equal length of time. From Luke and Freeman (1967).

longer measurable. Thus, an interaction between gibberellic acid and cytokinin in affecting stomatal opening seems plausible and should be further explored.

Sensitivity to applied gibberellic acid may also be acquired by a treatment that reduces the endogenous level of the hormone. In such a system the effect of added gibberellic acid on plant water balance could be assessed. A possible relevant system is described by Halevy and Kessler (1963). They reported that bean plants treated once through the soil with the growth retardants (2-chloroethyl)trimethyl ammonium chloride (CCC) and 2,4-dichlorobenzyltributyl phosphonium chloride (Phosphon) remained turgid and survived longer than control plants when irrigation was withheld. Under ample irrigation there were no differences in growth, in terms of fresh and dry weight, between Phosphon-treated and control plants. However, when irrigation was withheld to near the wilting point, the growth of the treated plants exceeded that of the untreated ones. This apparent increase in the drought tolerance of the plants by Phosphon or CCC might be related to interference of these compounds with biosynthesis of gibberellic acid. A reduced level of gibberellic acid in the treated plants could result in reduced stomatal opening and a lowered transpiration and thus also in a more positive water balance. If this sequence is indeed valid, it is predicted that applied gibberellic acid would increase stomatal opening and transpiration of plants treated with Phosphon and CCC.

The elucidation of the mode of action of cytokinin in promoting stomatal opening offers an intriguing challenge for hormonologists as well as for students of plant water relations. A proposed mechanism should take into consideration that stimulation of transpiration by kinetin can be observed also in the dark (Luke and Freeman, 1967; Pallas and Box, 1970). As mentioned above, the structural requirements for stimulation of transpiration by cytokinins are similar to those required for induction of cell division (Luke and Freeman, 1968). Furthermore, the proposed mechanism should account for the relatively rapid response of the tissue to the applied kinetin (Livne and Vaadia, 1965).

Meidner (1967) suggested that the effect of kinetin on stomatal resistance was partly indirect, since treatment of mature primary leaves of barley with $3 \times 10^{-6} M$ solutions of kinetin resulted in increased rates of net assimilation of CO_2. Since the effect of kinetin was apparent despite attempts to keep the concentration of CO_2 constant at the leaf environment, Meidner concluded that in addition to causing a reduction in the concentration of CO_2 inside the leaves, kinetin appeared to affect the stomatal mechanism directly. Pallas and Box (1970) proposed that the stomatal response to kinetin is osmopassive and turgorpassive: The osmotic potentials of the accessory, epidermal, and mesophyll cells of treated leaves would become less negative, presumably because of reduced pools of osmotically active compounds, such as sugars and amino or organic acids. Consequently, turgor would decrease and therefore guard cell turgidity would dominate and the stomata would open. Pallas and Box (1970) were indeed able to measure lower negative osmotic water potential and lower positive turgor pressure in excised oat leaves treated for 48 hr with $2.4 \times 10^{-6} M$ kinetin than in leaves kept in distilled water. The differences amounted to several atmospheres.

It is unlikely that such overall differences could be manifested in the kinetin-treated tissue within 1 hr or less—the time required to affect stomatal opening by kinetin (Fig. 1). Therefore, the effect of cytokinins in stimulating stomatal opening should be of a more direct nature. This conclusion is supported by the observation (Livne, 1966) that stomatal opening is stimulated by kinetin also in epidermal strips of leaves of responding species (*Nicotiana rustica* and *Senecio odoris*), with a dose response and a lag period similar to that found for intact tissue.

It has been advocated that light-dependent net uptake of potassium ions by guard cells is characterized by a sufficient rate and magnitude to account, in terms of osmotic values, for stomatal opening (Fujino, 1967; Fischer and Hsiao, 1968; Sawhney and Zelitch, 1969). A role of cytokinin in sustaining this process is possible if, for example, the hormone affects permeability of cell membranes to ions and/or water. Regulation of ATPase, which supports the efflux of K^+ from the guard cells (Fujino, 1967), is another possible site of action of cytokinin that can be explored experimentally.

C. ABSCISIC ACID

Although relatively little work has been devoted to the study of hormonal effects and water relations in trees, the first report on reduced transpiration by ABA was related to woody species. Little and Eidt (1968) concluded from their study of softwood and hardwood plants that a role may be assigned to ABA in regulating not only budbreak and cambial activity but also transpiration of trees. Mittelheuser and Van Steveninck (1969) showed that the reduction in transpiration by ABA in excised barley and wheat leaves was associated with stomatal closure.

Jones and Mansfield (1970) found that the stomatal response to ABA is rapid. When applied to detached tobacco leaves through the petiole at $10^{-4} M$, ABA caused almost complete closure of stomata within less than an hour. Stomatal closure induced by ABA is apparently not the result of an increase in the CO_2 concentration in the intercellular spaces, since this hormone effect was not reversed by CO_2-free air (Jones and Mansfield, 1970).

The effect of ABA in causing stomatal closure and in inhibiting transpiration has been documented in a number of plant species (Table II). No case has been reported of ineffectiveness of applied ABA in reducing transpiration in plants.

Since ABA is probably present in all higher plant species (Milborrow, 1969) it is of interest to assess whether the endogenous hormone plays a role in regulating stomatal opening. Support for such a role is available from several lines of work. As already mentioned, Wright (1969) showed an increase in ABA content in wilting wheat leaves, and indeed stomatal opening is commonly reduced in leaves undergoing water deficits. Mizrahi *et al.* (1970) reported that in tobacco plants subjected to osmotic or

TABLE II

INHIBITION OF TRANSPIRATION AND STOMATAL OPENING BY ABSCISIC ACID

Plant	Species	Effective concentration (M)	Stomatal closure	Reference
Balsam fir	Abies balsamea L.	1.5×10^{-6}–3.8×10^{-5}	ND[a]	Little and Eidt (1968)
White spruce	Picea glauca (Moench) Voss	1.5×10^{-6}–3.8×10^{-5}	ND	Little and Eidt (1968)
Wheat	Triticum vulgare Vill	3.8×10^{-7} and 3.8×10^{-6}	+	Mittelheuser and Van Steveninck (1969)
Barley	Hordeum vulgare L.	3.8×10^{-7} and 3.8×10^{-6}	+	Mittelheuser and Van Steveninck (1969)
Oats	Avena sativa L.		ND	Mittelheuser and Van Steveninck (1969)
Nasturtium	Tropeolum majus L.		ND	Mittelheuser and Van Steveninck (1969)
Tobacco	Nicotiana rustica L.	4×10^{-5}	ND	Mizrahi et al. (1970)
	Nicotiana tabacum L.	10^{-4}	+[b]	Jones and Mansfield (1970)
Xanthium	Xanthium pensylvanicum Wall	10^{-4}	+[b]	Jones and Mansfield (1970)
Tomato	Lycopersicon esculentum Mill.	3.8×10^{-6} and 3.8×10^{-5}	+	Tal and Imber (1970)

[a] ND, not determined.
[b] Only stomatal opening measured.

salinity root stress, transpiration rates decline while the content of ABA is increased.

These correlations are not unequivocal, since the distinction between cause and effect is yet to be clarified. More definite support comes from a genetic approach.

Tal (1966) described a tomato mutant, *flacca,* which wilts rapidly under water stress because of abnormal stomatal behavior. The stomata of the mutant remain open under conditions of darkness, wilting, plasmolysis of guard cells, and treatment with phenylmercuric acetate. Such conditions cause stomatal closure in the normal variety (Rheinlands Ruhm) from which the mutant was derived. Imber and Tal (1970) further showed that when ABA was applied to the mutant seedlings, either by a foliage spray (3.8×10^{-6} or $2.8 \times 10^{-5} M$) given once a day or in root medium ($3.8 \times 10^{-6} M$), a phenotypic reversion of the mutant took place within several days, including turgidity of the leaves. While guttation was completely absent in the control mutant plants, it appeared in the morning on the edges of the leaves of mutant plants treated with ABA, as in normal control plants. Furthermore, the rate of transpiration in leaves of treated mutant plants declined with increasing concentration of the hormone. However, the mutant phenotype fell short of full reversion with respect to transpiration rate. Decreased water loss was detected in both young leaves that developed during the hormonal treatment and in older leaves that had already been fully developed. It is of interest that this reversion is dependent on continued treatment with abscisic acid, since the treated mutant plants regained their wilty phenotype several days after the hormone spraying had ceased. This indicates that ABA undergoes a rapid turnover in these plants.

IV. HORMONES AND WATER BALANCE

A close coupling between changes of a regulator and a physiological process, with mutual effects, is a prerequisite for an effective regulation based on feedback mechanism. Furthermore, the availability of two or more regulators which affect the physiological process, each exerting an opposite effect, enables a fine adjustment of the outcome process.

These requirements appear to be met with respect to hormones and water balance.

The foregoing sections illustrated with some details the mutual effects of hormones and water balance. Water stress exerts a profound effect on hormonal distribution, particularly on the contents of cytokinins and ABA. As a consequence of water deficits, cytokinin activity is reduced whereas ABA is increased. These same hormones may greatly affect water balance.

An increase in abscisic acid reduces stomatal opening, while cytokinin exerts an opposite effect. Water loss through transpiration can thus be kept under hormonal control. Hormones may also affect the resistance of roots to water flow. Tal and Imber (1971a) showed that kinetin added to the root medium of tomato plants, either before or after decapitation, decreased the rate of exudation of stumps. The kinetin treatment also increased the difference between the osmotic pressure of the exudate and the root media. Auxin and ABA exerted an opposite effect, but the decreasing effect of added ABA on the difference in osmotic pressure was measurable only in mutant variety *flacca,* which is deficient in ABA. Due to the combined effects of the hormones on water uptake and on transpiration, Tal and Imber (1971a) concluded that kinetin would cause plant turgor to decrease, while an increase of turgor would result from elevation of ABA and IAA levels. A daily rhythm of hormones, reflecting the periodic changes of plant turgidity and affecting stomatal and root resistance to water flow, is highly probable. Furthermore, the autonomic diurnal fluctuations in the rate of exudation and root pressure (Vaadia, 1960) may be a manifestation of the unique interaction between hormonal regulation and water balance.

The combined effects of two types of stress imposed on a plant illustrate the involvement of hormones in the adaptation of plants to a changing environment. Such a case is presented by Mizrahi et al. (1972), by employing salinity stress and reduced root aeration. Cessation of root aeration to tobacco plants growing in culture solutions resulted in rapid wilting of the shoot. Plants growing in salinated solutions (0.1 M NaCl) for at least 2 days, retained turgor in the absence of aeration. This resistance to lack of aeration was found to be associated with increased concentration of ABA in leaves. Maximal values of ABA-like content in the leaves were reached by day 4 of salination and coincided with the attainment of maximal resistance to lack of aeration. Stopping aeration to plants which were root-fed for 48 hr with either ABA ($4 \times 10^{-6} M$) or kinetin ($5 \times 10^{-6} M$) resulted in water saturation deficits greatly below and above the control, respectively. Mizrahi et al. concluded that ABA is a major factor facilitating the adaptive response of plants to root stresses that impede water balance, whereas the ratio of cytokinins to ABA is implicated in directing the extent of the response.

Stomata commonly fail to resume regular opening after a period of wilting of leaves despite rapid recovery of leaf water content. This aftereffect, which may last for several days, has been amply documented in the literature for many plant species (Stålfelt, 1955, 1963; Iljin, 1957; Milthorpe and Spencer, 1957; Glover, 1959; Heath and Mansfield, 1962; Fischer et al., 1970). Fischer (1970) concluded that the major part of the temporary aftereffect was located in the guard cells, while only a minor part contributed by the mesophyll, possibly owing to an elevated CO_2 concentration in the stomatal cavity. Allaway and Mansfield (1970) independently reached a similar conclusion regarding the role of CO_2. By eliminating several other possibilities, Allaway and Mansfield concluded that the aftereffect was due to accumulation of an inhibitor (ABA?) following a period of water stress. Alternatively the aftereffect may be caused by deficiency of a substance which promotes stomatal opening (cytokinin?).

Meidner and Mansfield (1968) and Allaway and Mansfield (1970) considered the aftereffect to be a protective device that guards plants against excessive transpiration during periods when water is in short supply. They envisage a potential practical advantage of developing crop varieties which show pronounced aftereffects. The operation of an amplified endogenous mechanism for suppression of stomatal opening may thus provide a more promising approach for growing crops in semiarid regions than the use of chemical antitranspirants. The possible use of ABA as an antitranspirant has been pointed out (Milborrow, 1969; Jones and Mansfield, 1970). However, its high turnover complicates the practical application of ABA as an antitranspirant. Furthermore, Little and Eidt (1970) report some deleterious effect of ABA on wood production, since the effect of ABA on transpiration in balsam fir was accompanied also by reduced cambial activity.

A preparation of ABA which would result in slow release of the hormone in the plant may ensure a durable beneficial effect for water economy with minimal deleterious effects.

REFERENCES

Addicott, F. T., and Lyon, J. L. (1969). Physiology of abscisic acid and related substances. *Annu. Rev. Plant Physiol.* **20**, 139.

Allaway, W. G., and Mansfield, T. A. (1970). Experiment and observations on the aftereffect of wilting on stomata of *Rumex sanguirews. Can. J. Bot.* **48**, 513.

Allerup, S. (1964). Induced transpiration changes: Effects of some growth substances added to the root medium. *Physiol. Plant.* **17**, 899.

Andreenko, S. S., Potagou, N. G., and Kosulina, L. G. (1964). The effect of sap from maize plants grown at various pH levels on growth of carrot callus. *Dokl. Biochem.* **155**, 35.

Back, A., Bittner, S., and Richmond, A. E. (1972). The effect of abscisic acid on the metabolism of kinetin in detached leaves of *Rumex pulcher. J. Exp. Bot.* (in press).

Barnett, N. M., and Naylor, A. W. (1966). Amino acid and protein metabolism in Bermuda grass during water stress. *Plant Physiol.* **41**, 1222.

Ben-Zioni, A., Itai, C., and Vaadia, Y. (1967). Water and salt stress, kinetin and protein synthesis in tobacco leaves. *Plant Physiol.* **42**, 361.

Brandbury, D., and Ennis, W. (1952). Stomatal closure in kidney bean treated with ammonium 2,4-dichrophenoxyacetate. *Amer. J. Bot.* **39**, 324.

Brown, J. W. (1946). Effect of 2,4-dichlorophenoxyacetic acid on the water relations, the accumulation and distribution of solid matter and the respiration of bean plants. *Bot. Gaz.* **108**, 332.

Burrows, W. J., and Carr, D. J. (1969). Effect of flooding the root system of sunflower plants on the cytokinin content in the xylem sap. *Physiol. Plant.* **22**, 1105.

Darbyshire, B. (1971). Changes in indoleacetic acid oxidase activity associated with plant water potential. *Physiol. Plant.* **25**, 80.

Ferri, M. G., and Lex, A. (1948). Stomatal behaviour as influenced by treatment with β-naphoxyacetic acid. *Contrib. Boyce Thompson Inst.* **15**, 283.

Ferri, M. G., and Rachid, A. (1949). Further information on the stomatal behaviour influenced by treatment with hormone-like substances. *Ann. Acad. Brasil. Cienc.* **21**, 155.

Fischer, R. A. (1970). After effect of water stress on stomatal opening potential. II. Possible causes. *J. Exp. Bot.* **21**, 386.

Fischer, R. A., and Hsiao, T. C. (1968). Stomatal opening in isolated epidermal strips of *Vicia faba*. II. Responses to KCl concentration and the rate of potassium absorption. *Plant Physiol.* **43**, 1953.

Fischer, R. A., Hsiao, T. C., and Hagan, R. M. (1970). After effect of water stress on stomatal opening potential. I. Techniques and magnitude. *J. Exp. Bot.* **21**, 371.

Fujino, M. (1967). Role of adenosine triphosphate and adenosinetriphosphatase in stomatal movement. *Sci. Bull. Fac. Educ., Nagaski Univ.* **18**, 1.

Glover, J. (1959). The apparent behaviour of maize and sorghum stomata during and after drought. *J. Agr. Sci.* **53**, 412.

Gunning, B. E. S., and Burkley, W. K. (1963). Kinin-induced directed transport and senescence in detached oat leaves. *Nature (London)* **199**, 262.

Halevy, A. H., and Kessler, B. (1963). Increased tolerance of bean plants to soil drought by means of growth retarding substances. *Nature (London)* **197**, 310.

Heath, O. V. S., and Mansfield, T. A. (1962). A recording porometer with detachable cups operating on four separate leaves. *Proc. Roy. Soc., Ser. B* **156**, 1.

Iljin, W. S. (1957). Drought resistance in plants and physiological processes. *Annu. Rev. Plant Physiol.* **8**, 257.

Imber, D., and Tal, M. (1970). Phenotypic reversion of *flacca,* a wilty mutant of tomato by abscisic acid. *Science* **169**, 592.

Itai, C., and Vaadia, Y. (1965). Kinetin-like activity in root exudate of water-stressed sunflower plants. *Physiol. Plant.* **18**, 941.

Itai, C., and Vaadia, Y. (1971). Cytokinin activity in water-stressed shoots. *Plant Physiol.* **47**, 87.

Itai, C., Richmond, A. E., and Vaadia, Y. (1968). The role of root cytokinins during water and salinity stress. *Isr. J. Bot.* **17**, 187.

Johansen, S. (1954). Effect of Indole-acetic acid on stomata and photosynthesis. *Physiol. Plant.* **7**, 531.

Jones, R. J., and Mansfield, T. A. (1970). Suppression of stomatal opening in leaves treated with abscisic acid. *J. Exp. Bot.* **21**, 714.

Kende, H. (1965). Kinetinlike factors in the root exudate of sunflowers. *Proc. Nat. Acad. Sci. U. S.* **53**, 1302.

Kende, H. (1971). The cytokinins. *Int. Rev. Cytol.* (in press).

Koshimizu, K., Inui, M., Fukui, H., and Mitzui, T. (1968). Isolation of (+)-abscisyl-

β-D-glucopyraside from immature fruit of *Lupinus luteus*. *Agr. Biol. Chem.* **32**, 789.

Kozinka, V. (1967). Water uptake during rapid changes of transpiration induced by the presence of high concentration of growth substances in root medium. *Biol. Plant.* **9**, 222.

Kozinka, V. (1968). Water balance of plants during root application of high concentrations of growth substances. *Biol. Plant.* **10**, 398.

Kozinka, V. (1970). Inhibition of water uptake by high concentration of auxin-like substances. *Biol. Plant.* **12**, 180.

Krikon, J., Chorin, M., and Vaadia, Y. (1970). Unpublished data.

Kulaeva, O. H. (1962). The effect of roots on leaf metabolism in relation to the action of kinetin on leaves. *Sov. Plant Physiol.* **9**, 182.

Leopold, A. C., and Kawase, M. (1964). Benzyladenine effects on bean leaf growth and senescence. *Amer. J. Bot.* **51**, 294.

Little, C. H. A., and Eidt, D. C. (1968). Effect of abscisic acid on bud break and transpiration in woody species. *Nature (London)* **220**, 498.

Little, C. H. A., and Eidt., D. C. (1970). Relationship between transpiration and cambial activity in *Abies balsamea*. *Can. J. Bot.* **48**, 1027.

Livne, A. (1966). Unpublished observation.

Livne, A., and Vaadia, Y. (1965). Stimulation of transpiration rate in barley leaves by kinetin and gibberellic acid. *Physiol. Plant.* **18**, 658.

Livne, A., and Vaadia, Y. (1968). Unpublished data.

Lopushinsky, W. (1969). Stomatal closure in conifer seedlings in response to leaf moisture stress. *Bot. Gaz.* **130**, 250.

Luke, H. H., and Freeman, T. E. (1967). Rapid bioassay for phytokinins based on transpiration of excised oat leaves. *Nature (London)* **215**, 874.

Luke, H. H., and Freeman, T. E. (1968). Stimulation of transpiration by cytokinins. *Nature (London)* **217**, 873.

Mansfield, T. A. (1967). Stomatal behaviour following treatment with auxin-like substances and phenylmercuric acetate. *New Phytol.* **66**, 325.

Meidner, H. (1967). The effect of kinetin on stomatal opening and the rate of intake of carbon dioxide in mature primary leaves of barley. *J. Exp. Bot.* **18**, 556.

Meidner, H., and Mansfield, T. A. (1968). "Physiology of Stomata." McGraw-Hill, New York.

Milborrow, B. V. (1967). The identification of (+)-abscisin II [(+) dormin] in plants and measurements of its concentration. *Planta* **76**, 93.

Milborrow, B. V. (1969). The occurrence and function of abscisic acid in plants. *Sci. Prog., Oxf.* **57**, 533.

Milthorpe, F. L., and Spencer, E. J. (1957). Experimental studies of the factors controlling transpiration. III. The interrelationships between transpiration rate, stomatal movement, and leaf water content. *J. Exp. Bot.* **8**, 414.

Mittelheuser, C. J., and Van Steveninck, R. F. M. (1969). Stomatal closure and inhibition of transpiration induced by RS-abscisic acid. *Nature (London)* **221**, 281.

Mizrahi, Y., Blumenfeld, A., and Richmond, A. E. (1970). Abscisic acid and transpiration in leaves in relation to osmotic root stress. *Plant Physiol.* **46**, 169.

Mizrahi, Y., and Richmond, A. E. (1971). Personal communication.

Mizrahi, Y., Blumenfeld, A., and Richmond, A. E. (1972). The role of abscisic acid and salination in the adaptive response of plant to reduced root aeration. *Plant Cell Physiol.* **13**(1), in press.

Mothes, K., Engelbrecht, L., and Kulajewa, O. (1959). Uber die Wirkung des Kinetins auf Stickstoff-verteilung and Eiweisssynthese in isolierten Blättern. *Flora (Jena)* **147**, 445.

Osborne, D., and McCalla, B. R. (1961). Rapid bioassay for kinetin and kinins using senescing leaf tissue. *Plant Physiol.* **36**, 219.

Pallas, J. E., and Box, J. E. (1970). Explanation for the stomatal response of excised leaves to kinetin. *Nature (London)* **227**, 87.

Pegg, G. F., and Selman, I. W. (1959). An analysis of the growth response of young tomato plants to infection by *Verticillium albo-atrum*. II. The production of growth substances. *Ann. Appl. Biol.* **47**, 222.

Phillips, I. D. J. (1964). Root-shoot hormone relation. II. Changes in endogenous auxin concentration produced by flooding of the root system in *Helianthus annuus*. *Ann. Bot. (London)* [N. S.] **28**, 36.

Player, M. A. (1950). Effect of some growth regulating substances on the transpiration of *Zea mays* L. and *Ricinus Communis L. Plant Physiol.* **25**, 469.

Pohl, R. (1961). Wuchsstoffe und Wasseraufnahme. *In* "Handbuch de Pflanzenphysiologie" (W. Ruhland, ed.), Vol. 14, pp. 743–753. Springer-Verlag, Berlin and New York.

Reid, D. M., Crozier, A., and Harvey, B. M. R. (1969). The effect of flooding on the export of gibberellins from the root to the shoot. *Planta* **89**, 376.

Richmond, A. E., and Lang, A. (1957). Effect of kinetin on protein content and survival of detached Xanthium leaves. *Science* **125**, 650.

Sawhney, B. L., and Zelitch, I. (1969). Direct determination of potassium ion accumulation in guard cells in relation to stomatal opening in light. *Plant Physiol.* **44**, 1350.

Shah, C. B., and Loomis, R. S. (1965). Ribonucleic acid and protein metabolism in sugar beet during drought. *Physiol. Plant.* **18**, 240.

Stadelmann, E. J. (1969). Permeability of the plant cell. *Annu. Rev. Plant Physiol.* **20**, 585.

Stålfelt, M. G. (1955). The stomata as a hydrophotic regulator of the water deficit of the plant. *Physiol. Plant.* **8**, 572.

Stålfelt, M. G. (1963). Diurnal dark reactions in the stomatal movement. *Physiol. Plant.* **16**, 756.

Steadman, J. R., and Sequeira, L. (1970). Abscisic acid in tobacco plants. Tentative identification and its relation to stunting induced by *Pseudomonas solanancearum. Plant Physiol.* **45**, 691.

Takaoki, T. (1962). Relationships between plant hydration and respiration III. Water absorption of plants treated with auxins in culture solution. *J. Sci. Hiroshima Univ., Ser. B, Div. 2* **9**, 185.

Tal, M. (1966). Abnormal stomatal behavior in wilty mutants of tomato. *Plant Physiol.* **41**, 1387.

Tal, M., and Imber, D. (1970). Abnormal stomatal behavior and hormonal imbalance in *flacca,* a wilty mutant of tomato. II. Auxin and abscisic acid-like activity. *Plant Physiol.* **46**, 373.

Tal, M., and Imber, D. (1971a). Abnormal stomatal behavior and hormonal imbalance in *flacca,* a wilty mutant of tomato III. Hormonal effects on the water status in the plant. *Plant Physiol.* **47**, 849.

Tal, M., and Imber, D. (1971b). The effect of a prolonged 2,4-dichlorophenoxyacetic acid treatment on transpiration and stomatal distribution in tomato leaves. *Planta* **97**, 179.

Vaadia, Y. (1960). Autonomic diurnal fluctuations in rate of exudation and root pressure of decapitated sunflower plants. *Physiol. Plant.* **13**, 701.

Vaadia, Y., and Itai, C. (1968). Interrelationships of growth with reference to the distribution of growth substances. *In* "Root Growth" (W. J. Whittington, ed.), p. 65. Butterworth, London.

Vaadia, Y., Raney, F. C., and Hagan, R. M. (1961). Plant water deficits and physiological processes. *Annu. Rev. Plant Physiol.* **12**, 265.

von Guttenberg, H., and Beythien, A. (1951). Uber den Einfluss von Wirkstoffen auf die Wasserpermeabilitat des Protoplasmas. *Planta* **40**, 36.

Wareing, P. F., and Ryback, G. (1970). Abscisic acid: a newly discovered growth-regulating substance in plants. *Endeavour* **29**(107), 84.

Wright, S. T. C. (1969). An increase in the "Inhibitor-β" content of detached wheat leaves following a period of wilting. *Planta* **86**, 10.

Wright, S. T. C., and Hiron, R. W. P. (1969). (+)-Abscisic acid, the growth inhibitor induced in detached wheat leaves by a period of wilting. *Nature (London)* **224**, 719.

Zelitch, I. (1961). Biochemical control of stomatal opening in leaves. *Proc. Nat. Acad. Sci. U. S.* **47**, 1423.

Zelitch, I. (1963). The control and mechanisms of stomatal movement. *Conn., Agr. Exp. Sta., New Haven, Bull.* **664**, 18–36.

CHAPTER 9

PHYSIOLOGICAL BASIS AND PRACTICAL PROBLEMS OF REDUCING TRANSPIRATION

Alexandra Poljakoff-Mayber and J. Gale

DEPARTMENT OF BOTANY, THE HEBREW UNIVERSITY OF JERUSALEM,
JERUSALEM, ISRAEL

I. INTRODUCTION

A. PRACTICAL IMPORTANCE OF REDUCING TRANSPIRATION

Most of the water applied in irrigation is lost in the process of transpiration. It is therefore surprising that not more attention has been given to development of means for reducing this loss, although work in this field goes back many years (Miller *et al.,* 1950). Reduction of transpiration can be important both for decreasing irrigation water requirements and, under certain environmental conditions, for alleviation of water stress. Water stress develops in the plant when the rate of transpiration exceeds that of the water supply and transpiration capacity of the plant. This stress is expressed in a drop in plant water potential and turgor. Under such environmental conditions reduction of growth will occur, even when the soil is moist (Kozlowski, 1968).

Very few reviews and research reports on the physiological basis of antitranspirants have appeared in the last ten years (Laborde, 1964; Gale and Hagan, 1966; Waggoner and Zelitch, 1965; Waggoner, 1966a). In this chapter the theory behind the various methods for reduction of transpiration is discussed, as well as the practical problems involved.

B. Importance of Transpiration to Plants

Before an attempt can be made to reduce transpiration the question must be asked as to whether transpiration is a process that is essential to the plant's existence. The two most important ways by which plants may benefit from transpiration are expedition of mineral uptake and cooling of leaves.

Little progress has been made in elucidating the effects of transpiration on mineral uptake and transport since the reviews of Russell-Scott and Barber (1960) and Gale and Hagan (1966). In the absence of further evidence, the tentative conclusion that only a very low rate of transpiration is necessary for mineral transport may still be considered as valid.

Leaf temperatures depend on a delicate balance between net incoming radiation and factors of dissipation: mainly, sensible heat exchange, long wave radiation, and the latent heat of evaporation expended in transpiration.

The factors of dissipation are interdependent. For example, when transpiration is reduced by the presence of a resistance in the transpiration diffusion pathway, leaf temperatures will tend to rise. This in itself will raise again the rate of transpiration. However a rise in leaf temperature also increases the dark body radiation of the leaf and the sensible heat exchange with the surrounding air.

These considerations are expressed in the leaf energy balance equation [Eq. (1)]:

$$Q = R + S + LE$$
$$= K\epsilon T_1^4 + a(T_1 - T_a) + L(e_1 - e_a)/(r_1 + r_a) \tag{1}$$

where Q is net radiation entering the leaf in units such as $cal \cdot cm^{-2} \cdot min^{-1}$; R, S, and LE are dissipation terms for radiation, sensible heat, and evaporative cooling, respectively; K is a constant; ϵ is the black body emissivity of the leaf, which is usually close to 0.95; T_1 is the leaf temperature and T_a the air temperature in degrees Kelvin; a is a coefficient which depends on wind velocity and leaf dimensions; L is the heat of vaporization of water at temperature T_1 in cal·gm⁻¹; e_1 and e_a are the densities of water vapor in the leaf and in the air, respectively, in units of gm·cm⁻³; r_1, resistance to diffusion of water vapor from the leaf and r_a, resistance to water vapor transport in the leaf boundary layer, in units of min·cm⁻¹.

This leaf heat budget equation was first developed by Raschke (1960) and has been discussed in detail by Gates (1968). Inspection of Eq. (1) shows that all three components of heat dissipation are temperature dependent (e_1 present in LE also being an exponential function of temperature). It is also seen that the equation is transcendental. Consequently, there is no simple effect of the reduction of transpiration on leaf temperature. In the same way the effect of change in one of the factors determining the leaf energy balance on the rate of transpiration and on leaf temperature is not immediately apparent. However, the equation can be solved for any given combination of environmental parameters by numerical analysis.

Gates presents nomograms showing calculated rates of transpiration and leaf temperatures for a limited number of environmental conditions. These are presented as functions of some of the other environmental factors. Such nomograms, calculated from the energy balance equation, are very valuable for calculating the effect of transpiration on leaf temperatures and also for estimating the effect of an added leaf resistance (such as discussed in Section II,D) both on transpiration and leaf temperatures. It should be remembered that Gates' analysis was made for single leaves and not for an entire canopy; neither were changes considered in leaf resistance. Such changes in resistance may occur in response to changing conditions of water supply and demand. This will be further discussed in Section III.

In Table I are compiled some representative figures from Gates' nomograms which show the effect, on leaf temperature, of reduced transpiration, brought about by an increase of leaf diffusion resistance.

The data shown in Table I indicate that except for conditions of high insolation and relatively high humidity (NR 1.2 cal·cm⁻²·min⁻¹ and RH 50%) leaf temperatures were very close to air temperatures. This is in agreement with field measurements with a number of different plants made by Gale et al. (1965). Furthermore, the data of Table I show that a relatively large reduction of transpiration ($\sim$40%) did not result in more than an approximately 3°C rise in leaf temperature. With even a slight wind of 50 cm·sec⁻¹ (10 cm·sec⁻¹ being equivalent to conditions of essentially no wind or free convection) and with leaves of moderate size, a reduction of

TABLE I

CALCULATED RATES OF TRANSPIRATION AND OF LEAF TEMPERATURE UNDER VARYING ENVIRONMENTAL CONDITIONS, FOR DIFFERENT VALUES OF LEAF RESISTANCE TO WATER VAPOR DIFFUSION

Figure number (from Gates, 1968)[a]	Leaf dimensions (cm)	Total absorbed radiation (cal·cm⁻²·min⁻¹)	Environmental conditions				Leaf resistance to water vapor diffusion (sec·cm⁻¹)	Transpiration (gm·cm⁻²·min⁻¹ × 10⁶)	Leaf temperature (°C)
			Air temperature (°C)	Relative humidity (%)	Wind speed (cm·sec⁻¹)				
3	5 × 5	1.2	20	50	10	2	49	31.7
						5	29	35.0
					50	2	43	26.6
						5	22	28.7
1	5 × 5	1.2	40	50	10	2	64	43.0
						5	46	46.3
					50	2	70	41.0
						5	40	43.8
2	5 × 5	0.8	40	50	10	2	34	36.2
						5	23	38.2
					50	2	43	36.8
						5	27	38.6
9	1 × 1	1.2	40	20	100	4	61	40.5
						8	33	41.5
9	20 × 20	1.2	40	20	100	4	62	43.0
						8	41	47.0

[a] The figures given have been obtained by interpolation from nomograms presented by Gates (1968).

transpiration of 40% is predicted to produce at the most a 2–2.5°C rise in leaf temperature.

As pointed out by Gates and measured in some plants by Lange and Lange (1963) a complete stoppage of transpiration may, under certain very extreme conditions, produce a 10°C rise in leaf temperature. Such a rise, or even a 3–4° increase, may result in a detrimental rise in the respiration/photosynthesis ratio, and may even be lethal. However, as discussed by Gale and Hagan (1966), in many plants stomata tend to close during the hottest hours of the day, when cooling by transpiration would be most beneficial. Thus transpiration is reduced at the very time when its cooling effect would be most advantageous.

It may be concluded therefore that normally, a relatively large reduction in transpiration, of 40–50%, would not be detrimental to the plant. Furthermore, under conditions of excessive transpiration, reduction of the rate of transpiration would prevent the development of water deficits. This will be discussed in Section III.

C. The Driving Force and Pathway of Transpiration

The factors governing the loss of water from leaves have already been discussed by Tanner (1968) and Cowan and Milthorpe (1968) and have also been alluded to, in this chapter, in the LE component of the leaf energy balance [Eq. (1)]. Consequently, only a brief outline is given here.

Loss of water through transpiration is a passive process in which water moves from the leaves to the air, mainly by vapor diffusion and partly by turbulent transfer. The driving force is the vapor density gradient between the air of the substomatal cavities and the air around the leaf or above the leaf canopy. In its most simple form transpiration can be described by a diffusion equation of the form:

$$T = \frac{e_1 - e_a}{r_1 + r_a} \qquad (2)$$

where T is the transpiration flux from the leaf, r_1 is the resistance of the leaf to water loss, r_a is the boundary layer resistance to water vapor transfer in the immediate vicinity of the leaf, and e_1 and e_a are the water vapor densities in the leaf and the air, respectively (Gaastra, 1959).

The main pathway for water vapor diffusion from the leaf is through the stomata. The mesophyll wall and internal tortuosity resistances are usually, if not always, considered to be quite small (Slatyer, 1966; Gale et al., 1967; Jarvis and Slatyer, 1970). Consequently, the most important (and variable) resistance involved in water loss from leaves is that of the stomata.

Equation (2) is strictly an expression of the net-escaping tendency of water. The rate of transpiration, T, will be maintained only if there is a source of energy to replace the energy spent in evaporation. In the absence of such a source of energy, leaf temperature (and hence e_1) will soon drop. Furthermore, there must be an adequate supply of water; as noted in the introduction, the rate of transpiration must not exceed the water transport capacity of the plant roots and stems. If it does, leaf turgor will be lost and r_1 will increase.

Inspection of Eq. (2) shows that transpiration may be reduced in four ways: by reducing the net energy input and hence e_1; by increasing e_a, the humidity of the air near the leaf; by increasing r_1, the resistance of the leaf to water loss; and by increasing r_a, the resistance of the air near the leaf to vapor transport.

A first requirement of any method for reducing transpiration is that it should not interfere with plant growth. The reduction of plant transpiration is not in itself difficult to accomplish, but to do so without interference with photosynthesis and growth is the central problem of antitranspirant studies.

II. VARIOUS APPROACHES TO REDUCTION OF TRANSPIRATION

A number of methods have been studied as possibilities for reducing transpiration. The four most important are (*a*) increasing leaf reflectance, which reduces the net energy uptake; (*b*) windbreaks, which increase the air resistance to water vapor transfer; (*c*) enclosures in which the air humidity builds up, thus decreasing the leaf to air vapor density gradient; and (*d*) applying metabolically active materials which tend to close the stomata, or alternatively, coating of leaf surfaces with a plastic film. Both methods produce a physical barrier to diffusion, thus increasing the leaf resistance to water vapor loss.

A. MODIFICATION OF LEAF REFLECTANCE

1. Theory

Geographically, the areas of the world in which problems of water shortage are most critical, coincide with the regions in which the intensity of the sun's radiation often exceeds the maximum which can be used in photosynthesis. These regions lie between 10° and 35° latitude where radiation fluxes of 1.5 $cal \cdot cm^{-2} \cdot min^{-1}$ or more are common. Photosynthesis of exposed leaves is, however, usually saturated at 0.2–0.5 $cal \cdot cm^{-2} \cdot min^{-1}$. Furthermore, 60% of the solar radiation is of wavelength longer than 0.7 μ (Gates, 1962) and is not used in photosynthesis. However a considerable portion of this relatively long wave radiation is also absorbed by the leaves. Therefore in these regions (10°–35° latitude) the greater the average overall radiation the higher is the transpiration/photosynthesis ratio (de Wit, 1958). These considerations suggest two possibilities for reducing the transpiration/photosynthesis ratio: (a) application to plant leaves of a generally reflectant pigment, to decrease the net radiation load; (b) application of a pigment which ideally would selectively reflect all radiation below 0.4 and above 0.7 μ and would transmit the radiation between. Both methods should bring about lower leaf temperatures and hence lower e_1 [Eq. (1)].

Theoretical aspects of the effect of nonselective reflective materials on canopy transpiration have been studied by Seginer (1969), who calculated that about a 30% saving in irrigation water could be attained when natural albedo was increased from 0.25 to 0.4 of the incident shortwave radiation.

2. Practical Problems

Trunks of fruit trees are often whitewashed after pruning has exposed them to the direct rays of the sun. This is done to increase reflectance and to prevent overheating. In this case no consideration need be taken of the reduction of photosynthesis. Serr and Foot (1963) have reported the application of whitewash spray to walnuts, which reduced the temperature and increased the size of the nuts. However, this approach has not yet been tried extensively in the field for reducing transpiration.

A laboratory study of the effect of increased reflectance on leaf physiology was made by Abu-Khaled and collaborators (1970). They applied kaolinite to leaves of different plant species under laboratory conditions and made measurements of leaf temperatures, photosynthesis, and transpiration. A reduction in leaf temperature of 3–4°C and a decrease in transpiration of 22–28% were found. This was ascribed to increased leaf reflectivity. Photosynthesis was reduced only at low light intensities.

There are a number of problems, both practical and theoretical, which must be considered in the development of this method. The first is that pigments which reflect above 0.7 μ and transmit below this wavelength do not appear to exist at present (Chartrand, personal communication). However, we can envisage that such a pigment may be developed in the future, and that even nonselective reflective materials may be used. Even with an ideal pigment the following considerations will apply:

1. Any reflective material applied must be nontoxic, must stick and spread evenly on plant leaf surfaces, and must be stable and sufficiently permeable to gases so as not to interfere with photosynthesis and respiration.

2. Whenever there is even a slight breeze the temperature of normal thin leaves is very close to that of the air (Table I). Consequently, if as a result of decreased net shortwave radiation absorption the temperature of the leaf drops below air temperature, sensible heat exchange will add energy to the leaf. This will tend to offset the advantage of the increased reflectivity.

3. Although during the midday hours plants may receive levels of radiation which are supraoptimal for photosynthesis, this is not always so. For example, light is not supraoptimal during the early morning and evening periods, during much of the day in the winter season at middle latitudes, for much of the year at high latitudes, and when there is cloud cover.

4. At the same time that leaves at the top of the canopy are receiving supraoptimal levels of radiation (for photosynthesis), leaves at lower levels will be receiving much less. This, however, may not be such a difficult problem to overcome as light tends to decrease logarithmically from the top to the bottom of the canopy (Saeki, 1963) and a similar distribution of a reflective coating should be obtainable with a suitable spraying technique.

5. Although leaves of most temperate species of cultivated plants show light saturation for photosynthesis at relatively low light intensities, many semitropical and tropical species such as corn and sugar cane have very high saturation levels. The latter are plants which have the C_4-dicarboxylic carbon fixation pathway and show little photorespiration (Hatch and Slack, 1970). These plants would probably benefit only from nonspectrum-selective pigments during the midday hours. At other times of the day reduction of net radiation would almost always be detrimental to photosynthesis.

It appears, therefore, that nonselective reflective coatings would be mainly appropriate for crops of which the photosynthetic mechanism is saturated at relatively low light intensities.

B. WINDBREAKS

1. Theory

In advective situations where a significant amount of the energy taken up by the leaf originates from the air, windbreaks may be effective in decreasing the transpiration/photosynthesis ratio. This occurs in oases and in oasislike situations where large areas of land, upwind to the field, have only a sparse vegetation; the temperature of the air arriving at the plant is then higher than that of the leaves, which are being cooled by transpiration.

Lowering of the wind speed decreases the amount of advective heat brought to the leaves and increases the water vapor content of the air near the leaves [e_a, Eq. (2)]. Such an effect is predicted by the energy balance-resistance model of Rosenberg and Brown (1971) and has also been confirmed by them and others in practical field trials (Stoeckeler, 1962; Bouchet *et al.,* 1968; Lomas *et al.,* 1966).

It should be noted that a large quantity of water vapor is given off in transpiration, as compared to a relatively small quantity of carbon dioxide taken up in photosynthesis. Consequently the percentage change in the CO_2 concentration of the air, and in the air to leaf CO_2 gradient behind a windbreak, is small as compared to the change in water vapor concentration and to the leaf to air vapor gradient (Brown, 1969). As a result, photosynthesis benefits from the reduced stomatal resistance, often found in the sheltered plants, and is not seriously affected by reduced CO_2 levels. Thus, under advective conditions, windbreaks may increase the photosynthesis/transpiration ratio both by decreasing transpiration and increasing photosynthesis.

When advection is not a significant part of the heat budget of the leaf, the increased air temperature and decreased leaf diffusion resistance will tend to counterbalance the effect of the increased air resistance and humidity [r_a and e_a, Eq. (2)]. Under such conditions, there will be little effect of the windbreak on transpiration (Rosenberg and Brown, 1971).

2. Practical Problems

Problems associated with windbreaks are well known and will only be briefly mentioned. They are mainly economic and relate to the time required for establishment of live windbreaks (trees, hedges, tall grasses, etc.) and the cost of setting up and maintaining inert fences. The area of land taken up by the windbreak or affected by the roots of live windbreaks is often an important consideration. The particular climatic, crop, and economic situation will determine whether a windbreak is worthwhile and which type is most appropriate.

C. ENCLOSURES

1. Theory

Covered enclosures, such as plastic greenhouses, are theoretically very attractive for growing plants in sunny climates where water is at a premium. Enclosures present two possibilities, both of which are advantageous for plant growth.

1. Maintenance of high humidity in the air around the plant [e_a,

Eq. (2)]. This will reduce transpiration and also retard development of water stress.

2. Maintenance of a high concentration of CO_2 in the air during the hours of sunlight by the addition of this gas to the air. Under conditions of intense light the rate of photosynthesis is often limited by the low concentration of CO_2 normally present in the air (about 320 $\mu l/l$). The addition of CO_2 to the air will therefore raise the rate of photosynthesis [see Eq. (3), Section II,D)]. Furthermore, high concentrations of CO_2 tend to induce closure of stomata and thus reduce transpiration. Consequently CO_2 may be considered to be the ideal antitranspirant, as it operates favorably on both sides of the transpiration/photosynthesis ratio. Such an effect of CO_2 on both transpiration and photosynthesis has been demonstrated for corn (Moss *et al.,* 1961). Its practical application is, however, only possible in enclosures (Pallas, 1970).

2. Practical Problems

The main problem of enclosures in hot climates is the dissipation of unrequired heat during daytime hours, i.e., avoidance of the hothouse effect, by which longwave radiation is trapped in enclosures such as glass-covered greenhouses. The solution to this technological problem may be approached in a number of ways. First, the hothouse effect may be reduced by the use of thin, more transparent materials for the enclosure. Second, a cooling system can be installed based on either conventional fuel-consuming coolers, or internal heat-exchangers (storing heat during the day and releasing heat at night). Evaporative desert coolers may also be used; they have the double advantage of both cooling and moistening the air and, furthermore, could be operated with brackish or saline waters. Such systems have been installed at the University of Arizona (Hodges, 1967).

Another difficulty with this type of system is the prevention of plant diseases which tend to develop in the hot, humid atmosphere. However, the main problem is economic. Although all the engineering problems can be solved, enclosures which consume fuel for cooling or require heavy capital investment would be profitable only under very special circumstances.

D. Increase of Leaf Resistances

1. Theory

Coating leaves with an inert material to prevent water loss (increase of r_1) is the oldest recorded method for reducing transpiration (Theophrastus,

300 B.C.; Miller *et al.,* 1950). The surface of the leaves is also the most logical place at the plant level for increasing resistance to water movement and loss (Waggoner, 1966b). If resistance to water movement is increased at any position below the leaves (e.g., in the roots), leaf turgor is lost and photosynthesis reduced.

A common misconception in the literature of antitranspirants of more recent years was that certain film-forming polymer materials, which could be applied to leaves, such as polyethylene, are more permeable to carbon dioxide and oxygen than to water vapor (Gale and Hagan, 1966). However, comparison of the permeability coefficients of polymers to gases and vapors, when reduced to the same units, shows that this is not the case; even the most promising materials such as silicones and polyethylenes are more permeable to water vapor than to carbon dioxide by a factor of at least 4 (Waggoner, 1966b; Gale and Poljakoff-Mayber, 1967; Wooley, 1967). Some representative figures from data collated by Waggoner (1966b) are given in Table II.

A different method for increasing the resistance of leaves to water vapor loss is the application to the leaf of chemicals that induce stomatal closure. As noted in Section I,C, stomata form the main pathway for water vapor diffusion from the leaves. Consequently, closing stomata should be an effective means of reducing transpiration.

The fact that certain chemicals (insecticides or fungicides) reduce transpiration has been noted for many years (Blandy, 1957). Extensive study of compounds has been made in the last decade (Zelitch and Waggoner, 1962a,b; Pallas *et al.,* 1963, 1965; Pallas and Bertrand, 1966; Shimshi, 1963a,b; Waggoner *et al.,* 1964; Waggoner and Zelitch, 1965).

Under good growing conditions, including moderate to intense sunlight, the rate of carbon dioxide uptake, and hence photosynthesis, is determined

TABLE II

RESISTANCES OF MEMBRANES OF VARIOUS POLYMER MATERIALS TO GAS AND VAPOR PERMEATION[a,b]

Material	H_2O[c]	CO_2[c]
Cellulose acetate	1–7	720–2500
Hexadecanol monolayer	3	80
Polyvinyl chloride	4–36	$(12–25)10^3$
Polystyrene	10–22	330–1700
Polyethylene	52–620	200–3300
Cellophane	260–280	10^7
Saran	$(1–20)10^3$	$(5–100)10^5$

[a] From Waggoner (1966b), by permission of American Society of Agronomists.

[b] Membrane thickness, 1 μ.

[c] Per sec · cm^{-1}.

by the rate of diffusion of CO_2 from the air around the leaf to the site of fixation within the chloroplast. In the most simple analysis, this may be expressed as:

$$P = \frac{[CO_2]_{air} - [CO_2]_{chl}}{r'_a + r'_s + r'_m} \tag{3}$$

where P is the rate of uptake of CO_2; $[CO_2]_{air}$ and $[CO_2]_{chl}$ are the concentrations of CO_2 in the air and chloroplasts; r'_a and r'_s are the resistances to diffusion of CO_2 in the boundary layer and stomata, respectively; r'_m is an expression of the resistance to CO_2 fixation within the leaf, mainly in the liquid phase of the mesophyll cells (Gaastra, 1959).

Comparison of Eqs. (2) and (3) shows that the equation for photosynthesis contains an extra resistance, r'_m, not present in the equation for transpiration. If r'_m is of the same order of magnitude as r'_a and r'_s closure of stomata should bring about a larger percentage reduction of transpiration than of photosynthesis. As a result the transpiration/photosynthesis ratio will be decreased. Such an effect has been demonstrated by Zelitch and Waggoner (1962a) and by Slatyer and Bierhuizen (1964), who applied phenyl mercuric acetate to single leaves of cotton.

A number of theoretical considerations complicate this approach of increasing the leaf resistance: Plants appear to differ in the amount of their internal mesophyll resistance [r'_m, Eq. (3)]. Generally, it is found that plants in which CO_2 fixation takes place through the C_4-dicarboxylic pathway, and which lack photorespiration, show only very small values of r'_m. It follows from the above that if the ratio $r'_a + r'_s$ to r'_m is high, closure of stomata will reduce CO_2 uptake to almost the same degree as transpiration [compare Eqs. (2) and (3)]. There is also some evidence that mesophyll resistance to water vapor loss may not be negligible, as has been assumed in Eq. (2) (Gale et al., 1967; Jarvis and Slatyer, 1970).

A further theoretical difficulty arises from the complicated structure of the interface between the leaf surface and the air mass. Leaf resistances to water loss are rarely the same on the bottom and top surfaces of the leaves. The two surfaces of the leaves are parallel, but usually unequal conductors of water vapor and CO_2 (Gale and Poljakoff-Mayber, 1968; Moreshet et al., 1968). Consequently, partial closure or coverage of the stomata on one side of a leaf may have a different effect on the transpiration/photosynthesis ratio than the same degree of coverage applied to the other side (Waggoner, 1965).

The question of the relative magnitudes of the leaf and the air resistances must also be considered. Air resistance here includes both the boundary layer resistance in the immediate vicinity of the leaf and the resistance to water vapor transfer between the plant canopy and the air at some distance above the canopy. Earlier works considered this air resistance to be much larger than the plant resistance. The reason for this was that the plant canopy is formed of many layers of leaves. Each leaf and layer of leaves presents an individual parallel pathway for transpiration. The parallel sum of the resistances of these many conducting surfaces may therefore be relatively low in comparison with the resistance of the air in and around the canopy (Monteith, 1965). However, it was later shown that, especially in arid climates, leaf resistances comprise a considerable fraction of the total canopy resistance (Hunt et al., 1968; Turner, 1969). Therefore increasing leaf resistance should have a significant effect on transpiration. In fact in at least one field experiment, with bananas, an antitranspirant spray was found to reduce the extraction of water from the soil by approximately one third (Gale et al., 1964).

As discussed above, film-forming materials apparently do not present selectively permeable barriers to water vapor and carbon dioxide diffusion in the required direction (i.e., $P_{CO_2} > P_{H_2O}$). Instead they are from four to many thousand times more permeable to water vapor than to other gases (Table II). The question therefore arises why they do not prevent photosynthesis and growth almost entirely whenever transpiration is significantly reduced. Instead it is usually found that transpiration and photosynthesis are reduced to about the same degree. The answer appears to be that the film-forming materials are only covering part of the leaf. But where even a very thin film (of the order of $1-2$ μ) is present, it forms an essentially impermeable barrier to both CO_2 and water vapor. This hypothesis has been suggested independently by a number of workers (Waggoner, 1966a; Gale and Poljakoff-Mayber, 1967; Poljakoff-Mayber and Gale, 1967; Malcolm and Stolzy, 1968; Parkinson, 1970).

It follows that both inert films and physiologically active stomata-closing materials will operate in the same manner, by providing a partial, and essentially nonselective resistance to both water vapor and carbon dioxide. The ultimate effect of these treatments on transpiration and photosynthesis will depend on environmental conditions. This is discussed in Section III.

2. Practical Problems

The main practical problem with antitranspirants of the film-forming type is to formulate a material having selective permeability to gases in the required direction, i.e., $P_{CO_2} > P_{H_2O}$. There appears to be little chance of finding a normal polymer substance through which gases pass by diffusion that will meet this requirement (Stannett and Yasuda, 1964; Lebovits, 1966). This is because the solubilities and coefficients of diffusion of gases and vapors in polymer films are inversely proportional to their molecular weights and dimensions. The H_2O molecule is therefore basically more mobile than that of CO_2.

Even if a material which has a high permeability for CO_2 could be produced, other practical requirements of film-forming antitranspirant materials would still have to be met. These include nontoxicity, sticking and spreading properties, stability under the intense ultraviolet radiation of sunlight, elasticity, and durability. Finally, the cost of the material and its application must be less than the value of the water saved and/or the increased crop yield.

Stomatal-closing materials are generally cheaper to apply than are film-forming antitranspirants. They may be of particular value where rate of growth is of secondary interest and only conserving water is important. Such a situation occurs in water harvesting from forested watersheds. The application of phenyl mercuric acetate to a red pine forest has been shown to reduce the water use of the trees by about 10%. The concomitant 32% reduction of radial trunk growth may be considered acceptable if the trees are not being grown for timber (Waggoner, 1966a; Waggoner and Bravdo, 1967; Turner and Waggoner, 1968).

The main practical problem with antitranspirants of the stomatal-closing type is to obtain nontoxic materials which only affect the stomata and which are long-lasting. Presently available materials do not act specifically on stomata but are antimetabolites of sections of the respiration pathway (Zelitch, 1969). Materials of this nature, which also have a long residual action, or which contain toxic ions such as mercury, would not be acceptable for application to edible crops.

In addition to toxicity to humans and animals these metabolically active materials may also be harmful to plants. For example, decenyl-succinic acid (DSA), which has been suggested for use both as a stomatal-closing agent and as a material that increases root permeability to water, has been found to be toxic to beans. When DSA was applied via the roots, salt leaching from the roots was increased and root pressure was reduced (Newman and Kramer, 1966).

When DSA was applied as a spray to the tops of *Pinus resinosa* trees, Kozlowski and Clausen (1970) found that it was toxic. At those concentrations at which it affected needle water content, needles, buds, and stems were killed.

Recently abscisic acid was suggested by Jones and Mansfield (1970) as a nontoxic, naturally occurring stomatal-closing agent which could be used as an antitranspirant.

Antitranspirants may be of value for facilitating survival and growth of mesomorphic plants in regions of high evapotranspiration potential.

However, this possibility does not yet seem to have been specifically studied.

III. THE DEPENDENCE OF THE EFFECT OF STOMATA-CLOSING AND FILM-FORMING ANTITRANSPIRANTS ON ENVIRONMENTAL CONDITIONS

The effect of either stomata-closing or film-forming antitranspirants on plant transpiration, photosynthesis, and growth may be studied, both theoretically and practically, at three levels: the level of the single leaf, the entire plant, and the plant canopy. Studies so far made have shown how the effect of an antitranspirant will vary in both direction and degree, in accordance with environmental conditions (Gale, 1961; Gale and Hagan, 1966; Gale and Poljakoff-Mayber, 1967; Tomari, 1970). The complexity of this interaction increases from the single leaf to the canopy level.

A. INTERACTION BETWEEN PLANT, ANTITRANSPIRANT, AND ENVIRONMENT AT THE SINGLE LEAF LEVEL

As noted above, results of laboratory studies at the leaf level, in which the percentage decrease of photosynthesis was found to be the same as that for transpiration, led different workers to conclude that the various film-forming materials tried were all producing impervious and incomplete coatings (Section II,D). Evidently little gas and vapor were actually moving through the plastic films. The situation is almost the same for materials that close stomata. However, in this case, with partial stomatal closure, there may be a somewhat smaller decrease in photosynthesis than in transpiration. This will be so in those plants in which mesophyll resistance to CO_2 uptake is of the same order of magnitude as the other diffusion resistances [See Section II,D above and findings of Zelitch and Waggoner (1962a) and Slatyer and Bierhuizen (1964)]. Consequently, at the single leaf level, the effect of antitranspirants may be analyzed on the basis of partial coverage. Leaf conductivity to gases and vapors is directly proportional to leaf area. Conversely, leaf resistance to diffusion is the reciprocal of conductance. An example of these arithmetical relations is given in Table III.

The resistance figures given in Table III may be entered into the leaf energy balance equation [Eq. (1)] to give an estimate of the effect on leaf temperature and transpiration of different percentages of coating by an impervious antitranspirant. It is assumed that the plastic coating does not

TABLE III

RELATION BETWEEN PERCENTAGE COATING OF A LEAF BY AN IMPERVIOUS
ANTITRANSPIRANT AND CONDUCTIVITY AND DIFFUSION RESISTANCE[a]

Coating (%)	0	25	50	75
Conductivity (cm · sec^{-1})	0.6	0.45	0.3	0.15
Resistance (sec · cm^{-1})	1.66	2.22	3.33	6.66
Change in resistance as % of resistance in the uncoated leaf	—	+33	+100	+300

[a] The figures in this example are based on the arbitrary assumption that the untreated leaf with open stomata has a conductivity to water vapor of 0.6 cm · sec^{-1}. It is further assumed that there is no variation of stomatal width in response to the treatment.

directly affect the other factors of heat dissipation—radiation and sensible heat exchange.

The figures in Table III are strictly correct only if the coated patches of the leaf are very small. If, however, the coated patches are large, the entire pathway of diffusion from the leaf will be affected, i.e., both stomatal and boundary layer resistances. Asymmetric distribution of stomata on the two surfaces of the leaf is a further complication, which has been discussed by Waggoner (1965).

Gates' equation for the leaf energy budget (1968) may thus be used to estimate the effect of changes in leaf resistance, brought about by antitranspirants, on temperature and transpiration of single leaves. Figure 1 shows the results of such calculations made for a number of different values of leaf resistance and wind speeds. The other meteorological parameters are assumed to be constant and similar to those found at midday on a summer day, in a Mediterranean type of climate. In these calculations it has been assumed that only stomatal resistances have been affected by the antitranspirant treatment.

The calculations depicted in the nomogram (Fig. 1) indicate that under the conditions shown a relatively low percentage covering by an impervious antitranspirant, or even a small direct effect of a metabolic antitranspirant on the stomatal resistance may bring about a significant reduction in water loss. This is accompanied by only a very small increase in leaf temperature. Only when the air is almost still (wind speed, $V = 10$ cm · sec^{-1}) would there be a significant, and possibly harmful, rise in leaf temperature as a result of the reduced transpiration.

The data of Fig. 1 refer to conditions prevailing only at midday. Figure 2 depicts the calculated effect of antitranspirant treatments on leaf transpiration and temperature during an entire daytime period. The meteorological data used are average mid-August figures for Jerusalem. Calculations have been made only for that period of the day when light

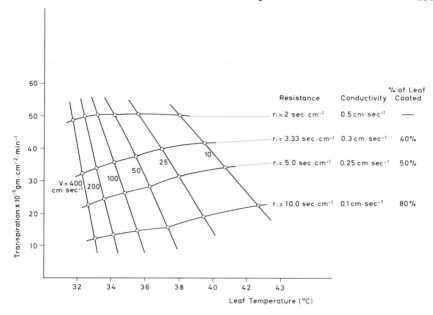

Fig. 1. Calculated leaf temperature and transpiration rates of single leaves under Mediterranean summer conditions, as affected by changes in leaf resistance to vapor diffusion and wind speed. Leaf dimensions, 5×5 cm; total absorbed radiation, 1.2 cal·cm^{-2}·min^{-1}; air temperature, 30°C; relative humidity, 50%; conductivity of leaf before treatment with antitranspirant, 0.5 cm·sec^{-1}.

intensity is considered to be sufficient for full stomatal opening. Radiation absorbed by the leaf was taken to be 85% of the total shortwave radiation, which varied between 0.41 and 1.39 cal·cm^{-2}·min^{-1}. It will be noted that there is a large range of air temperature and relative humidity and that wind velocities are moderate throughout the day.

The data depicted in Fig. 2B show that leaf temperatures remain within three degrees of air temperature at all hours of the day, and that the maximum rise in temperature, due to the treatment is 1.2°C. Before 8 A.M. and after 3 P.M. leaf temperatures are below air temperatures and the leaf is therefore gaining advective energy from sensible heat exchange with the air. Despite these changing conditions the increased leaf resistances are calculated to result in very significant reductions in transpiration at all hours of the day.

As, in practice, wind velocities do not fall below 200 cm·sec^{-1}, boundary layer resistance is low and the percentage effect of the treatments remains fairly constant throughout the day. Integration of the areas below the curves between 07:00 and 18:00 hours shows that the untreated

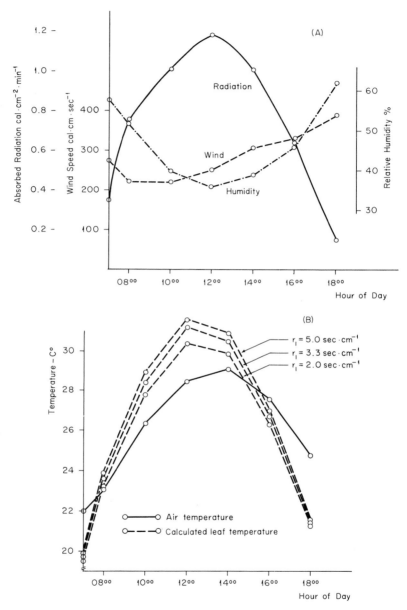

Fig. 2. Calculated effect of antitranspirant treatments on transpiration and leaf temperatures of a single leaf during the course of a typical mid-summer day in Jerusalem. For the purpose of these calculations total absorbed radiation was assumed to approximate 85% of the measured shortwave radiation. (A) Total

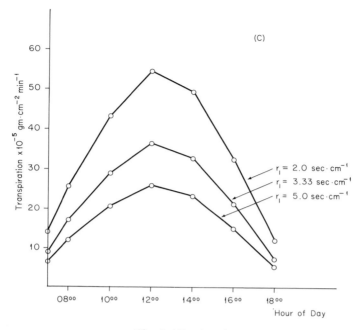

Fig. 2 (*Continued*)

absorbed radiation, relative humidity and wind speed. (B) Air temperature ($\bigcirc$—$\bigcirc$) and calculated leaf temperature ($\bigcirc$--$\bigcirc$). (C) Calculated rates of transpiration.

leaf ($r_1 = 2$ sec·cm^{-1}) loses 275 mg water·cm^{-2} and the leaves whose resistances are increased to 3.33 and to 5.0 sec·cm^{-1} lose 186 and 133 mg·cm^{-2}, respectively, a saving of 32.5 or 52%.

The effect of antitranspirants on photosynthesis, as dependent on environmental conditions at the leaf level, has not yet received the same attention as has transpiration. When it does the effect of change in leaf temperature, brought about by the antitranspirant treatments, on the relative rates of photosynthesis and respiration will have to be taken into account.

B. Interaction between Plant, Antitranspirant, and Environment at the Whole Plant Level

Gale and Hagan (1966) discussed how the effect of antitranspirant treatments on the transpiration and photosynthesis of whole plants changes under different environmental conditions. It was argued that when soil water is adequate and potential evapotranspiration moderate, presently available, nonselective, film-forming antitranspirants would reduce both transpiration and photosynthesis. On the other hand, when the evapo-transpiration demand exceeds the ability of either the soil to supply water or the plant to convey water from the soil to the leaf, leaf turgor of un-treated plants will be lowered; stomata will close and both photosynthesis and transpiration will be reduced. Under such conditions, the prior appli-cation of an antitranspirant which reduces the rate of transpiration will delay the buildup of water stress, turgor loss, and stomatal closure. As a result, photosynthesis and cumulative transpiration could be as great or even greater in a plant treated with an antitranspirant than in a nontreated plant.

This effect was first described by Gale (1961) and we have since re-peated these experiments under more closely controlled laboratory con-ditions. Results of two such experiments with potted bean and pine seedlings are given in Figs. 3 and 4. Bean (Fig. 3) is an example of a mesomorphic plant and pine (Fig. 4) of a xeromorphic plant. Full de-tails of experimental procedures have been given by Poljakoff-Mayber and Gale (1967).

The data presented in Fig. 3 show that the effect of the antitranspirant treatment was very dependent on the soil moisture and evaporation de-mand. Under moist soil conditions rates of photosynthesis and transpira-tion of treated plants were much lower than in controls, but under dry soil conditions the relative rates of gas exchange were reversed. Increasing the leaf to air vapor density gradient had the same effect as reducing soil moisture (as is particularly evident in the moist and dry soils) but to a lesser degree. The changes in the ratios of the photosynthesis and tran-spiration rates of the treated and control plants are due mainly to the re-duced CO_2 and vapor exchange of the controls under conditions of stress, as compared to the relative stability of the antitranspirant-treated plants.

The data of Fig. 4 show a complete reversal in the relative rates of photosynthesis and transpiration of treated and control seedlings, under medium-dry as compared to moist conditions. In this experiment air hu-midity (and hence evaporation demand) appears to have been a much smaller factor than the moisture of the soil in determining the relative effect of the antitranspirant. When the soil had been allowed to dry to a water potential of approximately −10 atm, no photosynthesis or transpira-tion could be detected in either the control or the antitranspirant-treated seedlings.

It is clear from the data shown in Figs. 3 and 4 that the results of treatment of an entire plant with an antitranspirant may vary considerably. The final effect on transpiration and photosynthesis will depend on the combination of the following factors: physical characteristics of the soil with regard to water content and conductance, distribution of plant roots,

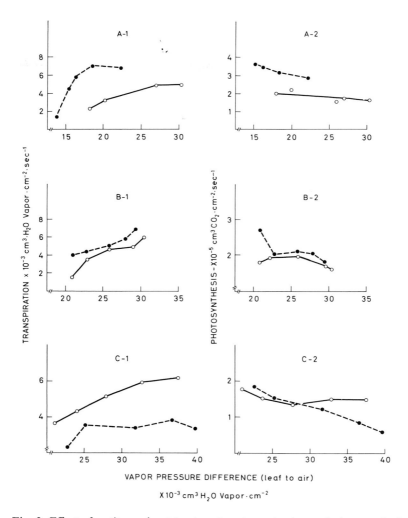

Fig. 3. Effect of antitranspirant treatment on transpiration and photosynthesis of potted bean plants under varying conditions of soil moisture and evaporation demand. Plants, *Phaseolus vulgaris;* antitranspirant, Machteshim Co. (Beersheba) TL-12 polyacrylate emulsion (10% dry weight) applied at least 2 days before experiment; light intensity, 0.4 cal·cm^{-2}·min^{-1} from a 1500 W quartzline lamp. A-1 and A-2, moist soil, ψ_w (water potential) $\sim$ −0.1 atm; B-1 and B-2, medium moisture, ψ_w $\sim$ −0.5 atm; C-1 and C-2, dry soil, ψ_w $\sim$ −9.0 atm. ●---●---●, controls; ○—○—○, antitranspirant treatment. Note: Changing scales on ordinates and abscissae.

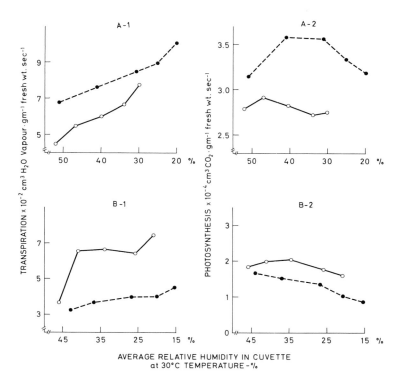

Fig. 4. Effect of antitranspirant treatment on transpiration and photosynthesis of potted pine seedlings under varying conditions of soil moisture and evaporation demand. Plants, one year old *Pinus halepensis* seedlings; antitranspirant and light intensity same as in Fig. 3. A-1 and A-2, moist soil, $\psi_w > -0.1$ atm; B-1 and B-2, medium-dry soil, $\psi_w \sim -4.0$ atm. ●- - -●- - -●, controls; ○—○—○, antitranspirant treatment. Note: Changing scales on ordinates and abscissae.

characteristics of the plant with regard to water transport, potential evapo-transpiration and the pattern of its daily variations, the physical characteristics of the antitranspirant, and finally the percentage of the leaves coated, or in the case of stomata-closing materials, the percentage closure of the stomata.

There may also be a number of important side effects from the use of antitranspirants. For example, the reduction of water stress and increase of turgor may lead to increased leaf area (Gale, 1962; Possingham *et al.*, 1969), and longer life of leaves (Gale *et al.*, 1964) in treated plants. This may compensate for the lower photosynthesis per unit leaf area but will also tend to increase total transpiration.

Water stress has often been considered to reduce photosynthesis and growth mainly by reducing leaf area and stomatal opening (Slatyer, 1971a,b). This is because under short-term laboratory experiments the photosynthetic apparatus itself—essentially the chloroplasts—has been found to be insensitive to low water potentials of the order generally found in stressed plants. However, when stress is considerable and wilting occurs, changes take place in cell fine structure (Nir et al., 1970), in the photochemical activity of the chloroplasts (Nir and Poljakoff-Mayber, 1967), and in the plant hormone balance (Benzioni et al., 1967; Itai et al., 1968). Furthermore, when even low levels of stress are allowed to prevail for long periods of time changes occur in the activity of various cellular enzymes (Vieira-da-Silva, 1971). The general effects of water deficits on physiological processes have been reviewed by Crafts (1968).

Treatment with an antitranspirant may help to maintain a relatively high plant water potential and turgor pressure in leaves throughout the day. Thus, many of the above cellular symptoms of water stress may be alleviated. This may be the case even when the cumulative daily transpiration is not reduced by the treatment.

C. Interaction between Plant, Antitranspirant, and Environment at the Level of the Canopy

The considerations relevant to the single plant are, in principle, operative at the level of the canopy. These include the soil water supply, plant characteristics, evapotranspiration demand, and the effect of the antitranspirant on the diffusion resistance of the leaves. The difficulty in assessing the effect of an antitranspirant under canopy conditions lies in a lack of knowledge of the microclimate prevailing around each plant and around each leaf at the different levels of the canopy. The microclimate within the canopy is determined by meteorological conditions outside the canopy and by the effect produced by the plants themselves. The overlapping of the leaves results in a closed canopy in which shade is increased, wind velocity lowered, air temperature changed, humidity increased, and CO_2 content of the air is lowered. Furthermore, under field conditions and particularly when the canopy is still open, a considerable amount of water may be lost by direct evaporation from the soil (Denmead, 1971).

As noted above (Section II,D) the total crop resistance per unit area of land is much less than the resistance per unit area of a single leaf (Monteith, 1965). The effectiveness of an antitranspirant treatment will therefore depend on the ratio between the overall canopy resistance originating in the leaves (r_c) and the overall air (or boundary layer) resistance of the canopy (r_a).

Early models of plant evaporation, such as the well-known Penman equation, were designed for moist temperate climates and assumed a very low r_c to r_a ratio. Later work (particularly that carried out in arid climates) showed a much larger contribution of r_c (Hunt et al., 1968; Waggoner and Reifsnyder, 1968; Turner, 1969). More recent evapotranspiration formulas have been modified to include plant resistance factors (Monteith, 1965).

Rosenberg and Brown (1971) have used such a modified equation of crop evapotranspiration to predict the effect of an antitranspirant on crop transpiration. Their calculations showed that antitranspirants would be effective mainly under hot, dry, advective conditions and even then transpiration would only be reduced at the most by some 13%. However, they assumed an antitranspirant which would increase leaf resistance to water vapor loss by just one third. As can be seen by reference to Table III, this is a very conservative estimate which is equal to only a 25% coating of the leaves. In practice 50% coating should be easily obtainable. If the film is impervious this would double leaf resistance to water loss.

IV. SUMMARY AND CONCLUSIONS

Each of the methods discussed for reducing transpiration is seen to have its own particular use and limitations.

Windbreaks will be of value in reducing transpiration, mainly in highly advective desert climates. Under such conditions other considerations, such as reduction of sand abrasion, may make them essential for successful plant cultivation.

Enclosures are theoretically very attractive in sunny climates for attaining more efficient use of water, for reducing damage caused by wind, and for making CO_2 fertilization practical. Small enclosures, such as plastic tunnels over vegetable beds, have recently come into widespread use, in spite of the considerable cost of construction (Enoch et al., 1970). Large enclosures in hot climates raise the problem of cooling, the cost of which may often be prohibitive.

Little work has yet been done on the use of reflective materials as antitranspirants. This may be a promising direction for research, particularly if a material reflecting above 0.7 μ and transmitting below this wavelength could be developed.

Metabolically active materials which reduce transpiration by inducing stomatal closure are of relatively low cost. They could be used in those situations where reduction of growth is a secondary consideration, such as with perennial ornamentals or in forest watersheds, as discussed above.

There remain three practical problems with materials of this nature which must be overcome: nonspecificity, short-term action, and toxicity.

The main practical problem with antitranspirants of the film-forming type is to obtain a material having the required physical characteristics. An ideal antitranspirant material would form a very thin, continuous, flexible, and durable film on the leaf surface. At the thickness obtaining on the leaf (about 0.1–1 μ) the film should be essentially impermeable to water vapor (H_2O diffusion resistance, ~ 2 sec·cm^{-1}·μ^{-1}) and sufficiently permeable to CO_2 so as not to greatly impair photosynthesis (CO_2 diffusion resistance, ~ 1 sec·cm^{-1}·μ^{-1}). Materials are presently available which have all the above characteristics except permeability to CO_2. However, even when such an ideal antitranspirant material is produced, the cost of treatment will probably limit its application to high-cash, essentially evergreen crops such as citrus or bananas.

A further practical advantage of film-forming materials is that they may be used as slow release agents for mineral nutrients (such as iron and zinc which are often supplied by foliar application) and also for insecticides and fungicides. Furthermore, the inert film may in itself form a mechanical barrier against salt from sprinkler-applied irrigation water (Malcolm et al., 1968) and against insects and fungal pathogens (King et al., 1960; Gale and Poljakoff-Mayber, 1962).

It is possible to envisage that a number of the above methods of reducing transpiration could be used in combination, for example: Crops grown behind windbreaks could also be treated with antitranspirant materials of the radiation-reflecting, stomata-closing, or film-forming type; enclosures may be covered with selectively reflective, pigmented films, and combinations of film-forming, stomata-closing, and reflective materials could be formulated.

Assuming that theoretically promising antitranspirants and reflective pigments are produced, they will still have to be tried out on different crops under various climatic conditions. As the number of different combinations of treatment, crop, climate, and economic situation are vast and the cost of field experiments is very high, there is particular importance to the development of predictive mathematical models for transpiration, photosynthesis, growth, and yield, as affected by antitranspirant treatments.

The benefit which may be obtained from such models is threefold: (a) the intellectual satisfaction of the orderly arrangement and integration of present knowledge; (b) the prediction of the most promising applications, which will reduce the number of field trials required; and (c) the designation of those areas in which knowledge is lacking and further research is required.

These models will have to be extended from the single leaf level (as shown for transpiration in Figs. 1 and 2) to the level of the entire crop. A complete model would have to be dynamic and preferably conceptual and not empiric. It would cover the entire period of crop growth, and include different rainfall and irrigation regimes. The principles behind such dynamic models have been described by de Wit and Brouwer (1971) and such models could no doubt be modified to include the effects of different types of antitranspirants.

REFERENCES

Abu-Khaled, A., Hagan, R. M., and Davenport, D. C. (1970). Effects of Kaolinite as a reflective antitranspirant, on leaf temperature, transpiration, photosynthesis, and water use efficiency. *Water Resour. Res.* 6, 280.

Ben-Zioni, A., Itai, C., and Vaadia, Y. (1967). Water and salt stress, kinetin and protein synthesis in tobacco leaves. *Plant Physiol.* 42, 361.

Blandy, R. V. (1957). The effect of certain fungicides on transpiration rates and crop yields. *Proc. Int. Congr. Crop Protect.* 4, 1513.

Bouchet, R. J., Guyot, G., and de Parcevaux, S. (1968). Augmentation de l'efficience de l'eau et amélioration des rendements par réduction de l'évapotranspiration potentielle au moyen de brise-vent. *Proc. UNESCO Conf. Methodes Agrometeorol. 1966* p. 167.

Brown, K. W. (1969). Unpublished Ph.D. Thesis, University of Nebraska (personal communication).

Cowan, I. R., and Milthorpe, F. L. (1968). Plant factors influencing the water status of plant tissues. *In* "Water Deficits and Plant Growth" (T. T. Kozlowski, ed.), Vol. 1, p. 37. Academic Press, New York.

Crafts, A. S. (1968). Water deficits and physiological processes. *In* "Water Deficits and Plant Growth" (T. T. Kozlowski, ed.), Vol. 2, p. 85. Academic Press, New York.

Denmead, O. T. (1971). Relative significance of soil and plant evaporation in estimating evapotranspiration. *Proc. UNESCO Symp. Plant Response to Climatic Factors, 1970* (in press).

de Wit, C. T. (1958). Transpiration and crop yields. *Versl. Landbouwk. Onderzoek.* 64.6, 1–88.

de Wit, C. T., and Brouwer, R. (1971). The simulation of photosynthesis systems. *Proc. IBP/PP Techn. Meet., 1969* (in press).

Enoch, H., Rylski, I., and Samish, Y. (1970). CO_2 enrichment to cucumber, lettuce and sweet pepper plants grown in low plastic tunnels in a subtropical climate. *Isr. J. Agr. Res.* 20, 63.

Gaastra, P. (1959). Photosynthesis of crop plants as influenced by light, carbon dioxide, temperature and stomatal diffusion resistance. *Meded. Landbouwhogesch. Wageningen* 59, 1–68.

Gale, J. (1961). Studies on plant antitranspirants. *Physiol. Plant.* 14, 777.

Gale, J. (1962). The influence of transpiration and its depression on plant economy. Ph.D. Thesis, Hebrew University, Jerusalem.

Gale, J., and Hagan, R. M. (1966). Plant antitranspirants. *Annu. Rev. Plant Physiol.* 17, 269.

Gale, J., and Poljakoff-Mayber, A. (1962). Prophylactic effect of a plant antitranspirant. *Phytopathology* **52**, 715.

Gale, J., and Poljakoff-Mayber, A. (1967). Plastic films on plants as antitranspirants. *Science* **156**, 650.

Gale, J., and Poljakoff-Mayber, A. (1968). Resistances to the diffusion of gas and vapor in leaves. *Physiol. Plant.* **21**, 1170.

Gale, J., Poljakoff-Mayber, A., Nir, I., and Kahane, I. (1964). Preliminary trials of the application of antitranspirants under field conditions to vines and bananas. *Aust. J. Agr. Res.* **15**, 929.

Gale, J., Poljakoff-Mayber, A., Nir, I., and Kahane, I. (1965). Effect of antitranspirant treatment on leaf temperatures. *Plant Cell Physiol.* **6**, 111.

Gale, J., Poljakoff-Mayber, A., and Kahane, I. (1967). The gas diffusion porometer technique and its application to the measurement of leaf mesophyll resistance. *Isr. J. Bot.* **16**, 187.

Gates, D. M. (1962). "Energy Exchange in the Biosphere." Harper, New York.

Gates, D. M. (1968). Transpiration and leaf temperature. *Annu. Rev. Plant Physiol.* **19**, 211.

Hatch, M. D., and Slack, C. R. (1970). Photosynthetic CO_2-fixation pathways. *Annu. Rev. Plant Physiol.* **21**, 141.

Hodges, C. (1967). "Controlled Environment Agriculture for Coastal Desert Areas," Res. Rep. University of Arizona, Tuscon.

Hunt, L. A., Impens, I. I., and Lemon, E. R. (1968). Estimates of the diffusion resistance of some large sunflower leaves in the field. *Plant Physiol.* **43**, 522.

Itai, C., Richmond, A., and Vaadia, Y. (1968). The role of cytokinins during water and salinity stress. *Isr. J. Bot.* **17**, 187.

Jarvis, P. G., and Slatyer, R. O. (1970). The role of the mesophyll cell wall in leaf transpiration. *Planta* **90**, 303.

Jones, R. J., and Mansfield, T. A. (1970). Suppression of stomatal opening in leaves treated with abscisic acid. *J. Exp. Bot.* **21**, 714.

King, J. R., Cohen, M., and Johnson, R. B. (1960). Film forming sprays on citrus in Florida. *Fl. Entomol.* **43**, 59.

Kozlowski, T. T. (1968). Introduction. *In* "Water Deficits and Plant Growth" (T. T. Kozlowski, ed.), Vol. 1, p. 1. Academic Press, New York.

Kozlowski, T. T., and Clausen, J. J. (1970). Effect of decenylsuccinic acid on needle moisture content and shoot growth of *Pinus resinosa. Can. J. Plant Sci.* **50**, 355.

Laborde, J. F. (1964). Some aspects of the control of plant transpiration. M.Sc. Thesis, University of California, Davis.

Lange, O. L., and Lange, R. (1963). Untersuchungen über Blatttemperaturen, Transpiration und Hitzeresistenz an Pflanzen mediterraner Standorte (Costa Brava, Spanien) *Flora (Jena)* **153**, 387.

Lebovits, A. (1966). Permeability of polymers to gases, vapors and liquids. *Mod. Plast.* **43**, 139.

Lomas, J., Jaakobi, D., Heth, D., Biton, A., and Zohar, J. (1966). The effect of an artificial windbreak on the microclimate, plant development and yields of peanuts in the Besor region. *Isr. Meteorol. Serv., Pub. Ser. C* No. 15.

Malcolm, C. V., and Stolzy, L. H. (1968). Effect and mode of action of latex and silicone coatings on shoot growth and water use by citrus. *Agr. J.* **60**, 598.

Malcolm, C. V., Stolzy, L. H., and Jensen, C. R. (1968). Effect of artificial leaf coatings on foliar chloride uptake during sprinkler irrigation. *Hilgardia* **39**, 69.

Miller, E. J., Gardner, V. R., Petering, H. G., Comar, C. L., and Neal, A. L. (1950). Studies on the development, preparation properties and applications of wax emulsions, for coating nursery stock and other plant materials. *Mich., Agr. Exp. Sta., Tech. Bull.* **218**, 1–78.

Monteith, J. L. (1965). Evaporation and environment. *Symp. Soc. Exp. Biol.* **19**, 205.

Moreshet, S., Koller, D., and Stanhill, G. (1968). The partitioning of resistances to gaseous diffusion in the leaf epidermis and the boundary layer. *Ann. Bot. (London)* [N. S.] **32**, 695.

Moss, D. N., Musgrave, R. B., and Lemon, E. R. (1961). Photosynthesis under field conditions. III. Some effects of light, carbon-dioxide, temperature and soil moisture on photosynthesis, respiration and transpiration of corn. *Crop Sci.* **1**, 83.

Newman, E. I., and Kramer, P. J. (1966). Effects of decenylsuccinic acid on the permeability and growth of bean roots. *Plant Physiol.* **41**, 606.

Nir, I., and Poljakoff-Mayber, A. (1967). Effect of water stress on the photochemical activity of chloroplasts. *Nature (London)* **213**, 418.

Nir, I., Klein, S., and Poljakoff-Mayber, A. (1970). Changes in fine structure of root cells from maize seedlings exposed to water stress. *Aust. J. Biol. Sci.* **23**, 489.

Pallas, J. E. (1970). Theoretical aspects of CO_2 enrichment. *Trans. ASAE (Amer. Soc. Agr. Eng.)* **13**, 240.

Pallas, J. E., and Bertrand, A. R. (1966). Research in plant transpiration: 1965. *U. S., Dep. Agr., Prod. Res. Rep.* **89**, 1–25.

Pallas, J. E., Harris, D. G., Elkins, C. B., and Bertrand, A. R. (1963). Research in plant transpiration: 1961. *U. S., Dep. Agr., Prod. Res. Rep.* **70**, 1–37.

Pallas, J. E., Bertrand, A. R., Harris, D. G., Elkins, C. B., and Parks, C. L. (1965). Research in plant transpiration: 1962. *U. S., Dep. Agr., Prod. Res. Rep.* **87**, 1–56.

Parkinson, K. J. (1970). The effect of silicone coatings on leaves. *J. Exp. Bot.* **21**, 333.

Poljakoff-Mayber, A., and Gale, J. (1967). "Effect of Plant Antitranspirants on Certain Physiological Processes of Forest Seedlings and Other Plant Material," Final Report to USDA on Project A10-FS-10, pp. 1–107. U. S. Dep. Agr., Washington, D. C.

Possingham, J. V., Kerridge, G. H., and Bottrill, D. E. (1969). Studies with antitranspirants on grapevines (*Vitis vinifera* var. sultana). *Aust. J. Ag. Res.* **20**, 57.

Raschke, K. (1960). Heat transfer between the plant and the environment. *Annu. Rev. Plant Physiol.* **11**, 111.

Rosenberg, N. J., and Brown, K. W. (1971). Measured and modeled effects of microclimate modification on evapotranspiration by irrigated crops in a region of strong sensible heat advection. *Proc. UNESCO Symp. Plant Response to Climatic Factors, 1970* (in press).

Russell-Scott, R., and Barber, D. A. (1960). The relationship between salt uptake and the absorption of water by intact plants. *Annu. Rev. Plant Physiol.* **11**, 127.

Saeki, T. (1963). Light relations in plant communities. *In* "Environmental Control of Plant Growth" (L. T. Evans, ed.), p. 79. Academic Press, New York.

Seginer, I. (1969). The effect of albedo on the evapotranspiration rate. *Agr. Meteorol.* **6**, 5.

Serr, E. F., and Foot, J. H. (1963). Effects of whitewash cover sprays on persian walnuts in California. *Proc. Amer. Soc. Hort. Sci.* **82**, 243.

Shimshi, D. (1963a). Effect of chemical closure of stomata on transpiration in varied soil and atmospheric environments. *Plant Physiol.* 38, 709.

Shimshi, D. (1963b). Effect of soil moisture and phenyl-mercuric acetate upon stomatal aperture, transpiration and photosynthesis. *Plant Physiol.* 38, 713.

Slatyer, R. O. (1966). Some physical aspects of internal control of leaf transpiration. *Agr. Meteorol.* 3, 281.

Slatyer, R. O. (1971a). The effect of internal water status on plant growth, development and yield. *Proc. UNESCO Symp. Plant Response to Climatic Factors, 1970* (in press).

Slatyer, R. O. (1971b). Effects of short periods of water stress on leaf photosynthesis. *Proc. UNESCO Symp. Plant Response to Climatic Factors, 1970* (in press).

Slatyer, R. O., and Bierhuizen, J. F. (1964). The influence of several transpiration suppressants on transpiration, photosynthesis and water use efficiency of cotton leaves. *Aust. J. Biol. Sci.* 17, 131.

Stannett, V. T., and Yasuda, H. (1964). Permeability. *In* "Crystalline Olefin Polymers" (R. A. V. Raff and K. W. Doak, eds.), Part II, 131. Wiley (Interscience), New York.

Stoeckeler, J. H. (1962). Shelterbelt influence on Great Plains field environment and crops. *U. S., Dep. Agr., Prod. Res. Rep.* 62, 1–26.

Tanner, C. B. (1968). Evaporation of water from plants and soil. *In* "Water Deficits and Plant Growth" (T. T. Kozlowski, ed.), Vol. 1, p. 74. Academic Press, New York.

Theophrastus (Greek). (300 B.C.). "Enquiry into Plants." Putnam, New York, 1916.

Tomari, I. (1970). Controlling transpiration and permeation of water through plant surfaces by multi-molecular film. *Bull. Nat. Inst. Agr. Sci., Ser. A* No. 17, pp. 1–102.

Turner, N. C. (1969). Stomatal resistance to transpiration in three contrasting canopies. *Crop Sci.* 9, 303.

Turner, N. C., and Waggoner, P. E. (1968). Effects of changing stomatal width in a Red Pine forest on soil water content, leaf water potential, bole diameter and growth. *Plant Physiol.* 43, 973.

Vieira-da-Silva, J. B. (1971). Influence de la secheresse sur la photosynthese et la croissance du cotonnier. *Proc. UNESCO Symp. Plant Response to Climatic Factors, 1970* (in press).

Waggoner, P. E. (1965). Relative effectiveness of closure in upper and lower stomatal openings. *Crop. Sci.* 5, 291.

Waggoner, P. E. (1966a). Transpiration of trees and chemicals that close stomata. *Proc. Int. Symp. Forest Hydrol., 1965* p. 483.

Waggoner, P. E. (1966b). Decreasing transpiration and the effect upon growth. *In* "Plant Environment and Efficient Water Use" (W. H. Pierre *et al.*, eds.), p. 49. Amer. Soc. Agron., Madison, Wisconsin.

Waggoner, P. E., and Bravdo, B. (1967). Stomata and the hydrologic cycle. *Proc. Nat. Acad. Sci. U. S.* 57, 1096.

Waggoner, P. E., and Reifsnyder, W. E. (1968). Simulation of the temperature, humidity and evaporation profiles in a leaf canopy. *J. Appl. Meteorol.* 7, 400.

Waggoner, P. E., and Zelitch, I. (1965). Transpiration and the stomata of leaves. *Science* 150, 1413.

Waggoner, P. E., Monteith, J. L., and Szeicz, G. (1964). Decreasing transpiration of field plants by chemical closure of stomata. *Nature (London)* 201, 97.

Woolley, J. T. (1967). Relative permeabilities of plastic films to water and carbon-dioxide. *Plant Physiol.* **42**, 641.

Zelitch, I. (1969). Stomatal control. *Annu. Rev. Plant Physiol.* **20**, 329.

Zelitch, I., and Waggoner, P. E. (1962a). Effect of chemical control of stomata on transpiration and photosynthesis. *Proc. Nat. Acad. Sci. U. S.* **48**, 1101.

Zelitch, I., and Waggoner, P. E. (1962b). Effect of chemical control of stomata on transpiration of intact plants. *Proc. Nat. Acad. Sci. U. S.* **48**, 1297.

CHAPTER 10

SOIL WATER CONSERVATION

D. Hillel and E. Rawitz

THE HEBREW UNIVERSITY OF JERUSALEM, FACULTY OF AGRICULTURE,
REHOVOT, ISRAEL

I. INTRODUCTION

Viewed on the global scale the term "water conservation" is meaningless, since the total quantity of the earth's water is constant, and circulates in a closed system. As a chemical substance water is generally indestructible as far as naturally occurring processes are concerned. However, when the subject of water conservation is looked at from a more local point of view, it is immediately obvious that in many locations an imbalance may exist between the supply and demand (or gain and loss) of water for considerable periods of time.

Precipitation, which is still the ultimate source of nearly all our fresh water, is not ideally distributed for the use of mankind in terms of time and place of occurrence. Although limited modification of local precipitation is apparently possible, the major components of the global hydrological cycle are as yet beyond our control. On the other hand, the disposition of water after it reaches the surface of the earth as precipitation is very much affected by human activity, though not always in a beneficial manner.

This chapter will deal with water conservation as a problem of management of available water resources, with emphasis on agricultural aspects of the problem. We have excluded consideration of water supply development as being outside the scope of our topic.

An important consideration in water management is that a very large portion of the water used for municipal, industrial, and agricultural purposes is nonessential in actual production processes. Admittedly, some controversy could arise concerning the definition of what is essential. For example, in order to produce a ton of grain, a corn crop may use 1000 or more tons of water in evapotranspiration, yet even at its most luxuriant stage the crop itself does not contain more than a few tons of water. Furthermore, in irrigated farming, actual water use often exceeds evapotranspiration, since some amount of water is lost during conveyance and application, and an additional amount may be lost through runoff from the soil surface and percolation beyond the rooting depth of the crop. Except for actual transpiration, all the above expenditures of water are both clearly nonessential and amenable to some measure of control. Even water expended on transpiration is not an immutable physiological necessity, although its control presents some formidable problems (see Chapter 9 of this volume).

What does happen to water as a consequence of its being used is some form of degradation, which usually affects its value for subsequent use. For example, water used for power generation in a hydroelectric plant loses energy, as does irrigation water that percolates through the root zone of a crop to the groundwater reservoir. If this water is to be recovered for reuse it will have to be pumped to the soil surface, and thus energy loss represents a loss in utility. Furthermore, as water percolates through the soil it generally acquires a higher concentration of soluble salts, which is equivalent to an additional loss of potential energy. Since soluble salts in the soil have a deleterious effect on either the soil or plants, increased salinity definitely affects the utility of water. Municipal and industrial effluent may be fit for agricultural use under certain conditions, whereas under different conditions it may become totally unusable because of pollution and its effect on soil and groundwater quality.

In this chapter we shall refer to water conservation problems mainly as they arise in semiarid and arid regions, since this is where these problems are most pronounced and have received the most attention. Under these conditions water generally constitutes the primary limiting factor for plant growth, and the major task of management is to ensure the supply, conservation, and efficient utilization of water in the field. Numerous methods have been proposed for this, yet the results are often baffling. Such is the complexity and variability of the soil–water–plant–atmosphere system that no single method of management can be expected a priori to apply equally or even similarly under different circumstances. Our best hope is not to search for universal prescriptions, but to understand the physical processes taking place in the habitat, and to learn how to modify them to best advantage in particular circumstances.

The purpose of this chapter is to elucidate these physical processes and to show their relevance to water conservation in the field. Field soil is the natural reservoir for water used by plants. However, since soil properties vary widely, not all soils are equally well suited for reception and storage of water. Different soils vary in the amount of water they can effectively store. Insufficient rainfall may result in inefficient utilization of soil storage capacity. Soils also differ in the rate at which they can absorb water. If rainfall cannot infiltrate the soil as rapidly as it reaches the surface, some of the water is lost by surface runoff and inadequate storage will result even when the quantity of rainfall is adequate in itself. Some of the water entering the soil may be lost by percolation beyond the depth of root penetration. Water stored in the root zone may be lost for crop production through direct evaporation from the soil surface, as well as through transpiration by weeds. In addition, water stored in the root zone and taken up by crop plants may be used at a low level of physiological efficiency owing to some other limiting factor, such as mineral deficiency, soil aeration, or atmospheric carbon dioxide. In the following sections the overall system will be considered first, after which the various component processes will be examined in greater detail.

II. THE FIELD WATER BALANCE

The field water balance, like a financial balance sheet, is an account of all quantities added to, subtracted from, or stored within the system. In its simplest form, it is merely a statement that, in a given body of soil, the difference between the input (W_{in}) and output (W_{out}) of water during a certain time interval is equal to the change in the amount of water stored in the soil (ΔW):

$$W_{in} - W_{out} = \Delta W$$

In order to itemize the accretions and depletions from the soil storage reservoir, one must examine the disposition of rain or irrigation water reaching a unit area of the soil surface during a given period of time. Some of this water infiltrates into the soil immediately. Water which does not infiltrate immediately accrues temporarily on the surface. Depending on the microrelief and slope of the soil, this water either will exit from the unit area as surface runoff, or will remain there stored as puddles in small depressions. Some of this water evaporates, while the remainder eventually infiltrates into the soil. Of the water infiltrated into the soil, some evaporates directly from the soil surface, some is taken up by plants for growth or for transpiration to the atmosphere, some may drain downwards beyond the root zone, and the remainder will accumulate and remain stored for a time within the root zone. Additional water may reach the area under consideration by runoff from a higher area, or by flow from an adjacent, wetter volume of soil.

To compute the water balance, it is first of all necessary to specify a relevant volume of soil. From a plant-ecological and agricultural point of view it is most appropriate to choose the rooting depth of the plant community or crop growing in the field.

The total water balance is expressed as follows:

$$[P + I] - [R + D + C + (E + T)] = \Delta S$$

where P is precipitation; I, irrigation; R, runoff (which generally represents a depletion but may be an accretion if runoff enters from some higher area); D, flow through the bottom of the root zone (which, if downward, is termed "drainage" and if upward is termed "capillary rise"); C, the amount of water incorporated into the plants; E, the direct evaporation from the soil surface; T, transpiration by plants; and ΔS, the change in root zone water storage.

The water balance is intimately connected with the energy balance (Lemon et al., 1963), since the processes comprising water transport naturally involve energy transformations. The soil water content affects the disposition of energy reaching the field. Likewise the fluxes of energy affect the state and movement of soil water.

While the field water balance is conceptually simple and readily understood, it is difficult to measure and even more difficult to control in actual practice. Generally, the largest component, and the one that is often the most difficult to measure directly, is evapotranspiration $(E + T)$. To be able to compute it from the water balance, very accurate measurements of all the other terms of the equation are needed, since the cumulative error of all measurements will appear as part of the computed value. In spite of some technical difficulties, it is generally possible to measure accretions by precipitation or irrigation with a sufficient degree of accuracy. Runoff is generally small from well-vegetated fields and may sometimes be taken as zero. Over a long period such as a whole year, the changes in the storage terms (water content of the crop and of the root zone) are likely to be small in relation to the total water balance. In this case the sum of rain and irrigation is approximately equal to the sum of evapotranspiration and drainage.

For shorter periods, however, the change in storage can be relatively large and must be measured, either by periodic sampling or by the use of such instruments as the neutron soil moisture meter. Heterogeneity of precipitation distribution and of soil hydraulic properties makes the evaluation of soil water storage in a field with satisfactory accuracy a formidable task. Measurement of the drainage component is even more difficult, as the flux to be measured is relatively slow and may be multidimensional, and the instrumentation required is delicate and not always reliable. A method for obtaining the drainage component of the field water balance by repeated measurements of the soil moisture and suction profiles was described by Rose and Stern (1967).

It is a common practice in irrigation to measure the total water content of the root zone just prior to an irrigation, and to apply the amount of water necessary to replenish the deficit which has developed since the previous irrigation. It is often assumed that this deficit is due entirely to evapotranspiration, thus disregarding the amount of water which may have entered or left through the bottom of the root zone. This flow is not always negligible, and may constitute a tenth or more of the total water balance (Hillel and Guron, 1970). While the error resulting from such an assumption may not have serious consequences for the irrigation management of a single field or farm, such an error cannot be neglected in evaluation of an irrigation project or the hydrological behavior of an area.

III. THE FIELD WATER CYCLE

The field water cycle includes the same processes that were discussed in the previous section, but here our concern is with the dynamics of the processes as they occur, sequentially or simultaneously, rather than with the total quantities of water involved over an extended period of time.

We shall begin with the process of water entry into the soil, called infiltration. If the soil is readily able to absorb the water as fast as it arrives, then the rate of water entry into the soil, called the infiltration rate, is controlled by and equal to the application rate. Under any given set of conditions, there is a maximum rate at which the soil can absorb water, and this limiting rate is often termed infiltration capacity, but we prefer the term "infiltrability" (Hillel, 1971).* This property is by no means a constant, even for a particular soil. It is, rather, a combined function of basic soil hydraulic characteristics and of initial moisture conditions within the profile. Infiltrability is generally a decreasing function of time during the infiltration process. This is due in the first place to the decrease

---

* The term "capacity" generally denotes a volume quantity (and thus is an extensive property) rather than a time-rate.

in hydraulic potential gradients which occurs as the soil is wetted more and more deeply. The frequent breakdown and compaction of the "open" structure at the surface also contributes to the decrease of soil infiltrability in time.

A number of equations, both empirical and theoretical, have been proposed to describe the infiltration function (e.g., Green and Ampt, 1911; Horton, 1940; Philip, 1957b,c) under different conditions. A common feature of these equations is that infiltrability is some function of $t^{-1/2}$. Depending on the conditions chosen for examination, fair to good experimental confirmation of the predicted function has been obtained on laboratory models (Bruce and Klute, 1956; Nielsen *et al.,* 1962). The laboratory models allow both simplification and good control of conditions. The situation in the field is much more complicated owing to soil heterogeneity both in terms of properties and transient conditions, and thus the discrepancy between predicted and observed data can be considerable.

The infiltrability of a soil is strongly influenced by the physical structure of the profile and its sequence of layers. In general, a profile that is uniform in depth will conduct water more rapidly than one that contains layers of different texture or structure. A particularly important factor affecting infiltrability is the physical condition of the surface, since it is through the surface that all the water entering the soil must pass. If the surface layer is compact, it can constitute a barrier which retards water entry and induces runoff and erosion. Such a compact layer, called a "crust," often forms under the beating and slaking action of raindrops and/or under repeated traffic by animals or machinery. The maintenance of an "open" (permeable) surface is a major task of soil management, no less in arid than in humid regions.

Soil infiltrability and rainfall regimen, in addition to the land surface configuration, combine to determine the quantity of runoff. In general, runoff constitutes only a small fraction of seasonal rainfall. However, catastrophic floods are produced when there occurs a combination of conditions such as a large, high intensity rainstorm over a large part of a watershed when soil infiltrability is low (due to high initial water content, compact soil, or frozen ground). The immediate concern of agriculture with runoff is loss of water storable in the root zone, and damage to the land by soil loss and gullies. In certain special cases, however, runoff water can be collected and utilized, so it is not necessarily a loss.

The water contained in the wetted part of the soil profile at the end of the infiltration process does not remain immobile. Even if this process were to culminate in a state of internal equilibrium, evaporation of water from the soil surface would soon create an upward potential gradient and consequently upward flux. However, the soil water potential at the end of infiltration is usually not uniform throughout the profile, and as a consequence water in the lower part of the wetted profile continues to move downward. This process is termed redistribution. The rate of flow and the volume of water involved in redistribution also depend on soil physical properties and on soil moisture conditions in the profile, the latter being in turn determined both by antecedent (preinfiltration) moisture and by the rate and amount of water infiltrated. Like the infiltration process, redistribution proceeds at a decreasing rate with time, and generally does not reach static equilibrium within a time interval of practical significance (W. R. Gardner *et al.*, 1970a,b). Within a few days, the rate of flow can become so low as to be considered negligible. However, this is an arbitrary criterion, since what is negligible in one case may be considerable in other cases.

The postinfiltration soil moisture profile and the rate and extent of redistribution affect the rate of water and nutrient loss from the profile, and hence the efficiency of the soil as a dynamic reservoir for the plant.

The continued movement of water within and beyond the root zone cannot be evaluated by measurements of water content alone. Even if the water content at a given depth stays constant, we cannot conclude that the water there is immobile, since it might be moving steadily without causing any moisture changes in the conducting layer (Rose and Stern, 1967). Tensiometric measurements can, however, indicate the direction and magnitude of the hydraulic gradient through the profile (Richards, 1965), and allow us to compute the fluxes on the basis of the hydraulic conductivity versus water content function for a particular soil (Watson, 1966; van Bavel *et al.,* 1968a,b). As was mentioned above, these measurements are difficult to carry out and to apply to the calculation of the deep percolation component from a large and heterogeneous area. More direct measurements of this component at a point will become possible with the perfection of soil moisture flux meters (Cary, 1968), which are, however, still in the preliminary development stage.

Water that penetrates an appreciable distance beyond the lower root-zone boundary during infiltration or redistribution is considered as a loss in the water balance of a particular field. It is not always a total loss, however. If the area is underlain by an exploitable aquifer, the water "lost" to the field by deep percolation can be credited to groundwater recharge. In the presence of a shallow water table, water can enter the root zone from below by capillary rise. Such a process is possible even in the absence of a water table, though to a lesser degree. Upward movement of water into the root zone from wetter layers below may in some cases account for an appreciable portion of the total water use by plants (Hillel and Guron, 1970).

The nature of the evaporation process from the soil surface is determined first of all by the source of the water. If there is a relatively shallow groundwater table, it is possible that a quasi-steady-state upward water flux will be established.* As in the case of profile-controlled infiltration, there is a maximal rate at which a given soil profile can conduct water from a water table to the surface of evaporation. This limiting rate depends upon the soil's hydraulic transmission properties as well as on depth of the water table (W. R. Gardner, 1958).

If the initially wet soil profile is not underlain by a shallow water table, the soil will gradually dry as evaporation proceeds. As the surface layer begins to dry out, suction gradients are established which draw from the moister layers below. For a time, this increase in gradient compensates for decrease in hydraulic conductivity of the drying layers and thus the evaporation rate can remain constant and is determined by meteorological conditions. Eventually, however, the evaporative flux must begin to decrease as the surface becomes air-dry. The drying process and its control will be discussed in greater detail in Section VI.

When the soil surface is covered by active vegetation, the major abstraction of soil water is by transpiration, a process in which plants transmit water from soil to atmosphere in response to the meteorological demand (commonly termed the "evaporativity").

In an arid environment, situations may develop in which the plants cannot draw soil moisture fast enough to satisfy the meteorologically imposed demand, and plants must limit the transpiration stream if they are to avoid dehydration. Plants can generally accomplish this, in a limited way, by closing their stomata. However, for this limitation of transpiration the plants must pay, sooner or later, in reduced growth, since the same stomatal openings which transpire water also serve for uptake of carbon dioxide needed in photosynthesis. Furthermore, reduced transpiration affects the energy balance of leaves and can result in undesirable and possibly excessive warming of plants (Gates, 1962).

IV. INFILTRATION CONTROL

Effective utilization of precipitation depends in the first place upon the capability of soil to absorb water that arrives at the soil surface. In attempting to increase infiltration, two factors are important: One is to maintain a high value of infiltrability, the other to ensure eventual infiltration even where application rate may temporarily exceed infiltrability.

---

* Since evaporation at the surface follows a diurnal pattern, true steady-state flow cannot prevail in the upper layer of soil, but these fluctuations of flow rate are usually damped out with depth.

The infiltrability of a soil is a function of more or less permanent soil properties, and of transient soil conditions. The relevant soil properties are hydraulic conductivity–suction–water content–depth relations, which are affected by soil texture and type of clay minerals, soil structure, and the constitution of the soil profile. These properties affect not only the infiltrability of a soil, but also the rate of internal drainage and hence the storage of water in the soil. Thus these properties also affect the antecedent soil moisture status (i.e., the antecedent soil water content and its energy potential distribution within the profile), which is a transient condition.

Most amenable to direct modification are those soil properties affected by vegetation, tillage, mulching, chemical amendments, and depth of water table.

Unfortunately, land management practices all too often tend to impair rather than improve the condition of the soil. Beneficial control of infiltration can be exercised through improvement and stabilization of soil aggregation, creation of an "open" surface, and shaping of the land to detain runoff and provide sufficient pocket storage over the soil surface.

Of all management operations, soil tillage has the greatest immediate effect on soil structure. As commonly carried out, a sequence of tillage operations can break up a massive soil layer into large clods and subsequently further break up the clods into smaller crumbs, and possibly even grind the soil into a powder. The surface soil may be inverted in the process, thus also incorporating undecomposed organic material (plant residues) into the soil. The number of large pores is generally increased by tillage, and infiltrability is thus increased, at least for a short time. However, tillage may also tend to break down the smaller soil aggregates (thus increasing their vulnerability to slaking and erosion by rain) and to compact the soil layer immediately below the plow layer, forming a so-called "plow sole." Thus the immediately beneficial effects of tillage are sometimes accompanied by long-term deleterious effects. As a consequence of structural breakdown a dispersed crust may form on the soil surface. Such a crust, particularly in conjunction with a dense plow sole, or with traffic-caused compaction, will result in lowered infiltrability, thus requiring more frequent and deeper tillage in order to achieve the same temporary improvement.

To avoid this vicious circle of excessive tillage and traffic in the field, a number of management systems have been developed, including "stubble mulching," "minimum tillage," and "precision tillage."* The principle underlying these approaches is to develop management practices based on minimal mechanical disturbance of the soil consistent with requirements

---

* See Papers on Zero-Tillage [*Neth. J. Agr. Sci.* **18**, No. 4 (1970)].

of seedbed preparation, soil conservation, and control of weeds, pests, and diseases. As such, these practices are not concise or universal formulas but necessarily vary from place to place according to specific conditions. Particularly noteworthy is the method of precision tillage, which aims at permanent division of a row-cropped field into parallel zones: seeding rows, which are brought to a fine-structured condition and loosened seasonally; coarse-structured ("open") interrow areas, treated for maximal intake and minimal evaporation of water; and travel tracks, which alone bear the traffic of tractors and machinery and are permanently compacted. The travel tracks are confined, as far as possible, to a small percentage of the field. Random traffic and compaction of the planting rows are thus avoided, and only a small fraction of the surface is pulverized to the extent necessary for seedbed preparation (Meyer and Mannering, 1961).

Soil structure can be stabilized by decomposition products of organic matter and by synthetic soil conditioners (Greenland *et al.*, 1962; Strickling, 1957). The effectiveness of organic matter in improving infiltrability has often been demonstrated, but is difficult to apply in practice, mainly for economic reasons. Organic materials are usually scarce, and their transport expensive. Cultivation of a green-manure crop eliminates a cash crop from the rotation, while incorporation of crop residues is sometimes more difficult or expensive than burning or removal of these residues. Synthetic soil conditioners are still too expensive for widespread use. In semiarid climates, especially where irrigation is practiced, organic matter decomposes so rapidly in the soil that it is extremely difficult to achieve a lasting increase of organic matter content. For all these reasons, maintenance and improvement of soil structure by addition of organic matter in the open field are seldom practical on a large scale.

A number of measures are, however, available for proper management of soil structure. The most important of these is proper execution and timing of tillage. As mentioned, tillage operations should be kept to a minimum consistent with adequate seedbed preparation and field sanitation. Manipulation of an excessively wet soil can cause undue compaction, while that of an excessively dry soil can result in irreversible breakdown of aggregates. The choice of optimal moisture for tillage is a problem to be resolved for each soil individually. The timing and type of tillage should expose the soil to minimum damage by raindrop impact. This can be done either by leaving a fallow field covered with plant residues or by creating a rough surface of large clods which are broken down in final seedbed preparation just prior to sowing.

Surface soil structure can also be protected by a variety of mulching materials, natural or artificial. Some of these mulches, including industrial wastes, paper or perforated plastic sheets, sprayable oils and waxes, gravel, etc., can have additional desirable effects such as thermal insulation, low conductivity to the upward flux of vapor or liquid, and high albedo, all of which can aid in water conservation (e.g., Holmes *et al.,* 1961; Bond and Willis, 1969).

Where the surface layer of the soil is compact, infiltration can be improved by opening crevices or slits and filling them with chopped straw to maintain their stability over a number of years. This method, known as "vertical mulching," has been practiced successfully in orchards as well as in field crop management (Swartzendruber, 1964).

A special problem of infiltration is that of alkali soils, which have a high content of sodium in the cation exchange complex. This monovalent cation causes the clay to disperse. Sodic soils slake and become muddy when wet, and form a massive and tight structure when dry. Such soils have extremely low infiltrability, and generally also suffer from poor internal drainage and impaired aeration. Their characteristically high pH may also have a specifically toxic effect on plants. Such soils can be reclaimed by displacing the adsorbed sodium by calcium and leaching the excess sodium from the soil. In order to accomplish this it is necessary to provide for a large excess of soluble calcium in the soil and for adequate drainage. The application of gypsum (calcium sulfate) is a common way of both lowering the pH and supplying calcium to the soil. Artificial drainage installations may be necessary. While such reclamation is both expensive and slow, dramatic improvement in soil infiltrability can be achieved by successful reclamation.

Even with high infiltrability, it is still to be expected that some rainfall will occur at an intensity exceeding the infiltrability. In order to prevent losses of water and soil owing to runoff and erosion, a number of procedures have been devised that increase the total eventual infiltration by providing for temporary storage of water over the surface of the land. Various types of terraces, contour furrows, lister furrows, infiltration beds, and pitting methods used in conjunction with appropriate tillage and cropping practices make it possible to either store excess water over the soil surface until it infiltrates, or to let it flow in a controlled manner to the point of utilization or disposal (e.g., Zingg and Hauser, 1959).

Another special case of infiltration control is the improvement of water intake in hydrophobic soils, a condition apparently caused by certain types of organic material (DeBano and Letey, 1969).

V. DRAINAGE

Drainage practices are generally undertaken to remove excess water from either the soil surface or soil profile in order to eliminate an impediment to farming operations or to crop growth. The majority of land areas requiring drainage are located in humid regions, but large tracts of such land also occur in arid regions, principally in river valleys. The occurrence of excess water may appear to be but little connected with water conservation, except that water that impairs agricultural production is not being used efficiently. Where watersheds do not have an outlet to the sea, particularly in arid regions, surface and subsurface flow may collect in the lowest part of such basins forming salt marshes, playas, and saline inland seas. Drainage in the immediate vicinity of such "sinks" may not be possible, since the prime requirement of any drainage project is a lower point to which the drainage water can be discharged. Since the soils in river valleys are often potentially very productive, the drainage of these lands and elimination of phreatophytic vegetation can prevent wasteful evapotranspiration and reclaim valuable land for agriculture.

The reclamation of saline and alkaline soils requires the leaching of soluble salts out of the soil profile, and replacement of sodium with calcium in the soil's cation exchange complex. These measures require that the soil be well drained.

Failure to consider the need for drainage can lead to disastrous failures of irrigation projects. As soon as irrigation is expanded, either by pumping water from a local river, or worse yet, by import of additional water from outside the region, the danger of a rising water table is created. As the water table nears the soil surface, the evaporation rate increases and, since groundwater in arid regions often contains appreciable concentrations of salts, the surface soil can soon become saline. Once this happens, only major drainage and reclamation measures can render the land productive again.

Removal of excess water from the land may involve both surface and subsurface drainage. Surface drainage is required to remove accumulations of water on the land, to guide and control the flow of runoff in order to prevent erosion, and to prevent excessive amounts of water from entering the soil. Surface drainage is the simpler problem to solve. It is also less commonly a problem, especially in dry regions.

In order for a surface drainage system to function properly the land surface must be sufficiently smooth for water to flow over it at a controlled rate to a collection point. Surface depressions can be drained by leading a ditch from each depression, but when the land surface has many small depressions, land leveling is required. Where irrigation is practiced, surface drainage may also be necessary for removal of possible irrigation excesses. From the point of local water concentration the surface runoff must be conveyed in ditches or grassed waterways to the point of disposal. Such ditches require maintenance if they are to function properly in time of need.

The main operation is weed control and bank stabilization. Accumulated trash may be brought down to the ditches by a heavy flow, and this should be removed in order to prevent clogging of the system during the next rainfall.

Subsurface drainage systems may consist of underground pipelines, generally of open-jointed sections of tile, concrete, or asbestos–cement pipes, enclosed in a pad of gravel. Subsurface drainage is also carried out by means of deep ditches, dug to below the water table. In other cases, mole drains are used. These are tunnels formed in the soil by passage of a bullet-shaped implement attached to a subsoiler shank pulled by a large tractor. Mole drains are relatively short-lived in many soils. Since the slope of the drain is in effect determined by the shape of the land surface over which the tractor travels, mole drains are effective only if the surface is properly graded ahead of time.

Drainage occurs primarily under saturated conditions, and the flow can be described mathematically by the appropriate combination of Darcy's law and the law of continuity.*

Though based fundamentally on simple laws, the design of drainage systems can be very complicated in practice. The reason for this is that soil is not a uniform material. Soil properties can vary considerably with depth at a point, as well as laterally from point to point. Layering is a very common feature of soil profiles, and each layer will generally have different hydraulic properties. Furthermore, the hydraulic conductivity may also be different for horizontal and vertical flow within each layer, a condition called anisotropy. The situation may be so complex as to preclude analytical solution, and in such cases one may resort to solution by electrical analog or by numerical methods.

All calculated solutions of drainage problems must be based upon reliable information on soil properties which must be obtained from the field itself. This is a fairly laborious and uncertain task at any given point,

---

* Darcy's law states that velocity of flow is proportional to the hydraulic gradient. The coefficient of proportionality is called the hydraulic conductivity. The one-dimensional form is:

$$v = - k(dh/dx)$$

where v is the flow velocity, h is the hydraulic head and dh/dx is its gradient in x-direction, k is the hydraulic conductivity.

The law of continuity is simply an expression of the law of conservation of matter. The equation in one dimension is:

$$(\partial v/\partial x) = - (\partial \theta/\partial t)$$

where $\partial v/\partial x$ is the change in flow velocity with distance and $-\partial\theta/\partial t$ is the decrease of volumetric water content with time.

and ever so much more for an area where soil is heterogeneous. The undetected presence of sand and gravel lenses, rock outcrops, subterranean channels, and impermeable layers can change the effective flow rates by several orders of magnitude relative to flow where such irregularities are absent.

Experience indicates that even where a great effort is made to obtain reliable data, drainage designs all too frequently result in failure. Much of what is known about practical drainage design was acquired by trial-and-error experience, and where such knowledge is available for a particular set of conditions, it is well to utilize it at least in conjunction with the theory-based approach. The first drainage network to be installed in a new area always involves risk, and it should be so designed that modification (e.g., of variables such as spacing between drains) is subsequently possible. A comprehensive monograph on drainage was published by the American Society of Agronomy (Luthin, 1957).

VI. LIMITING BARE SOIL EVAPORATION

In the absence of plants, as well as in exposed areas between plants, evaporation from the soil surface constitutes an important water loss mechanism.

Soil water evaporation may be controlled both by atmospheric evaporativity or by the profile's own transmission properties. Typically, following a wetting, the initial evaporation rate (while the soil is relatively moist and its water transmission rate is not limiting) is constant and dictated by the external evaporativity acting on the soil surface. Sooner or later, however, if evaporation persists and the soil surface dries, the ability of the soil profile to transmit water to the surface falls below evaporativity of the atmosphere, and the actual evaporation rate drops—sometimes quite abruptly (Lemon, 1956).

The length of time the initial stage of drying can persist depends upon intensity of the meteorological factors which determine atmospheric evaporativity, as well as upon conductive properties of the soil itself. When external evaporativity is relatively slight, the initial constant-rate stage of drying is slower and can persist longer. Under similar external conditions, the first stage of drying will be sustained longer in a clayey than in a sandy soil, since the clayey soil retains higher water content and conductivity as suction develops in the upper zone of the profile (Hillel, 1968).

In principle, the evaporation flux from the soil surface can be modified in three basic ways: by controlling energy supply to the site of evaporation (e.g., modifying the albedo through color or structure of the soil surface, etc.); by reducing the potential gradient, or the force driving water through the profile (e.g., lowering the water table if present, or warming the surface so as to set up a downward-acting thermal gradient); or by decreasing the conductivity or diffusivity of the profile (e.g., tillage and mulching practices).

The actual choice of means for reduction of evaporation depends on the stage of the process one wishes to regulate: whether it be the first stage, in which the effect of meteorological conditions on the soil surface dominates the process; or the second stage, in which the rate of water supply to the surface, determined by the transmitting properties of the profile, becomes the rate-limiting factor. Methods designed to affect the first stage cannot *a priori* be expected to serve during the second stage, and vice versa.

Covering or mulching the surface with plant residues or with reflective materials can reduce the intensity with which external factors, such as radiation and wind, act upon the surface (Hanks *et al.*, 1961; Bond and Willis, 1969). Thus, such surface treatments can retard evaporation during the initial stage of drying. A similar effect can result from application of materials which lower the vapor pressure of water (Law, 1964), or from shallow cultivations designed to produce a "dust mulch" at the surface (Army *et al.*, 1961). Retardation of evaporation during the first stage can provide the plants with a greater opportunity to utilize the moisture of the uppermost soil layers, an effect which can be vital during the germination and establishment phases of plant growth. The retardation of initial evaporation can also enhance the process of internal drainage, and thus allow more water to migrate downward into the deeper parts of the profile, where it is conserved longer and is less likely to be lost by evaporation (Hillel, 1968).

During the second stage of drying, the effect of surface treatments is likely to be only slight and reduction of the evaporation rate and of water loss in the long run will depend on decreasing the diffusivity or conductivity of the soil profile in depth. Deep tillage, for instance, by increasing the range of variation of diffusivity with changing water content, can decrease diffusivity at the lower water contents and thus reduce the rate at which the soil can transmit water toward the surface during the second stage of the drying process. However, the second-stage rate is usually much lower than during the first stage and it may be questionable whether it is worthwhile to invest in the control of evaporation in this stage.

An irrigation regime having an excessively high irrigation frequency can cause the soil surface to remain wet and the first stage of evaporation to persist most of the time, resulting in a maximum rate of water loss. Water loss by evaporation from a single deep irrigation is generally smaller than from several shallow ones with the same total amount of water. Most irrigation regimes have such an irrigation frequency that first-stage drying is the dominant factor contributing to evaporation losses.

Tillage for seedbed preparation, weed control, or other purposes is perhaps the most ubiquitous of soil management practices. Soil disturbance by tillage implements generally results in opening and loosening of the tilled layer. This often enhances drying of the tilled layer, but might also reduce water movement from the layers below. Tillage might thus produce either a larger or smaller total loss than would have occurred in the undisturbed soil, the net effect depending on duration of the process, as well as on depth, degree, and frequency of tillage. The long-term effect also depends on subsequent rainfall and its influence in reconsolidating the tilled layer. In many cases, failure to conserve water by shallow cultivation is due to the simple fact that most soils cannot be cultivated when wet. Once the soil surface becomes dry enough to permit cultivation, the first stage of evaporation is generally over, so that surface modification can no longer help (since it is the profile in depth which now begins to control the process).

Marshall and Gurr (1966) reported that the zone of evaporation from a moist soil is a layer about 1 cm thick at the top of the moist zone. This zone may thicken somewhat as the drying effect progresses more deeply into the soil, though H. R. Gardner and Hanks (1966) found that it remains about 1 cm thick even as its location descends into the soil. Work of W. R. Gardner (1959) and of Hanks and Gardner (1965) suggests that the total water loss during the first stage of drying may be reduced if the water diffusivity in the wet range is reduced. Only if this is achieved to an appreciable depth does it affect the second stage of drying as well. Accordingly, a shallow aggregated layer or other mulch at the soil surface can be expected to affect the evaporation process primarily during its initial stage. An important qualification is that the profile and surface conditions desired must be achieved prior to water application, since after wetting no mechanical treatments can be performed in practice until the surface dries, by which time much of the water loss has already taken place.

Due to indiscriminate application of experimental findings by F. H. King in Wisconsin during the latter part of the nineteenth century there arose the misconception that soil water could always be conserved by cultivation after every rain. It was assumed that working the soil surface layer into a finely pulverized condition could help to break the continuity of capillary tubes through which water was presumed to flow from lower layers to the soil surface. As a consequence, expensive operations were adopted which in many cases not only failed to conserve water, but also damaged soil structure instead of improving it. The finely pulverized soil tended to slake and crust after each cultivation because of raindrop impact.

There are two situations in which cultivation appears to be necessary for water conservation. The first is in control of growth of weeds, which represent an uneconomical use of soil moisture. In the last decade, increasing use of chemical herbicides, which can be applied by spraying from the air, seemed to obviate the need for cultivating wet soils for weed control. However, recent recognition of the danger of environmental pollution may curtail the widespread application of some of these chemicals.

The second situation requiring tillage is the case of soils which, because of appreciable swelling and shrinking, as a consequence of wetting and drying, tend to crack markedly during the dry seasons. These cracks become secondary evaporation surfaces penetrating inside the soil, causing it to desiccate deeply and to an extreme degree. Johnston and Hill (1944) found in a field study that soil water content decreased around cracks to a distance of 7.5 cm on each side of the crack and to a depth of 30–45 cm. Adams and Hanks (1964) and Adams et al. (1969) demonstrated with laboratory and wind tunnel models the dependence of evaporation rate on wind velocity and crack width. In a recent laboratory study, Selim and Kirkham (1970) investigated the effect of crack width, radiation intensity, and wind velocity on soil evaporation. They found that cracks caused a 5–10% decrease of volumetric soil water content within a period of 12 days. Lateral water movement toward a crack was detected at distances of 5–14 cm from the crack wall. Evaporation from bare soil was increased 12–30% depending upon the conditions of the experiment.

Timely cultivations can prevent the development of such cracks or even obliterate cracks after they have begun to form. The quality of tillage, i.e., the degree of pulverization and clod size distribution and porosity of the tilled layer are also important. Coarse tillage leaves large cavities among the clods into which air currents can penetrate and increase vapor outflow (Hillel, 1968). Tillage that is too fine may pulverize the soil unduly and result in a compact surface that would defeat the aim of reducing diffusivity. Each set of soil and meteorological conditions may require its own optimum depth, timing, and quality of tillage to effect maximal reduction of evaporation losses.

In general the deeper the water is stored in the soil, the more slowly will it be removed by evapotranspiration. Thus, depth of water penetration is an important consideration in reduction of evaporation losses. Soil texture and profile stratification are the principal properties controlling distribution of water storage in the soil. Coarse-textured soils usually drain more rapidly than fine-textured ones and thus retain less water for plant use per unit volume of soil after equal elapsed time intervals. This may be an advantage in arid regions where the small amount of precipitation water is transmitted to a greater depth in the soil profile, where it is less subject to evaporation. Provided this water did not penetrate below the rooting depth of the plants, more of it will be available for plant use. Under arid conditions, therefore, coarse-textured soils can constitute a moister habitat than fine-textured ones (Hillel and Tadmor, 1962). In contrast, under humid climatic conditions, it is the finer-textured soils which hold more water in the root zone for a longer time and can provide more effective moisture storage for plant growth.

In principle, the best condition from the standpoint of both evaporation and infiltration control is to have a coarse layer of large pores at the surface overlying a finer textured or structured profile. The coarse surface layer would conduct the water readily at saturation during infiltration but would restrict liquid flow as suction develops during evaporation and the soil with the large pores desaturates and its hydraulic conductivity decreases drastically. Since vapor diffusion in dry soils is generally much slower than liquid flow in moist soils (typically by several orders of magnitude) rapid drying of the surface few millimeters may reduce subsequent evaporation.

The effectiveness of vegetative mulches may be limited unless they are sufficiently thick, because their high porosity permits rapid diffusion and air currents (Hanks and Woodruff, 1958). The initial evaporation rate under a mulch is usually reduced, but for extended periods a mulch may keep the soil surface moister and thus produce no net saving of water. Gravel and clod mulches, on the other hand, can effectively reduce evaporation, both in the short and in the long run because the immediate surface dries rapidly and such materials are dense enough to act as diffusion barriers. Since mulches of crop residues tend to be more reflective than most soils, they may result in cooler surface temperatures. Consequently, early plant growth under mulches is often retarded.

Since evaporation is an energy-consuming process, treatments designed to decrease the energy flux to the soil can reduce evaporation, particularly during its initial stage. Color, slope aspect, and surface roughness may alter the energy absorption of the soil surface (Hanks et al., 1961). Soils of light color reflect a higher percentage of incident radiation than do darker colored ones. As soil organic matter and water content increase, colors darken and reflectance decreases. Artificial treatments, such as application of chalk powder over dark soils, are sometimes effective in reducing direct soil water evaporation (Stanhill, 1965; Hillel, 1968). Slope aspect and degree alter the energy absorption rate per unit area. In the Northern Hemisphere, north-facing slopes receive a lower radiation intensity than south-facing slopes. Thus, ridging the soil can have an effect on evaporation. In addition to the radiation effect, there can also be a differential effect of wind on the upwind versus downwind sides of the ridges.

VII. IRRIGATION AND WATER-USE EFFICIENCY

Any concept of efficiency is a measure of the output obtainable from a given input. Irrigation or water-use efficiency can be defined in different ways, however, depending upon the point of view or the emphasis that one wishes to place on certain aspects of the problem.

One might, for example, define an agronomic or economic irrigation efficiency as the marketable yield or financial return per amount of water applied or money invested in the water supply. The decision of whether to base this calculation on volume of water applied or money invested is an illustration of point of view brought to the problem. The choice of the one base or the other might depend on relative cost of the water itself and of the investment in its conveyance and application.

A different measure of efficiency is the plant water-use efficiency. This can be defined as the amount of plant tissue or dry matter produced per unit volume of water taken up by the plant from the soil. Since most of the water absorbed by plants in the field (in arid regions—99% or more) is transpired while generally only a small fraction is retained for growth, the plant water-use efficiency as defined is approximately the reciprocal of what is known as the "transpiration ratio" in classic plant physiology texts.

What we shall refer to as the technical efficiency is what is most commonly understood as irrigation efficiency by water supply and irrigation engineers. It is generally defined as the net amount of water added to the root zone divided by the amount of water taken from some source, multiplied by 100 to yield an answer in percent:

$$E_a = 100 \, (W_r/W_a)$$

where E_a is the application efficiency, W_r is the water added to the root zone, and W_a is the volume of water supplied at the source. Since the source of supply may differ for different users, it is possible to define the technical irrigation efficiency in terms of a complex regional project efficiency, or an individual farm, or a specific field. In each case, however, the difference between the net amount of water accrued in the root zone and the total amount of water withdrawn from the source represents losses.

When we consider the efficiency of an entire irrigation project, the water input represents the volume of water made available at the (often distant) source, and the water actually applied to the fields represents the output. The losses that may take place in the process of delivering water from a project's source to individual farms are such that occur during storage and conveyance of the water. Thus, if the water supply of a project is stored in a surface reservoir, there may occur losses due to evaporation of water from the open water surface, percolation of water through the bottom or sides of the reservoir, evaporation and leakage from conveyance systems such as canals and pipelines, and, possibly, transpiration by useless riparian vegetation. The evapotranspiration and evaporation losses are not recoverable as a rule within the boundaries of any single project. On the other hand, water that escapes the storage or conveyance system by downward seepage may reach the groundwater and may reappear as surface flow further downstream or may be recoverable by pumping from wells. It is, nevertheless, common practice to consider water which leaks out of a storage and conveyance system as a project loss. Recovery of such losses for subsequent use usually entails not only additional expense, but also a deterioration of water quality.

Control of seepage and of riparian phreatophytes should not be a major problem in well-designed pipeline or lined canal conveyance systems, although it is a major item of maintenance in the case of unlined earth ditches and along the shores of open reservoirs. In the latter case it is often desirable or necessary to have these banks vegetated in order to protect them from erosion and to prevent accelerated silting-up of the reservoir itself. The vegetative cover of reservoir banks where needed should at least be effective in controlling soil erosion. It would of course be preferable if this vegetation also had some additional economic or aesthetic value.

Evaporation from the surface of open reservoirs is a major loss of water from storage facilities in arid regions. Not only does evaporation cause water loss, but since surface water in dry regions generally contains appreciable amounts of soluble salts, evaporation also causes a deterioration of water quality. It has proved difficult so far, but not impossible, to control evaporation from open reservoirs. One important factor which can be taken into consideration in the planning stage is cross-sectional shape of the reservoir. The most favorable shape is that which exposes the minimum amount of surface per unit volume of water stored. Hence, a deep reservoir is obviously preferable to a shallow one. Many attempts have been made to decrease open water evaporation by formation of a monomolecular layer of various fatty alcohols and similar substances on the water surface. Similar attempts to limit surface evaporation have been based on the use of low-density solids (pellets or rafts) which float on the water and form a physical barrier to vapor movement.* These approaches have been thoroughly documented in the literature (e.g., Myers, 1970); however, the overall result of such efforts is that the use of surface barriers is still not universally practical for both technical and financial reasons.

---

* For the best effect, such solids should have high reflectivity so as to dissipate as much as possible of the radiant energy which would otherwise heat and vaporize the water.

Monomolecular layers are only effective in decreasing evaporation if they form a continuous cover over the entire water surface both in space and in time. However, monomolecular layers are very vulnerable to break-up by wind and wave action, solar radiation, and bacterial activity. Solid barriers, such as foam-plastic rafts or plastic films, are still quite expensive. One solution to the problem is to avoid surface reservoirs altogether, but this is only feasible where groundwater recharge and subsequent recovery are possible. Seepage control in surface reservoirs is also a problem (Myers, 1970).

Some large-scale irrigation projects operate in an inherently inefficient way. Thus, in surface-irrigation schemes one or a few farms may be allocated large flows representing the entire discharge of a lateral canal for a specified period of time. Since water is delivered to the consumer only at the specified time, and charges may be based not on measured delivery but on calculated delivery during that time (based on known capacities of the conveyance network), customers often tend to accept and apply the entire amount of water made available to them whether it will be used effectively or not. This often results in overirrigation, which is not only wasteful of water but may also result in project-wide difficulties owing to return flow, waterlogging of soils, leaching of nutrients, and raising of the regional groundwater table (consequently requiring expensive drainage). Although it is difficult to arrive at reliable statistics, it has been estimated that the average efficiency of irrigation water use in the western United States is around 50%. Since it is definitely known that individual irrigations can be applied with a field efficiency of 90% and better, it appears that many of the losses encountered in practice could be eliminated by proper planning and management.

The factors determining farm irrigation efficiency (as distinct from field application efficiency) are very similar to those determining project efficiency. Since surface storage of irrigation water on irrigated farms is not common practice, the main water losses are connected with conveyance and distribution rather than with storage. Where open and unlined distribution ditches are used, ditch leakage and evapotranspiration by riparian phreatophytes are again the major causes of water loss. Even pipeline distribution systems do not always prevent losses, contrary to what might be assumed. Leaky joints because of poor workmanship, corrosion, or obsolescence are only one of the possible causes of water loss. Another cause is leaky turnout devices such as valves, owing to poor maintenance, faulty operation, or mechanical damage by farm implements. Sometimes this damage is not immediately obvious, and if a pipeline is constantly under pressure, a cracked or damaged riser may fail during nighttime hours when no one is in attendance. Such a failure of course not only causes loss of water, but erosion and waterlogging of the field soil as well.

Besides these unintentional losses, there are occasional wastes which might almost be called deliberate. By this we mean management practices which lead to overirrigation, not necessarily because of technical problems or lack of knowledge, but simply because it may appear more convenient or possibly more economical in the short run to waste water than to apply proper management practices. Such wasteful practice typically occurs when the price of irrigation water is lower than the cost of labor or of automated equipment necessary to avoid overirrigation. It is clear both from theory and from experience that such wastage is against long-term interests not only of the entire project but also of the individual farmer, since it might cause eventual water shortage, soil waterlogging, impeded drainage, salinization, and loss of nutrients.

In the final analysis irrigation efficiency is determined in the field, since it is here that the ultimate purpose of an irrigation project either succeeds or fails. This topic may be divided into two aspects: the technical aspect of applying irrigation water and the agronomic aspect of water utilization.

In addition to various losses resulting from leakage of conveyance systems, we often encounter losses by surface runoff originating from individual fields. In irrigated areas which receive appreciable rainfall as well, surface runoff of rainwater is difficult to prevent. Unavoidable rainwater runoff must be guided and conveyed in a controlled manner to a point of exit from the field in order to avoid flooding and erosion damage.

Field surface runoff as a result of application of irrigation water ideally should not occur, particularly when sprinkler irrigation is practiced. Sprinkler irrigation systems should and can be designed so that the water application rate will at no time exceed soil infiltrability. Even in those cases where this condition is not met at the outset owing to improper design, it is often possible to modify the irrigation equipment in order to achieve this (e.g., by replacing the exchangeable sprinkler nozzles with a smaller size nozzle of lower discharge).

In contrast to sprinkler irrigation, it is often virtually impossible to achieve uniform water distribution over the land surface in gravity irrigation systems without having some runoff from the field ("tail water"). Some systems are deliberately designed to produce some tail water, while in others it appears practically impossible during operation to irrigate large areas at reasonable labor efficiency without the occurrence of a least some runoff. Only when provision is made to collect both irrigation and rainwater surpluses at the lower end of the field and guide them as controlled return flow to the farm or project conveyance system, can this water be considered as anything but a loss.

Evaporation can cause a major loss of water during irrigation application. Evaporation losses associated with water application may be attributed to evaporation from water droplets during their flight from the sprinkler to the ground, to evaporation from the open water surface of border checks or furrows, evaporation from wetted crop canopies during and after irrigation, evaporation from wet soil immediately after the irrigation, and evaporation from small sprinkler drops carried by wind drift away from the target area. At least for the case of a densely growing crop which effectively covers most or all of the soil surface, it is not at all certain that evaporation losses from spray, open water surface, canopy interception, and even from wet soil should be considered a loss, since it might merely replace the transpiration which would occur otherwise during the same period.

When a sprinkler irrigation is applied during excessively windy conditions, and particularly if the sprinkler is operated at too high a pressure, which causes excessive break-up of the irrigation stream into very small droplets, some of the water is carried by the wind outside the target area, and may thus be lost.

From the foregoing material it is seen that no really universal definition of such an apparently simple matter as technical irrigation efficiency is possible since it depends very much on the particular circumstances at hand. What may technically be considered a loss from a single irrigation in the fall, may very well be prudent storage of water in the root zone for future use if applied during the early stages of a crop in the spring. An intended beneficial leaching of excess soluble salts from the soil profile may result in the raising of the groundwater table and unintentional salinization of the soil if drainage problems were not taken into account. It is therefore difficult to provide a set of universal recommendations as to the optimal practice from the point of view of irrigation efficiency. A successful application of the principles of water conservation to the process of irrigation depends rather on a proper understanding of basic principles and processes and application of these to the particular conditions at hand.

When the soil surface is covered by a close-growing crop, the major abstraction of soil water is through transpiration. In the open field, there is rather little that can be done about transpiration of a crop while maintaining conditions required for high yields. Even though transpiration may not be physiologically essential for growth, it might be said that under open field conditions it is the aim of water conservation to make sufficient water available for transpiration by crop plants. Attempts have been made to decrease field transpiration by the application of various chemicals, known as antitranspirants, to the plant canopy, and by controlling wind movement above the crop by windbreaks. The use of antitranspirants has not proved itself to be economically sound so far, since decrease of transpiration has generally been accompanied by decreased yield (Waggoner, 1966). The effect of windbreaks on transpiration is not clear, since conflicting results have been obtained. Under certain conditions, a windbreak may actually decrease water use efficiency by the crop, since it might act as a heat trap. Evapotranspiration can be reduced appreciably when plants are grown under cover (e.g., in a greenhouse). The physical barrier of the enclosure decreases the vapor pressure gradient between the canopy and the atmosphere, thus reducing water flux. If a greenhouse is operated as a closed system, whatever water condenses on the underside of the enclosure can be collected and reused. Far from impeding growth, the overall control practicable in greenhouses leads to yields far superior to those obtainable in open field culture. However, the major problem here is heat dissipation during the summer season. To characterize the relation of total seasonal water use by a crop to its dry matter production, early investigators (e.g., Briggs and Shantz, 1912) introduced the term "transpiration ratio." This index was found to depend primarily on the climate, and to range up to several hundred (or even one thousand or more) in arid regions. Later, the inverse concept of "water use efficiency" came into use, defined (Viets, 1962) as the ratio of the dry matter production to the amount of water transpired by a crop in the field.

Transpiration can be limited either by the supply of water to be evaporated (by precipitation or irrigation), or by the supply of energy needed for vaporization, the latter determined mainly by such climatic factors as radiation, temperature, and humidity. When the supply of water is plentiful, it is the climate, and particularly the net radiation, that determines water use (Penman, 1956). When water is limiting, the assimilation rate, plant growth, and consequently crop yield are all related quantitatively to the water supply, but this relationship may not be a simple one. An interesting comprehensive review and analysis of this relationship in temperate and arid climates was published some years ago by de Wit (1958). He found that in climates with a large percentage of bright sunshine (i.e., arid regions) a relation

$$Y_d = mE_a/E_0$$

exists between total dry matter yield (Y_d) and the ratio of actual transpiration (E_a) and free water evaporation (E_0). In climates with a small percentage of bright sunshine (i.e., temperate regions) the relation

$$Y_d = nE_a$$

was found (that is to say, dry-matter production is proportional to transpiration and hence water use efficiency is constant). These relations were found for both container-grown and field-grown plants, and the values of constants m and n were reported to be characteristic of the crops. These relations obviously cannot hold as potential evapotranspiration is approached and water ceases to be the limiting factor in plant growth. Beyond the point where transpiration reaches its climatic limit, the promise of increasing production lies in obviating the other possible limiting factors, such as light intensity, carbon dioxide concentration in the air, nutrients in the soil, etc. Hillel and Guron (1970) reported that water use efficiency of an irrigated corn crop (in terms of total dry matter yield per unit quality of evapotranspiration) increased twofold as ET was increased by 30% by the wetter treatment (in which actual ET was approximately equal to the potential ET). The wettest treatment yielded 2.3 times as much grain as the driest treatment, while consuming only 1.3 times as much water.

The presence of a limiting factor not connected with water may affect the physiological efficiency of water use by the crop and also the economic efficiency of irrigation. Thus plant diseases and pests as well as insufficient soil fertility may cause a serious depression of yield without being accompanied by proportional saving of irrigation water and other production costs. In the wider sense, therefore, the presence of limiting factors to crop production certainly affects the overall efficiency of field water use.

Often, water resources for irrigation are scarcer than land resources. It is therefore of utmost importance to choose for irrigation development only that land which is most suited for the purpose. Such land should have, in addition to location and proper topographical features, a productive soil with the desirable combination of physical and chemical properties. Only thus can the yields obtained justify the investment in irrigation installations. These desirable properties include adequate soil depth, stable structure, high infiltrability (though an extremely high infiltrability is not desirable where surface irrigation is practiced as it might prevent even distribution of water in the irrigation furrow or border), good aeration, high fertility, and freedom from excess salts. Water applied to poor land, on the other hand, is used inefficiently and may be largely wasted.

Proper planning of farm irrigation systems, by which we mean adjusting the system to the specific conditions of topography, soil, crop, and climate, has a great bearing on water use efficiency.* The choice of the method of irrigation should depend upon knowledge of soil properties. Optimal irrigation frequency and quantity depend upon soil, crop, and climate relationships. An excellent review of the subject of programing irrigation for greater efficiency was recently given by Jensen (1972).

---

* In a water-scarce area, the aim of the farmer should be to get as high a return as possible from the limited amount of irrigation water available to him. Attainment of such an aim involves optimal integration of all the farm operations, including crop rotation, fertilization, salinity and alkalinity control, and others.

VIII. RUNOFF INDUCEMENT AND UTILIZATION

Uncontrolled runoff is never desirable and since its control requires additional investment, in most situations it is indeed preferable that no runoff should occur from the land. In areas where natural rainfall is on the borderline of being insufficient for dryland farming, it is particularly important that as much of the precipitation as possible infiltrate into the soil. However, in many arid regions large tracts of land remain unused owing to insufficient or unstable rainfall, unarable soils (too shallow, stony, or saline) or unsuitable topography. Rain falling on such land is almost invariably lost. Often rainfall amounts per storm are insufficient to either recharge the ground water or to support an economically valuable crop. This water generally infiltrates to a shallow depth only, and is quickly lost either by direct evaporation or by transpiration of ephemeral vegetation, usually weeds. Some of this water may also appear as runoff causing destructive flash floods typical of the desert, although natural runoff seldom exceeds about 10% of annual precipitation (Myers, 1967). If total rainfall amount is appreciable but mostly occurring in such small amounts as to make control and collection of the runoff water an uneconomic proposition, both the water that infiltrates and that appearing as runoff is lost. However, when it is feasible, the possibility of controlling and increasing the amount of surface runoff obtainable from such lands can be of great importance, particularly where water is scarce and the runoff thus obtained can augment the water supply of the region. It should be pointed out that there are areas receiving about 500 mm of rainfall annually, not too far from most, if not all, of the cities of the world's semiarid and arid regions (Myers, 1967). Water obtained from such areas could be delivered by gravity to the point of use often in existing storage and conveyance facilities.

The collection and utilization of runoff have been practiced by desert peoples since antiquity. In large areas of the Middle East it is still possible today to find functional ancient systems of terraced stream beds which were used for farming, whose water supply was often augmented by runoff water collected on the adjoining hillsides and brought to the fields by means of sophisticated conveyance systems (Evenari and Koller, 1956; Evenari et al., 1958; Lowdermilk, 1954). There is evidence that the builders of these systems understood not only the inadequacy of natural runoff but also perfected techniques of inducing soil crusting and increased runoff. Noticing apparently that the soil had a natural tendency to crust, which was impeded, however, by a protective cover of stones (desert pavement), this natural gravel mulch was collected from the soil surface into geometrically arranged mounds of stone, thereby exposing the intervening

soil surface to the impact of raindrops, which caused a dense and impermeable crust (Tadmor *et al.,* 1957). This method of decreasing soil infiltrability was both laborious and of temporary effectiveness since erosion
of the unstable soil crust eventually exposed additional stones on the soil
surface. As a consequence, the ancient runoff farmers required a water-
contributing area 20 times or more larger than the area to which water was
brought for utilization. Similar techniques were used to collect runoff water
in cisterns both in the open countryside and within the limits of fairly
large cities in the desert.

Modern technology holds the promise of more effective runoff inducement than was obtainable in ancient times. It is possible to increase runoff
severalfold, thus reducing the ratio of contributing to receiving area by
means of mechanical treatment (stone clearing, smoothing, and compaction) as well as by a variety of chemical treatments to seal and stabilize the
surface. Rendering the soil surface water repellent or impermeable may
be achieved either by the application of mechanical barriers to water
movement, such as plastic, aluminum foil, concrete, or sheet metal, or by
artificial formation of an impervious soil crust with the aid of various
chemicals, such as sodic salts and various petroleum distillation byproducts
(Myers, 1961, 1967, 1970; Hillel, 1967). While Myers' initial efforts
were directed primarily at the collection of runoff for watering of stock
and wildlife, he considers that there are possibilities for irrigation use if
methods can be further improved.

Where rainfall is insufficient but the soil is otherwise arable, it is possible to utilize the land in a system of "runoff farming" where part of the
land surface is shaped and treated appropriately for runoff inducement
while another portion of the land receives the runoff water so produced.
The irrigation may be carried out either immediately with runoff water
flowing directly off the contributing plots to the receiving plots, or water
can be stored for future use in reservoirs. Although the former method
is less expensive, it is also less flexible since the irrigation is applied immediately following precipitation, regardless of whether the root zone storage
has been depleted sufficiently to effectively store all the water applied.

Several approaches have been tried in respect to size and arrangement
of the contributing area in relation to the water-receiving area. At one
extreme entire small watersheds can be treated so as to make them less
pervious with the water being concentrated at the outflow of the basin and
conveyed to an irrigated field. On a smaller scale alternate strips are
treated on a slope where runoff strips are sealed and stabilized so as to
contribute their share of rainfall as runoff to adjacent lower lying "runon
strips" in which the crops can be grown (Hillel, 1967). Microwatersheds
are the third approach possible where each single contributing area serves

either a single tree or row of plants (Shanan *et al.*, 1970; Aase and Kemper, 1968; Fairbourn and Kemper, 1971). Experience with these runoff farming systems is still too limited to pass judgment on their relative effectiveness, but it appears that different approaches might be required for various conditions of topography, climate, soil, and crop.

A number of technological and engineering problems still require solution before the economic performance of runoff farming can be fairly examined. However, even in the preliminary stages of this development it has been shown that relatively cheap treatment based on application of sodium chloride or of petroleum byproducts can double or triple runoff yields, depending on soil properties and rainfall intensity distribution during a particular season. Furthermore, the threshold of rain (as defined by amount and intensity required to produce runoff) is appreciably lower for treated areas than for untreated ones.

As an example of what can be achieved with runoff inducement, let us consider the growth of wheat in the semiarid parts of Israel. The crop, which is generally sown in the late fall or early winter, can only produce a reasonable yield provided that precipitation during the growing season will total 300–500 mm of rainfall. Drought seasons are frequent, however, and growers face the possibility of a crop failure roughly one season out of three. Let us consider the operation of a runoff-farming system during a season with only 150 mm of rainfall. In order to be conservative, we shall assume a runoff yield of 50% from a treated area. Under such conditions the ratio between the water-contributing and water-receiving area of 2:1 will double the amount of water available to the runoff-receiving area for a total of 300 mm, while a ratio of 4:1 would supply a total amount of 450 mm to the growing area, well within the range required for satisfactory yields. Looked at in another way, such a treatment could provide the equivalent of a Mediterranean-type climate with roughly 500 mm of annual water supply for 20–25% of the land. It appears likely that satisfactory yields from a portion of the land are capable of producing a larger economic return than low and uncertain yields from the entire area.

Water produced by runoff inducement also represents a potential municipal water supply. The minimum essential per capita amount of water required for domestic use is about 100 l per day (Phelps, 1948), or about 35 m^3 per year. If we again assume a 50% runoff yield from precipitation and a mean annual precipitation of 150 mm, a catchment area of somewhat less than 500 m^2 is capable of supplying the annual domestic needs of one consumer. If a city were to rely for its entire domestic water supply on runoff inducement, an area of 500 hectares could supply a population of 10,000. Considering that such a system would no doubt also utilize runoff from paved areas, which yield virtually 100% of precipitation as runoff (again taking an example from ancient cities in the desert), it is seen that the area which would have to be treated for runoff production is of manageable size. The value of the water produced is probably out of proportion to any alternative land use of such land in the vicinity of a city in an arid climate.

The above examples demonstrate that the marginally utilized and inadequate precipitation falling in semiarid climates nevertheless represents appreciable amounts of water which, if properly collected and concentrated, represent an important potential water source. There are indications that current technology is already adequate to make this water source competitive with other unconventional sources of water, especially if it is kept in mind that, as can be seen from Myers' statistics (1967), consumers in many localities are already paying more for water than the present cost of water produced by runoff inducement where conditions are favorable for this.

While water from unconventional sources is likely to remain more expensive than what most consumers are accustomed to pay today, current trends in population and water consumption patterns make it a safe prediction that where water is really needed, users will pay what it costs.

REFERENCES

Aase, J. K., and Kemper, W. D. (1968). Effect of ground color and microwatersheds on corn growth. *J. Soil Water Conserv.* 23, 60–62.

Adams, J. E., and Hanks, R. J. (1964). Evaporation from soil shrinkage cracks. *Soil Sci. Soc. Amer., Proc.* 28, 281–284.

Adams, J. E., Ritchie, J. T., Burnett, E., and Fryrear, D. W. (1969). Evaporation from a simulated soil shrinkage crack. *Soil Sci. Soc. Amer., Proc.* 33, 609–613.

Army, T. J., Weise, F. A., and Hanks, R. J. (1961). Effects of tillage and chemical weed control practices on soil moisture losses during the fallow period. *Soil Sci. Soc. Amer., Proc.* 25, 410–413.

Bond, J. J., and Willis, W. O. (1969). Soil water evaporation: Surface residue rate and placement effects. *Soil Sci. Soc. Amer., Proc.* 33, 445–448.

Briggs, L. J., and Shantz, H. L. (1912). The relative wilting coefficient for different plants. *Bot. Gaz. (Chicago)* 53, 229–235.

Bruce, R. R., and Klute, A. (1956). The measurement of soil water diffusivity. *Soil Sci. Soc. Amer., Proc.* 20, 458–462.

Cary, J. W. (1968). An instrument for *in situ* measurement of soil moisture flow and suction. *Soil Sci. Soc. Amer., Proc.* 32, 3–5.

DeBano, L. F., and Letey, J. (1969). Water repellent soils. *Proc. Symp. Water Repellent Soils,* University of California, Riverside.

de Wit, C. T. (1958). Transpiration and crop yields. *Versl. Landbouwk. Onderz.* No. 64.6, p. 88.

Evenari, M., and Koller, D. (1956). Masters of the desert. *Sci. Amer.* 194, 39–45.

Evenari, M., Aharoni, Y., Shanan, L., and Tadmor, N. H. (1958). The ancient agriculture of the Negev. III. Early beginnings. *Isr. Explor. J.* 8, 231–268.

Fairbourn, M. L., and Kemper, W. D. (1971). Microwatersheds and ground color for sugarbeet production. *Agron. J.* 63, 101–104.

Gardner, H. R., and Hanks, R. J. (1966). Evaluation of the evaporation in soil by measurement of heat flux. *Soil Sci. Soc. Amer., Proc.* 30, 425–428.

Gardner, W. R. (1958). Some steady state solutions of the unsaturated moisture flow equation with application to evaporation from a water table. *Soil Sci.* 85, 228–232.

Gardner, W. R. (1959). Solutions of the flow equation for the drying of soils and other porous media. *Soil Sci. Soc. Amer., Proc.* 23, 183–187.

Gardner, W. R., Hillel, D., and Benyamini, Y. (1970a). Post irrigation movement of soil water. I. Redistribution. *Water Resour. Res.* 6, 851–861.

Gardner, W. R., Hillel, D., and Benyamini, Y. (1970b). II. Simultaneous redistribution and evaporation. *Water Resour. Res.* 6, 1148–1153.

Gates, D. M. (1962). "Energy Exchange in the Biosphere," Biol. Monogr. Harper, New York.

Green, W. H., and Ampt, G. A. (1911). Studies on soil physics. I. Flow of air and water through soils. *J. Agr. Sci.* 4, 1–24.

Greenland, D. J., Lindstrom, G. R., and Quirk, J. P. (1962). Organic materials which stabilize natural soil aggregates. *Soil Sci. Soc. Amer., Proc.* 26, 336–371.

Hanks, R. J., and Gardner, H. R. (1965). Influence of different diffusivity-water content relations on evaporation of water from soils. *Soil Sci. Soc. Amer., Proc.* 29, 495–498.

Hanks, R. J., and Woodruff, N. P. (1958). Influence of wind on water vapor transfer through soil, gravel and straw mulches. *Soil Sci.* 86, 160–164.

Hanks, R. J., Bower, S. B., and Boyd, L. D. (1961). Influence of soil surface conditions on the net radiation, soil temperature and evaporation. *Soil Sci.* 91, 233–239.

Hillel, D. (1967). "Runoff Inducement in Arid Lands," Final Tech. Rep. U. S. Dep. Agr., Volcani Inst. Agr. Res., Rehovot, Israel.

Hillel, D. (1968). "Soil Water Evaporation and Means of Minimizing It," Rep. U. S. Dep. Agr., Hebrew University of Jerusalem, Israel.

Hillel, D. (1971). "Soil and Water: Physical Principles and Processes." Academic Press, New York.

Hillel, D., and Guron, Y. (1970). "The Application of Radiation Techniques in Water Use Efficiency Studies," Res. Rep. IAEA, Vienna.

Hillel, D., and Tadmor, N. H. (1962). Water regime and vegetation in the Negev Highlands of Israel. *Ecology* 43, 33–41.

Holmes, S. W., Green, E. L., and Gurr, C. G. (1961). The evaporation of water from bare soils with different tilths. *Trans. Int. Congr. Soil Sci., 7th, 1960* Vol. 1, pp. 188–194.

Horton, R. E. (1940). Approach toward a physical interpretation of infiltration capacity. *Soil Sci. Soc. Amer., Proc.* 5, 339–417.

Jensen, M. E. (1972). Programming irrigation for greater efficiency. *In* "Optimizing the Soil Physical Environment Toward Greater Crop Yields" (D. Hillel, ed.). Academic Press, New York (in press).

Johnston, J. R., and Hill, H. O. (1944). A study of the shrinking and swelling properties of rendzina soils. *Soil Sci. Soc. Amer., Proc.* 9, 24–29.

Law, J. P. (1964). Effect of fatty alcohol and a nonionic surfactant on soil moisture evaporation in a controlled environment. *Soil Sci. Soc. Amer., Proc.* **28**, 695–699.

Lemon, E. R. (1956). The potentialities for decreasing soil moisture evaporation loss. *Soil. Sci. Soc. Amer., Proc.* **20**, 120–125.

Lemon, E. R., Shinn, J. H., and Stoller, J. H. (1963). Experimental determination of the energy balance. *U. S., Dep. Agr., Progr. Rep. No.* **71**.

Lowdermilk, W. C. (1954). The use of flood water by the Nabataeans and the Byzantines. *Isr. Explor. J.* **4**, 50–51.

Luthin, J. N., ed. (1957). "Drainage of Agricultural Lands," Monogr. No. 7. Amer. Soc. Agron., Madison, Wisconsin.

Marshall, T. J., and Gurr, C. G. (1966). Movement of water and chlorides in relatively dry soil. *Soil Sci.* **77**, 147–152.

Meyer, L. D., and Mannering, J. V. (1961). Minimum tillage for corn: Its effects on infiltration and erosion. *Agr. Eng.* **42**, 72–75, 86, and 87.

Myers, L. E. (1961). Waterproofing soil to collect precipitation. *J. Soil Water Conserv. India* **16**, 281–282.

Myers, L. E. (1967). New water supplies from precipitation harvesting. *Int. Conf. Water Peace,* Washington, D. C.

Myers, L. E. (1970). Opportunities for water salvage. *Civil Engi. (New York)* 41–44.

Nielsen, D. R., Biggar, J. W., and Davidson, J. M. (1962). Experimental consideration of diffusion analysis in unsaturated flow problems. *Soil Sci. Soc. Amer., Proc.* **26**, 107–112.

Penman, H. L. (1956). Evaporation: An introductory survey. *Neth. J. Agr. Sci.* **4**, 9–29.

Phelps, E. E. (1948). "Public Health Engineering." Wiley, New York.

Philip, J. R. (1957). Evaporation, moisture and heat fields in the soil. *J. Meteorol.* **14**, 354–366.

Philip, J. R. (1957b). Numerical solution of equations of the diffusion type with diffusivity concentration-dependent. 2. *Aust. J. Phys.* **10**, 29–42.

Philip, J. R. (1957c). The physical principles of soil water movement during the irrigation cycle. *Proc. Int. Congr. Irrig. Drainage, 3rd* Vol. 8, pp. 125–128 and 154.

Richards, S. J. (1965). Soil suction measurements with tensiometers. *In* "Methods of Soil Analysis," Monogr. No. 9. pp. 153–163. Amer. Soc. Agron., Madison, Wisconsin.

Rose, C. W., and Stern, W. R. (1967). The drainage component of the water balance equation. *Aust. J. Soil Res.* **3**, 95–100.

Selim, H. M., and Kirkham, D. (1970). Soil temperature and water content changes during drying as influenced by cracks: A laboratory experiment. *Soil Sci. Soc. Amer., Proc.* **34**, 565–569.

Shanan, L., Tadmor, N. H., Evenari, M., and Reiniger, P. (1970). Runoff farming in the desert. III. Microcatchments for improvement of desert range. *Agron. J.* **62**, 445–449.

Stanhill, G. (1965). Observation on the reduction of soil temperature. *Agr. Meteorol.* **2**, 197–203.

Strickling, E. (1957). Effect of cropping systems and VAMA on soil aggregation, organic matter and crop yields. *Soil Sci.* **84**, 489–498.

Swartzendruber, D. (1964). Seepage into stratified water-saturated soil from idealized vertical mulch channels. *Soil Sci. Soc. Amer., Proc.* **28**, 314–317.

Tadmor, N. H., Evenari, M., Shanan, L., and Hillel, D. (1957). The ancient desert agriculture of the Negev. I. Gravel mounds and heaps in the Shivta area. *Isr. J. Agr. Res.* **8**, 127–151.

van Bavel, C. H. M., Stirk, G. B., and Brust, K. J. (1968a). Hydraulic properties of a clay loam soil and the field measurement of water uptake by roots. I. Interpretation of water content and pressure profiles. *Soil Sci. Soc. Amer., Proc.* **32**, 310–317.

van Bavel, C. H. M., Brust, K. J., and Stirk, G. B. (1968b). Hydraulic properties of a clay loam soil and the field measurement of water uptake by roots. II. The water balance of the root zone. *Soil Sci. Soc. Amer., Proc.* **32**, 317–321.

Viets, F. G., Jr. (1962). Fertilizers and the efficient use of water. *Advan. Agron.* **14**, 228–261.

Waggoner, P. E. (1966). Decreasing transpiration and the effect upon growth. *In* "Plant Environment and Efficient Water Use" (W. H. Pierre *et al.*, eds.), pp. 49–72. Amer. Soc. Agron., Madison, Wisconsin.

Watson, K. K. (1966). An instantaneous profile method for determining the hydraulic conductivity of unsaturated porous materials. *Water Resour. Res.* **2**, 709–715.

Zingg, A. W., and Hauser, V. L. (1959). Terrace benching to save potential runoff for semiarid land. *Agron. J.* **51**, 289–292.

AUTHOR INDEX

Numbers in italics refer to the pages on which the complete references are listed.

SUBJECT INDEX

A

ABA, 258, 259, 263, 267–279, 290
Abortion, 242
Abscisic acid, *see* ABA
Abscission, 53, 95, 103, 114, 256
  buds, 95
  flowers, 95
  fruits, 96, 103
  leaves, 40, 95, 242
Abscission tissue, 50
Abscission zone, 50, 51, 102, 114, 117
Absorption
  minerals, 217, 218, 220, 224–227, 232,
    236, 267, 278
  water, 2, 23, 31, 42, 46, 56, 57, 75, 95,
    125, 217, 219–221, 255, 260, 261,
    296, 309, 311, 325
Absorption lag, 39, 56, 57
Acetone, 196, 199
N-Acetylglutamic acid, 193
N-Acetylglucosamine, 200
Accessory cells, 33, 266
Acid phosphatase, 140, 250
Acrylamide gel, 249
F-actin, 200
Active transport, 38
Adenine, 245
Adenine–RNA complex, 245
Adenosine triphosphate, *see* ATP
Adhesion, 47, 50, 51
Adsorption, 74, 155, 156, 158, 159, 194,
  205, 231
Advection, 284, 285, 300
Aecidia, 46
Aecidiospores, 46
Aeration, 66–68, 71–73, 83, 104, 257,
  270, 309, 331
Aging, *see also* Senescence
  fruits, 112

  leaves, 31, 37, 243, 256
  seeds, 67
Aggregation, 141, 150, 165, 203, 315, 316
Air resistance, 288, 289, 299, *see also*
  Resistance
Alanine, 199, 247, 249
Albedo, 317
Albumin, 140
Alcohol dehydrogenase, 208
Alcohols, 111, 133, 326
Aldehydes, 111
Alkalinity, 318
Alternating temperature, 70, *see also*
  Temperature
Amides, 166, 207, 245
Amino acid(s), 108, 137, 157, 166, 183,
  193, 199, 200, 206, 241, 243–249,
  266, *see also* individual amino acids
D-Amino acid oxidase, 208, 209
α-Aminobutyrate, 247
Amino group, 155
p-Aminophenylalanine, 205
Ammonia, 245
Ammonification, 220
Amylase, 161, 179, 180, 196, 208, 249,
  250
Amylopectin 1,6-glucosidase, 196
Anions, 198, 200
Anisotropy, 319
Annual ring, 56
Annulus, 47, 48
Anther, 49, 96
Anthesis, 49
Anthocyanin, 164
Antidesiccants, *see* Antitranspirants
Antimetabolites, 290
Antitranspirants, 99, 100, 271, 286–302,
  329, *see also* Transpiration
  reduction

354